John Haldeman Cooper

A Treatise on the Use of Belting for the Transmission of Power

Second Edition

John Haldeman Cooper

A Treatise on the Use of Belting for the Transmission of Power
Second Edition

ISBN/EAN: 9783744713795

Printed in Europe, USA, Canada, Australia, Japan

Cover: Foto ©berggeist007 / pixelio.de

More available books at **www.hansebooks.com**

"COOPER ON BELTING."

A Treatise on the Use of Belting for the Transmission of Power. With numerous illustrations of approved and actual methods of arranging Main Driving and Quarter Twist Belts, and of Belt Fastenings. Examples and Rules in great number for exhibiting and calculating the size and driving power of Belts. Plain, Particular, and Practical Directions for the Treatment, Care, and Management of Belts. Descriptions of many varieties of Beltings, together with chapters on the Transmission of Power by Ropes; by Iron and Wood Frictional Gearing; on the Strength of Belting Leather; and on the Experimental Investigations of Morin, Briggs, and others for determining the Friction of Belts under different tensions, which are presented clearly and fully, with the text and tables unabridged.

By JOHN H. COOPER, M. E.

Second Ed. One vol., demi octavo. Cloth, $3.50.

The Publishers will send copies by mail, postage prepaid, on receipt of price.

WHAT THEY SAY OF IT.

" It contains a great deal of much-needed information."—BROWN & ALLEN, N. Y.

" A useful and instructive volume, typographically creditable."—JAS. CHRISTIE, Philadelphia.

" It collects in the simplest manner the opinions of practical and theoretical men."—R. BRIGGS, M. E., Philadelphia.

" It contains everything that need be said on this important subject."—H. HOWSON, M. E., Philadelphia.

" We confidently welcome it as the standard treatise on belting."—POLYTECHNIC REVIEW, Philadelphia.

" You have studied *clearness* instead of *mystification*."—J. C. TRAUTWINE, C. E., Phllada.

" More to be found in your book upon this subject than in all the world beside."—Prof. J. E. SWEET, Cornell University.

" A thorough and complete treatise on the subject of belting."—SCIENTIFIC AMERICAN, N. Y.

" There is also much original matter never before printed."—HERALD AND FREE PRESS, Norristown, Pa.

" This work is exhaustive in character, and creditable to author and publishers."—AM. R. R. JOURNAL, N. Y.

" Fully illustrated in every respect, and a most valuable contribution to technical literature."—LEFFEL'S MILLING NEWS, Springfield, O.

" A complete treatise, embracing every variety of transmitting power by belts and ropes."—J. W. NYSTROM, M. E., Philadelphia.

" Written in the plainest language; easiest book to understand I ever read."—G. V. CRIPPS, Philadelphia.

" An encyclopedia, eighty illustrations, and numerous tables of great value."—N. AMERICAN, Philadelphia.

" Comprehensive, practical work, the careful study of which would save millions of dollars annually."—E. S. WICKLIN, Millwright, Wis.

" This work has a good index; use of belting explained in clear language."—PRESS, Philadelphia.

" An eminently practical work, subject treated with fulness and perspicuity."—PRINTERS' CIRCULAR, Philadelphia.

" The mass of facts and figures presented cover every point of theory and practice. It includes information from every available source; a valuable assistant."—N. W. LUMBERMAN, Chicago, Ill.

" I consider it a most valuable contribution to technical literature."—Prof. A. BEARDSLEY, Swarthmore College, Pa.

" A very admirable and exhaustive treatise."—Hon. ELLIS SPEAR, Commissioner of Patents, Washington, D. C.

" 'Use of Belting' supplies a want long felt by all mechanical engineers."—TAWS & HARTMAN, Engineers, Philadelphia.

" The need for such a book as this has long been manifest."—VAN NOSTRAND'S ECLECTIC MAG., N. Y.

" The most complete collection of rules, tables, and statistics upon the use of belts now in print."—JOURNAL OF FRANKLIN INSTITUTE, Philadelphia.

" No intelligent man can read your book carefully without informing himself pretty thoroughly as to what can actually be done with belting."—SAM'L WEBBER, M. E., Manchester, N. H.

A TREATISE

ON THE

USE OF BELTING

FOR THE

TRANSMISSION OF POWER.

BY

JOHN H. COOPER.
MECHANICAL ENGINEER.

SECOND EDITION.

Fully Illustrated.

PHILADELPHIA:
E. CLAXTON & COMPANY,
LONDON: E. & F. N. SPON, 16 CHARING CROSS.
1883.

Entered, according to Act of Congress, in the year 1877, by

CLAXTON, REMSEN & HAFFELFINGER,

in the Office of the Librarian of Congress, at Washington.

81,119

DORNAN, PRINTER,
PHILADELPHIA.

PREFACE

TO THE SECOND EDITION.

THE demand for a second edition of *Use of Belting* offers occasion to me to thank my readers and critics for the very favorable opinions entertained regarding my work, and for the good words expressed in its behalf.

Of the many full and favorable notices concerning the book, there occurs one statement—friendly and true, I believe—but about which I deem necessary certain explanations in justification of the object I had in view when placing this work before the public.

The statement referred to exists in a paper on a "Formula for the Horse-Power of Leather Belts," read at the Hartford meeting of the American Society of Mechanical Engineers, held in May, 1881, of which the following is the substance and text: In "a recent book on the *Use of Belting*, I found a large and valuable collection of matter on the subject, but no attempt was made to assimilate the same."

I am pleased to know that the book is found on perusal by intelligent readers, not only to be full of valuable matter, but that it also supplies many helps in the shape of experimental data and practical working of belts.

As already stated in *Use of Belting*, my whole aim in producing it was to present the largest array of facts and figures consistent with accuracy and in harmony with best use, and *not* to reduce the particulars of performance to one rule or statement. A square inch of paper holds area sufficient for a horse-power rule of belt performance, but a large book is necessary to tell the whole story of the use of belting.

Assimilation is impossible where the modifying circumstances exist in such great number and variety as they do under which belts act.

If we assume that the various published rules for ascertaining the horse-power of a belt represent the expressed assimilations of the data collated by the author of each, we shall have a considerable group of fairly formulated statements; but how near to the average truth they will approach can best be ascertained by reference to the rules for belts already given in *Use of Belting*.

In the making of many rules, too much stress is placed upon niceties of mathematical expression, to the utter disregard of the many disturbing conditions of practice and the persistent facts of working. What does it avail with formulated fine-spun, if, for example, the length and tightness of belts, the angle at which they are running, and the character of the adhesives applied, are left out of consideration by the rule, when we know that any one of these conditions will greatly affect the driving power of a belt?

PREFACE TO THE SECOND EDITION.

Adhesion is a greater fulcrum to the driving power than friction; each is subject to its own laws; the law of this has been largely investigated and relied upon, but for that there seems to be no written law or certainty of result, yet when friction fails, the rule is—*apply an adhesive.*

I believe the largest presentation of facts and conditions affords the best material for study of the subject. I have no fears for good proportions and best running conditions in the hands of the intelligent mechanic, if he is provided with the results of reliable experiments, and studies them in connection with the circumstances under which they were made.

On page ix of the Introduction, I have given an outline of reliable data for the guidance of the practical engineer who wishes to have knowledge of a fair average of belt strength, and of the force which belts are capable of transmitting in the ordinary way of using them; and on page xi I have given some of the conditions to be fulfilled in the use of belts in order to get best service from them.

It is safe to say that no better assimilation of belt data can be made than that already presented in these pages. More conditions and variety of conditions, when discovered, will be added to this considerable list; but the last deduction, when all the facts are in, will give to the practical man for use a simple statement, something like this: *a flat leather belt of the usual make, one inch in width, will transmit a force of say 55 lbs. at any ordinary velocity, running freely on smooth-faced pulleys.*

An easily made experiment will prove that such a belt will transmit a force of 100 lbs., or even more, but for long unfailing service the amount named is not far from the best practice. One can easily see and understand that a unit of this character, like a known tensile strength of good wrought-iron per square inch of section, is a factor of almost universal application, and is in harmony with the use of belting.

There are theoretical crotchets in many books; nevertheless, rules for belting should be "practical as far as possible, theoretical as far as necessary," and yet it will ever be true on test, that each material gives its own figures, which can only be known to those who lay a hand to the practical side of the subject as well as a head to the principles which govern the movements.

Mention is made of a difference between Adhesion and Friction. It may be said, in general, that friction is in proportion to the weight, but independent of the surface exposed to the sliding action; adhesion, on the other hand, is in proportion to the extent of surface and to the nature of the adhesive.

Either or both of these may act with running belts. For many reasons, therefore, closer observation and comparisons of them with more extended experimentation are much needed to fully explain the many anomalies of belt driving. There is a repeated outcropping in the public prints of misapprehension concerning these irregularities, and I hope that conclusive tests will be fully made in the near future.

JOHN H. COOPER.

LOS ANGELES, CAL., *May*, 1883.

ERRATA.—See page 310.

PREFACE

TO THE FIRST EDITION.

SOME of the facts and figures given in this treatise were originally collated for private use, and were printed in the "Journal of the Franklin Institute." They are now rearranged, greatly enlarged, and are offered to the public with the hope that they may, at least, spare the reader's time in making searches, and help him to extend his knowledge of the subject treated.

In publishing these materials, which have been selected from the best available sources, foreign and American, my object is to acquaint the reader with existing written rules, together with the particulars of working examples, in order that he may study the subject for himself.

I fully believe that the *useful* part of every division of mechanical science may be expressed in simple language, and therefore propose, in the following pages, to tell just what all need to know about belting, in a way that every one will clearly understand.

The first five chapters, which make up the greater part of this treatise, may be said to fitly represent the practice of the workshop and factory.

The sixth chapter, presenting the text and results entire of Mr. Briggs's elaborate essay, with Mr. Towne's reliable experiments, forms in itself a theoretical and practical record of the tension of leather belts of the highest permanent value.

I have given in the seventh chapter, by the aid of an efficient translator, a literal and, I believe, an exact reproduction of the subject-matter and experiments of Morin on the variation of the tension of leather belts, with every *figure*, *formula*, and *tabular statement* exactly as it exists in the original, repeating, also, the measures in mètres and the weights in kilogrammes. As Morin is frequently quoted in detached portions, I felt it to be right, as it is in keeping with the work I had undertaken, to at least attempt a full and complete translation of his words and meaning into plain English.

The eighth chapter is devoted to the transmission of power by ropes, beginning with the particulars and practice of wire-rope driving, by the Roeblings, selected from printed matter, full of good and standard data, generously offered by them, embracing also an article from the French of A. Achard, which, like Morin's, contains some mathematical nuts for the curious to crack, besides the opened kernels of fact, all of which may prove both pleasant and profitable to find.

Finally, in chapter ninth, Mr. Wicklin's liberality has enabled me to lay before the reader a particular account, in unmistakably clear language, of his method of transmitting power by the friction of rolling contact.

I have endeavored, in all parts of this work, to discover and name the author of every statement made; and, in addition to this, I offer my thanks to Messrs. Alexander Bros., Philadelphia; Messrs. D. Appleton & Co., New York; Mr. H. C. Baird, Philadelphia; Messrs. Brown & Allen, New York; Mr. R. H. Buel, New York; Mr. James Christie, of Pencoyd Iron Works, Philadelphia; Messrs. Hoyt Bros., New York; the "Journal of the Franklin Institute," Philadelphia; "The Polytechnic Review," Philadelphia; Profs. Morton and Thurston, of the Stevens Institute, Hoboken; Mr. J. W. Nystrom, Philadelphia; Mr. Samuel Webber, N. H.; and to many others, whose names are appended to their contributions, for favorable opinions kindly expressed, and for the privilege, freely given, of making extracts from their published works.

JOHN H. COOPER.

135 Wister Street,
GERMANTOWN, PHILA.,
May, 1877.

CONTENTS.

	PAGE
INTRODUCTION	v

CHAPTER I.
RULES AND DATA FOR BELTING 17

CHAPTER II.
METHODS OF BELT TRANSMISSION 154

CHAPTER III.
CEMENTS, ADHESIVES, AND FASTENINGS. 181

CHAPTER IV.
VARIETIES OF BELTING 190

CHAPTER V.
STRENGTH OF BELTING LEATHER 210

CHAPTER VI.
EXPERIMENTS OF BRIGGS AND TOWNE ON LEATHER BELTS . . 214

CHAPTER VII.
EXPERIMENTS ON THE TENSION OF BELTS, BY A. MORIN . . . 232

CHAPTER VIII.
ROPE TRANSMISSION OF POWER 253

CHAPTER IX.
FRICTIONAL GEARING 288

INTRODUCTION.

EXPLANATIONS.

IN order to lessen labor I have introduced two abbreviations;—Rpm for *revolutions per minute*, and Fpm for *feet per minute*, which I have found useful in the every-day pocket-book jottings of business.

The Greek letter π, as in all mathematical formulæ, stands for 3.1416, denoting the circumference of a circle whose diameter is 1. On page 49 is a formula in which 6.28 occurs. This is meant to be 2π, or 3.1416×2. The signification of all other Greek letters is given in the articles where they are used.

In Chap. 7 the comma (,) is used for the same purpose that we employ the period (.); that is, to divide the decimals from the whole numbers; and the period is not used at all to denote any operation to be performed in the formulæ; while in Art. 181 the comma is used for the decimal point, and the period for the sign of multiplication. See formula at head of page 266, where the numbers are separated by both periods and commas.

Three dots, thus, ($\therefore$) stand for *therefore*.
The angular character ($<$) means *less than*, and the same reversed ($>$) means *greater than*. Thus $A<B$ is the same as saying A is less than B, and $C>D$ is the same as saying C is greater than D.

The word logarithms is thus abbreviated: *log.*

The words "No. 1 and following," on page 242, refer to articles in Morin's memoir of experiments on the friction of journals.

INTRODUCTION.

The *metre* is equal to 39.37043 inches, or it is equal to 3.2808693 feet; or, to give the length of a metre in terms of English measure, and in a way easily remembered, it is quite near to call it equal to 3 feet 3 inches and $\frac{3}{8}$ths of an inch.

The metre is divided into one thousand equal parts, each part called a millimetre. Therefore, to express a measure by this system, this form is chosen: $3^m,425$ — a distance equal to 3 metres and 425 millimetres, or 3 metres and 425 thousandths of a metre. The following table gives all the minor metric denominations and their equivalents in English measure.

	METRES.	INCHES.	FEET.
1 millimetre	$\frac{1}{1000}$	0.03937	
1 centimetre	$\frac{1}{100}$	0.39370	
1 decimetre	$\frac{1}{10}$	3.93704	
1 metre	1	39.37043	3.2808693

The *gramme* is equal to 15.43234874 grains, and the kilogramme is 1000 grammes, or it is equal to $\dfrac{15.43234874 \times 1000}{7000} = 2.20462125$ lbs., and to express kilogrammes and fractions, the following form is used: $56^k,824$ — meaning 56 kilogrammes, and 824 thousandths of a kilogramme, or 56.824 grammes.

The unit of measure for *man-power*, established by Morin, is the equivalent of 50 lbs. raised 1 foot in a second of time, and that of a *horse-power*, established by Watt, is the equivalent of 550 lbs. raised 1 foot in a second, which is deduced from the original data given herewith:

"Mr. Watt made some experiments on the strong horses employed by the brewers in London, and found that a horse of that kind, walking at the rate of 2½ miles per hour, could draw 150 pounds avoirdupois, by means of a rope passing over a pulley, so as to raise up that weight, with vertical motion, at the rate of 220 feet per minute. This exertion of mechanical power is equal to 33,000 pounds raised vertically through a space of one foot per minute, and he denominated it a horse-power, to serve for a measure of the power exerted by his steam-engines."—*John Farey on the Steam-Engine, London, 1827.*

The following equation will make this clear:

$$\frac{5280 \times 2\frac{1}{2} \times 150}{60} = 33,000 \text{ lbs., raised 1 foot in a minute.}$$

This unit of measure is employed in this treatise, and is the form of expression used in this country, because it is customary to give the speeds of machines in *feet*, or *revolutions per minute*. In Europe, the unit of time for the expression of speeds is the *second*, not the minute.

Units for Horse-Power, from Nystrom.

Horse-Power.		Foot-lbs. per Second.
English...	33,000 lbs., raised 1 foot per minute	550.
French....	75 kilogrammes, raised 1 mètre per second*...	542.47
German...	....	582.25
Swedish...	....	542.06
Russian...	....	550.

* Equal to 0.986 of an English horse-power.

The letters employed in the opening formulæ of Morin's essay are explained on pages 235–7, the same being used in the formulæ relating to the friction of journals, given in his earlier articles.

Morin's work is published under the authority of the Institute of France, which in itself is the highest recommendation of its intrinsic value. This is also abundantly proven by the universal employment of his deductions by subsequent writers, and by the invariable coincidence of results with those obtained from all later experimentation, so that it needs no introduction to the reader other than to say the translation here given is an exact reproduction of the original.

The author of Art. 58 is the same referred to in the preface, and I call attention to his clear and fully detailed experiments on a double leather belt, as well as to the mention of those of larger dimensions employed in rolling mills.

On page 165 the word *rigger* is used, and in this case it is clearly synonymous with small pulley or drum. Rigger, in "Weale's Dictionary of Terms of Art," is defined "a wheel with a flat or slightly curved rim, moved by a leather band." Box, in his "Treatise on Mill-Work," repeatedly says "rigger or pulley," as if they were one and the same; but I do not find the word in use by Abel, Baker, Buchanan, Fairbairn, Rankine, or "The Engineer and Machinist's Assistant."

Buchanan says, on page xx. of his preface: "Arkwright used iron bevel-wheels and band-pulleys in the cotton spinning mills at Cromford and Belper, in 1775."

The term *bays*, used on pages 137–8, may be defined as the spaces in a ceiling or roof of a building marked by the rafters or beams, or by the buttresses or pilasters of the walls.

The figures given in Arts. 68 and 96 coincide exactly with those in Art. 163, which leads one to believe they were derived from Mr. Towne's published experiments.

The reader will be surprised and amused at the different results to which the rules here assembled lead him, and he may find it a difficult task to reconcile them to what in his view may be considered a fair average performance of a belt; but as they are exhibited exactly as I found them, no apology is offered for their repetition word for word, and herein, indeed, consists the true value of this collection, in which will be found in one volume a great variety of data gathered in from many.

I give below what may be termed *the philosophy of the belt*, in which the principles of action are presented, and have added the *mechanics of belting*, introducing dimensions which may not be reliable in practice for definite rules, because of the greatly varied circumstances under which belts are used; yet when certain attainable conditions are fulfilled, belts will transmit forces commensurate with the area of pulley contact. The statements made on pages 9–11 are founded upon experiment, backed up by good authority, and may be depended upon for general practice.

The Philosophy of Belting.

"Motion communicated by cords, bands, or straps, is remarkably smooth, and free from noise and vibration, and on this account, as well as from the extreme simplicity of the method, it is always preferred to every other, unless the motion is required to be conveyed in an exact ratio.

"As the communication of motion between the wheels and bands is entirely maintained by the frictional adhesion between them, it may happen that it may occasionally fail through the band sliding on the pulley. This, if not excessive, is an advantageous property of the contrivance, because it enables the machinery to give way when unusual obstructions or resistances are opposed to it, and so prevents breakage and accident. For example, if the pulley to which motion is communicated were to be suddenly stopped, the driving pulley, instead of receiving the shock and transmitting it to the whole of the machinery in connection with it, would slip round until the friction of the band upon the two pulleys had gradually destroyed its motion. But if motion is to be transmitted in an exact proportion, for example, such as is required in clock work, where the hour hand must make one exact revolution while the minute hand revolves exactly 12 times, bands are inapplicable; for supposing it practicable to make the pulleys in so precise a manner that their diameters should bear the exact proportion required, which it is not, this liability to slip would be fatal.

"But in all that large class of machinery in which an exact ratio is not required to be maintained in the communication of rotation, endless bands are always employed, and are capable of transmitting great forces."—*Prof. Willis's Mechanism.*

The Practical Mechanics of Belting.

Force Required to Break Belts.

The mean ultimate strength of a single belt one inch wide and $\frac{1}{4}$ inch thick, made of good leather, may be taken at 1000 lbs. in the solid part; and ordinary leather belts will average $\frac{3}{4}$ of this amount. (See pages 14, 212, and 213.)

The strength of fastenings varies according to the loss of section by the rivet and lace holes, and by the tenacity of the cements and lace leather used. The weakening effect of the several methods of joining the ends of belts is such that no more than 200 lbs. per inch of width can be depended upon for ultimate strain. (See Art. 163.)

We do not have a sufficient number of tests of other belting materials to compare fairly with leather, and can therefore only refer the reader to the figures on page 213.

Force Transmitted by Belts.

Each inch of width of good leather belting will transmit a force of 55 lbs., and this may be depended upon for continuous service when the belt is properly surfaced on smooth pulleys running at high speeds. (See Arts. 1, 15, and 40.)

Taking the average thickness of belts at $\frac{1}{6}$ of an inch, this would give a strain to the square inch of section equal to 330 lbs. (See Art. 1.)

A strain of 55 lbs. to the inch of width is equal to a surface velocity of 50 square feet per minute per horse-power, which is safe practice for single belts in good condition; and on double belts a strain proportioned to the thickness may be used; in all cases there must be ample contact with the pulleys. (See Arts. 1, 15, and 103.)

To facilitate the conversion of belt strains into surface velocity, the following table is given:

x   INTRODUCTION.

Table for Converting the Strain upon a Belt into its Surface Velocity per Horse-Power, and Vice Versa.

These figures were obtained by dividing 33,000 by 12 times the strain in lbs., to which one inch of width of belt may be subjected.

Example.

$$\frac{33,000}{10 \times 12} = 275.$$

Lbs. of Strain on each Inch of Width of Belt.	Square Feet of Belt per Min. per Horse-Power.	Lbs. of Strain on each Inch of Width of Belt.	Square Feet of Belt per Min. per Horse-Power.	Lbs. of Strain on each Inch of Width of Belt.	Square Feet of Belt per Min. per Horse-Power.
10	275.	41	67.0731	71	38.7323
11	250.	42	65.4761	72	38.1944
12	229.1666	43	63.9534	73	37.6712
13	211.5384	44	62.5	74	37.1621
14	196.4285	45	61.1111	75	36.5853
15	183.3333	46	59.7826	76	36.105
16	171.875	47	58.5106	77	35.6371
17	161.7647	48	57.2916	78	35.1812
18	152.7777	49	56.1224	79	34.7368
19	144.7368	50	55.	80	34.375
20	137.5	51	53.9215	81	33.9506
21	130.9523	52	52.8846	82	33.5365
22	125.	53	51.8867	83	33.1325
23	119.5652	54	50.9259	84	32.738
24	114.5833	55	50.	85	32.3529
25	110.	56	49.1071	86	31.9767
26	105.7692	57	48.2456	87	31.6091
27	101.8518	58	47.4137	88	31.25
28	98.2142	59	46.6101	89	30.8988
29	94.8275	60	45.8333	90	30.5555
30	91.6666	61	45.0819	91	30.2197
31	88.7096	62	44.3548	92	29.8913
32	85.9375	63	43.6507	93	29.5698
33	83.3333	64	42.9687	94	29.2553
34	80.8823	65	42.3076	95	28.9473
35	78.5714	66	41.6666	96	28.6458
36	76.3888	67	41.0447	97	28.3505
37	74.3243	68	40.4411	98	28.0612
38	72.3684	69	39.855	99	27.7777
39	70.5128	70	39.2857	100	27.5
40	68.75				

Conditions to be Fulfilled in the Use of Belts.

Special conditions of successful practice, within which is embraced the driving capacity measurable by the area of contact, and modified by the state of the pulley and belt surfaces, and adhesive used, which must not permit the belt to slip or stick; the proper material for, and treatment of belts; the utmost contact or arc of enrolment on the pulley; the proportion of diameter of pulley, and length and width of belt for best running; the least rounding of the pulley faces, and the greatest smoothness obtainable; the hair side of leather belts always to pulleys, as it is the smoother side, for what is lost in contact must be made up in strain, and because the stronger fibres lie nearest the flesh side, and should be preserved; the amount of adhesion or traction developed by the tension employed; the fastenings, which should be of the best; the disposition of the laps such, that the motion of driving will run *with*, not *against*, them; the employment of large pulleys, high speeds, and light belts; the careful putting on and skilful joining of belts; the running of them, slack as possible, in the upper fold or strip; the avoidance of tightness by excessive strain or binders, and of lateral straining, as in some quarter twist methods; the introduction of fly-wheels, or like devices, for rendering the work of the belt uniform; the increase of driving capacity, for overcoming occasional resistances and starting frictions; the uniformity of belt section, and weight and texture of material, the straightness of edges for smoother running at high speeds; the employment of gum belts for elevators, or to run in moist or hot situations, or where uniformity of section, without joints, is desirable — in fact, in many places where leather is generally used, always avoiding twists, all devices rubbing the edges, and the contact with any solvent of gum; the adoption of leather covering for pulleys, by which 33 per cent. of adhesion is gained; the securing of strips to the outer edges of single belts to increase adhesion; the running of a belt atop of another to make it drive; increasing the speed of a belt, which may be as high as a mile in a minute and be safe and advantageous. The introduction of the devices for augmenting the tractive pull of belts; the utilization of belts for imparting and arresting motion; the substitution of " wrapping connectors " for gear, as in twist belt arrangements, which do not over-strain the fibres; these, and a multitude of other conditions, involving the essential elements of best practice, will be found on examination to be accepted and used by the numerous authorities quoted, and to which the reader is directed and assisted by a complete index, arranged especially for ready reference.

Vulcanized Rubber Belts, by the N. Y. Belting and Packing Co.

In Art. 148, I have given in detail many facts relating to the belting fabricated by this company; but since that was written, belts of larger proportions, and of more perfect workmanship, if such be possible, have been produced. As an evidence of the extent to which their belting is employed, "they manufactured in the year 1875 over 1,200,000 feet of various widths of belting, and some idea may be formed of the magnitude of their industrial operations when we mention one order lately filled amounting to $45,000, which included many large and long belts, and also one driving belt for the elevator of the New York Central and Hudson River Railroad, which measured 48 inches in width, 330 feet in length, and weighed 4000 pounds," thus outdoing their "Champion Belt," described on page 197, and producing a belt *larger* than "*the largest belt in the world,*" of which account is given in Art. 76.

In order to fairly present the scientific improvements developed by this Company, and the mechanical perfection to which their processes of manufacture have been brought, I beg leave to offer in part the report of L. C. de Montanville and E. Sternhein, agents of the French Government to the Centennial, not omitting the grand compliment they have paid us in the following words, which should be written in letters of gold:

"The Philadelphia exhibition was the greatest manifestation of industrial work, in all its ramifications, ever offered to mankind," and then proceed to requote the "praiseworthy terms in which Messrs. Kuhlmann, Jr., and Dietz-Monin, of France, and Mr. De Wilde, of Belgium, spoke of the manufactured products exhibited by this establishment." They said:

"For some years past India-rubber, which has passed through so many new processes, has formed the basis of several large industries, which manufacture an infinite variety of articles useful in domestic economy, as well as various instruments of undoubted value, not only to surgery, but also to the physical, chemical, and mechanical arts, and to navigation. Up to the present time, we have observed no more important progress in this manufacture than that of the introduction of Oxide of Lead, Zinc, and Sulphur in the vulcanization of rubber, thus rendering it capable of being used for mechanical purposes. What we especially noticed in the goods manufactured by the Company which is the present subject of our special attention, was that their rubber fabrics retained their flexibility, and all their essential qualities, without the slightest alteration, even when subjected to the most opposite temperatures. We cannot too highly praise its special excellencies—namely, *its resistance to the action of chemical agents, the perfect smoothness of its surface, and the evenness of its vulcanization.* The tests that we made of the different bands manufactured by this Company, and destined for use as machine belting, showed us conclusively that, in addition to the above-mentioned good qualities, it possessed *great tenacity and large power of resistance under great pressure.* For these important reasons, *we should prefer its use for driving machinery even to the best and thickest leather belting.* The mixture of Oxide of Lead, Zinc, and Sulphur used in combination with the rubber during its process of vulcanization, obviates all tendency to its becoming hardened or rigid by extreme cold, or to its softening and becoming porous under the influence of heat. These probabilities, which were serious obstacles to the general use of vulcanized rubber for such purposes, are, in our opinion, completely overcome by the use of Oxide of Lead, Zinc, and Sulphur. We cannot do better than follow these remarks by giving verbatim, and without further comment, the very flattering terms in which the jury of Philadelphia spoke of the New York Belting and Packing Company. Their official report reads as follows:

PHILADELPHIA, December 20th, 1876.
REPORT ON AWARDS.

Product, *Rubber Belting.*
Name and address of Exhibitor, *New York Belting and Packing Company, New York City.*

The undersigned, having examined the product herein described, respectfully recommends the same to the United States Centennial Commission for Award, for the following reasons, viz.:

The belting is of various widths to 48 inches, of thickness from three to five ply, of length to 320 feet. Its strength, as determined by experiment under direction of Capt. Albert: A three-ply three inch belt gave way at 3,000 (three thousand) pounds. In adhesion, a six inch belt, with a weight of fifty pounds at either end, over a 15¾ inches exterior diameter smooth, cast-iron fixed pulley, slipped at 70 (seventy) pounds. The thickness of the belt was three-ply—$\frac{7}{8}$ of an inch. Commended for adhesion, strength, smooth finish, and care in workmanship and curing.

E. N. HORSFORD.

UNCLASSIFIED FIGURES AND NOTES.

From "Tables, Rules, and Data, by D. K. Clark," London, 1877.

Average Tension.

"Dr. Hartig found, from the results of experiments made by him in a woollen mill, that the tension of the driving belts varied from 30 lbs. to 532 lbs. per square inch of section, and that it averaged 273 lbs. per square inch."

Average Working Strain.

"An average working strength of 300 lbs. per square inch of section of leather belts may be accepted for purposes of calculation."

Surface Velocity.

"The performances of belts may be compared by calculating the number of square feet of belt-surface passed over either pulley per minute per horse-power; involving the elements of working stress and velocity. It is found by multiplying the velocity in feet per minute by the breadth of the belt in feet, and dividing the product by the horse-power transmitted."

Rule for Horse-Power of a Belt.

"M. Claudel gives the following empirical formula, in common use, for finding the breadth of a leather belt enveloping half the circumference of a pulley."

Altering the measures: $b = c \dfrac{H}{v}$:

In which $b =$ breadth of belt in inches.
$H =$ horse-power.
$v =$ the speed of the belt in feet per second.
$c =$ a constant $= 26$ for upright shafts and
20 for horizontal shafts.

"M. Claudel instances the common experience that a belt 3¼ inches broad, moving at a velocity of 9 feet per second, can very well transmit one horse-power with ordinary tension, and without over-straining, working on turned and smooth pulleys of equal diameter.

"This example, if adopted as a basis, would give a coefficient $c = 29$. The working tension is only about 20 lbs. per inch wide. At the same time, the values given by the empirical formula are little more than those deducible from the data of M. Morin."

"Mr. Kirkaldy's Tests of Norris & Co.'s Belting, gave the following Results for Ultimate Tensile Strength.

Size.	English Belting.	Helvetia Belting.	English per Inch of Width.	Helvetia per Inch of Width.
12 inch double	14,861 lbs.	17,622 lbs.	1238	1469
7 " "	6,193 "	11,089 "	884	1584
6 " "	5,603 "	10,456 "	934	1743
4 " "	4,365 "	6,207 "	1091	1552
2 " "	2,942 "	4,237 "	1471	2118
10 " single	8,846 "	11,888 "	885	1189
5 " "	4,060 "	5,426 "	812	1085
4 " "	3,248 "	3,948 "	812	987
3½ " "	3,007 "	3,377 "	859	965

Tightening Pulleys.

The tightening pulley is applied to belts for increasing their adhesion to the pulleys; and as this is liable to fail first on the smaller pulley, it is usual to place them on the slack side of the belt, nearer this pulley, in order to increase adhesion as well as the area of contact, which it effectually does in this position; but it also increases the friction of driving, in proportion to the thrusting of the same from the line of its natural curvature. If placed nearest the larger pulley, it would increase the adhesion where such is not wanted, and would, at the same time, diminish the area of contact on the smaller pulley by pulling the belt away from it, thus lessening the very effect which it was applied to remedy. It would, however, increase adhesion by augmenting the tension of the belt; but this in turn would add to the resistance to be overcome, by creating additional friction in the shaft bearings.

Effect of Speed on Slack Belts.

In making close calculations for belt adhesion, care must be taken to note the degrees of contact of the belt on the pulley, as the adhesion increases in a greater ratio than the area of contact (Art. 97), especially when the upper fold is very slack and the speed high, in which case the belt throws itself against the face of the pulley it is approaching with much force, increasing the area of contact, as well as the pressure on the pulley.

This banking of the current of the belt against the face of the pulley augments adhesion, and in a great measure compensates for the loss of contact due to centrifugal force.

Belting of Intestines.

" Belting is made in America from the entrails of sheep, which average (the entrails — not the sheep) some fifty-five feet in length. They are thoroughly cleaned, and subjected for some days to the action of brine, and are then wound upon bobbins, after which the process is the same as making common rope. If a flat belt is required, a loom is employed, and the strands are woven together. A $\frac{3}{4}$-inch rope thus made will stand a strain of seven tons, and is guaranteed to last ten years; the best hemp rope of same thickness has a life of about three years."

To Measure for Belting.

To find the length and course of a belt, apply a tape-line or string to the pulleys where the belt goes, and then measure the length of the string by a two-feet rule, which a mechanic should always have at hand; and when such measure cannot be made, make a drawing, full size or to scale, and step dividers around the course of the belt. By means of such drawings, the places where the belt passes floors and the like can also be found. These methods are more trusty and more convenient than any system of calculations, however tabulated, formulated, or prepared.

Pulleys of Paper and Raw-Hide.

Pulleys may be made of paper and of raw-hide in the same manner as directed in Art. 61 for leather pulleys. Raw-hide has superior qualities for resisting wear; it may therefore be very successfully used for shaft bearings or for the hubs of loose pulleys.

Paper also suits well for pulley covers.

A TREATISE

ON THE

USE OF BELTING.

CHAPTER I.

RULES FOR ASCERTAINING THE DRIVING POWER OF BELTS, AND FACTS AND FIGURES RELATING THERETO.

Mr. Samuel Webber, C. E., of Manchester, N. H., who has much practical acquaintance with mill work and belting, says:

1. "I have had a working rule for many years, which was given to me by an old and experienced machinist, and may be thus expressed: *Ordinary leather belting one inch wide, having a velocity of* 600 *Fpm, will transmit one horse-power.*

"Practice has shown me the safety of this rule, and for which, and for its extension to all cases, I have sought a reason and a formula.

"After an examination of the text-books, I found that Morin's data gave me the clue to the truth of this rule, and that it was supported by other good authority. Morin says: 'Belts designed for continuous service may be made to bear a tension of 0.551 lbs. per .0000107 square feet, or .00155 square inches of section, which enables us to determine their breadth according to the thickness.' This is equal to 355 lbs. per square inch of belting leather, and is also equal to from $\frac{1}{8}$ to $\frac{1}{12}$ the breaking strength of the same, as given by Rankine and other authorities.

"From this I see my way to a simple formula: Substituting 330 lbs. for 355 lbs. per square inch, I strike the component part of a horse-power, and deduce the following: one square inch of belting, at a velocity of 100 Fpm, will transmit one horse-power with safety, and from these data get this rule:

"The denominator of the fraction expressing the thickness of the belt in inches gives the velocity in hundreds of feet per minute at which each inch of width will transmit one horse-power, that is: $\frac{1}{6} = 6 \times 100 = 600$ ft. : $\frac{1}{3} = 3 \times 100 = 300$ ft., and so on.

"Now, $\frac{1}{6}$ inch being about the ordinary thickness of a single belt, this shows me why my old 'rule of thumb' proved right, and a careful examination of many of the large belts running in our New England cotton-mills within the last year or two confirms my opinion as to the safety of the rule.

"This gives a strain of 55 lbs. per inch, and a belt speed of 50 square feet of surface per minute per horse-power, as safe, ordinary practice for single belts; and I find the same velocity, with a strain proportioned to the thickness, works perfectly well with double belts. This, however, is applicable where there is a sufficient holding surface on the smaller pulley; if the arc of contact be small, a wider belt will be necessary, and I am not yet able to formulate a rule for belt contact. The nearest approach I have made to it yet is to allow 10 to 12 square inches of pulley surface in contact with belt for each horse-power.

"It is generally conceded that the friction of a belt passing half around a pulley, is equal to one-half the strain on the belt; or that an inch belt, at 600 Fpm, with a strain of 55 lbs., would give a traction of 27.5 lbs. and require a pulley which would give 1200 lineal feet per minute of surface contact, to obtain the one horse-power to which the belt would be equal.

"Morin, in his 'Mechanics,' gives, as the result of actual trials with a loaded belt over a wooden drum, an average friction of 57 per cent., which would be increased by using a pulley covered with leather; and a polished iron pulley, with a smooth, flexible belt, may, I think, be depended on in actual use for 50 per cent.

"The friction of a belt varies with the arc of the circle of the pulley with which it is in contact, and is only half as great on $\frac{1}{4}$ of a pulley as on $\frac{1}{2}$ of one; so that double the surface in square inches will be required to transmit the same power in the former case that would be needed in the latter.

"Carrying out these rules, it will be easily seen that where high speed is to be obtained by the use of small pulleys, a much greater width of belt is necessary to get the frictional surface than is called for by the strength of the leather; and it will be found that, for circular saws, cotton-pickers, spinning-frames, etc., a wider belt is needed than is due to the actual power transmitted. Take, for instance, a spinning-frame with a 7-inch pulley, 900 Rpm, or 1650 feet belt velocity, and requiring 1½ horse-power. One inch of belt at that speed would transmit 2½ horse-power, but the contact surface of the pulley would not be over 10 inches in length, and, by the above rules, calls for a 3-inch belt, which is the standard size for that purpose. Looms, and other machines which are constantly stopped and started, also require wider belts, to stand the wear and tear of 'shipping,' than would be needed for the power. Let me cite, as an example, an instance of a main belt running in this city 24 inches wide, double, transmitting 160 horse-power at 3200 Fpm, to a pulley 4 feet 10 inches diameter. Taking my formula for double belts, it would be:

$$W = \frac{160 \times 3660}{3200 \times 7.58} = 24.14 \text{ inches wide.}$$

"This belt has run four years without repairs, and looks likely to run forty more. According to my rules for the strength of leather, it would transmit 192 horse-power, but to do it the smaller pulley should be 5 feet 9.6 inches diameter instead of 4 feet 10 inches.

"My formula for single belts is:

$$W = \frac{HP \times 5500}{velocity \times contact \text{ } in \text{ } ft.},$$

and for double belts:

$$W = \frac{HP \times 3660}{velocity \times contact \text{ } in \text{ } ft.}.$$

"The tendency with us now is to use large pulleys, high surface speeds, and light belts. In my rules for double belts I have assumed ¼ inch in thickness and 82½ lbs. strain; but if the belt be, as many are, ⅜ inch thick, it would, of course, bear from 110 to 120 lbs., and 300 Fpm would give one horse-power per inch.

"One other point I would also mention: The better friction is not the only reason for putting the 'grain' side of the leather next the pulley; it is harder, and not so elastic and fibrous as the flesh side; will wear better on the surface of the pulley, while it will crack and break if exposed to the expansion and contraction to which the outside of the belt is continually subjected."

Tests of Shafting, by SAMUEL WEBBER, C. E., Manchester, N. H., 1874.

DATE.	PLACE.	Length.	Diameter.	Weight.	Weight of Pulleys.	Total Weight.	No. of Bear'gs.	Rev. per Minute.	Ft. Lbs.	Horse-Power.	Coeff. Friction.	REMARKS.
				LBS.	LBS.	LBS.						CONTINUOUS OILING. [Kerosene oils mixed.
April, 1871	Amoskeag Mills	8 ft. 6 in.	2⅛ inc.	101	577	678	2	216	49	.089	.0336	Single Counter. Dreyfuss Pat. Oiler. Sperm and
"	"	34	"	404	1,974	2,378	8	216	196	.357	.0413	4 Count's like above. Connected with Belts. Same
"	"	114	"	1,366	1,859	3,225	15	216	325	.590	.0500	Single line. Oilers as above. [Oils mixed.
"	"	228	"	2,732	3,617	6,349	30	216	650	1.181	.0501	2 Lines like above connected. Oils, etc., same.
"	"	342	"	4,098	5,331	9,429	47	216	1022	1.858	.0552	3 " " " " "
"	"	16	2⅜ inc.	2,427	2,988	5,415	25	216	378	.687	.0338	Single line. Oils, etc., same.
"	"	178	2⅛ "									
"	"	10 ft. 4 in.	4⅛ "									
"	"	80	2⅞ "									
"	"	32	2⅝ "	3,910	5,393	9,303	26	210	873	1.587	.0334	Single line. Oils, etc., same.
"	"	48	2⅜ "									
"	"	32	2⅛ "									
July, 1871	"	10	2⅝ "									ORDINARY OILING.
"	"	48	2⅜ "	1,289	1,456	2,745	12	150	275	.499	.0640	Single line. Oiled in ordinary way, daily. Tallow in Boxes, as safeguard in case of Heating.
"	"	32	2⅛ "									
"	"	Similar line			1,006	2,295	12	150	217	.394	.0610	" " " "
"	"	34	2⅜ "									
"	"	32	2⅜ "	1,484	1,736	3,220	13	150	291	.537	.059	" " " "
"	"	32	2⅛ "									
"	"	10 ft. 4 in.	2⅞ "	2,336	2,999	5,335	24	211	793	1.442	.0759	Single line. Had been oiled A. M., test at 11 A. M.
"	"	176	2⅛ "									
"	"	Similar line		2,336	2,999	5,335	24	211	679	1.234	.0650	" " Taken just after oiling.
Jan., 1872	Whittenton Mills	Ab't 200 ft.	2¼ "	2,700*	....	....	.....	155	314	.571	....	
Feb., 1872	Langdon Mills	24 ft. 10 in.	2⅛ "	295	350	645	4	210	143	.260	.114	" " Sprung in centre by pull of Belt.
Mar., 1872	Haydensville	42	2⅛ "	428	....	....	5	120	147	.267	....	
April 1, '72	Salmon Falls, N. H.	9	4 "	3,151	2,354	5,805	31	211	857	1.558	.0714	" " Tallow in Boxes, taken at noon; had been oiled early in A. M.
		231	2⅛ "									
April 2, '72	"	Same shaft		3,151	2,354	5,805	31	211	619	1.120	.0516	Tallow removed from Boxes, and sponge saturated in oil substituted. Time of testing as before.
May, 1872	Rockport, Mass.	9 ft. 10 in.	3 "									
	"	32	2⅜ "									
	"	32	2⅛ "	1,987	2,000*	3,987	30	185	376	.685	.0568	Oiled daily in usual manner.
		96	1⅞ "									
May, 1872	Masconom't M'l Newburyport	10 ft. 8 in.	3 "									
		24	2½ "									
		127 ft. 1 in.	2¼ "	3,554	4,268	7,882	37	245	1028	1.870	.0585	Oiled daily in usual manner.
		64	2 "									
		64 ft. 3 in.	1¾ "									
Nov., 1872	Paterson, N. J.	100	1⅝ "	unknown	unknown	unknown	11	170	184	.335	....	Dreyfuss Oilers.
Nov., 1873	Granite Mills, Fall River	Ab't 200 ft.	2 "	"	"	"	......	200	302	.549	....	" "

* Estimated.

2. "**The Operative Mechanic and British Machinist,**" by John Nicholson, Esq., C. E., American ed., 1826, makes mention of "The *fast and loose pulley*, which is represented in Fig. 1. B is a pulley firmly fixed on the axle A, and C a pulley with a bush, so that it can revolve upon the axle A without communicating motion to it. This contrivance is remarkable for its beautiful simplicity, as the axle A can be thrown in and out of gear at pleasure, without the least shock, by simply passing a strap from the one pulley to the other."

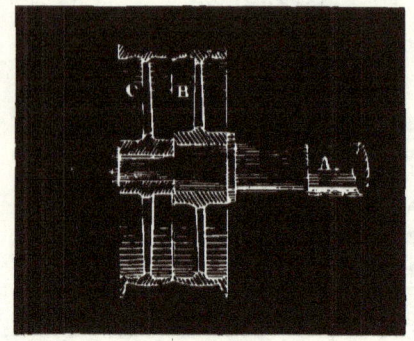

Fig. 1.

The above work also describes the grooved sliding clutch pulley and the "tightening roller," or pulley for increasing the tension of a belt to the tightness required for driving.

From "The Encyclopædia of Arts, Manufactures, and Machinery," by Barlow & Babbage, London, 1848.

3. "In the numerous contrivances for the purpose of engaging and disengaging machinery, the particular object aimed at has been to communicate motion without a shock; as, in consequence of the inertia of bodies, or their disposition to remain in the state in which they are, the parts of Machinery, when acted upon too suddenly by a moving power, are liable to fracture and derangement.

"The inventions for this purpose may be divided into two classes, viz.: when the motion is communicated by bands, belts, or chains, and when it is communicated by wheel work; the former generally possesses the advantage of bringing on the motion more gradually, although the application to large Machinery is attended with inconvenience."

Fig. 2 represents a contrivance termed the fast and loose pulley, which is remarkable for its simplicity. It is attended with no shock, and is considered the most perfect method yet invented for the purpose where it can be applied.

It consists simply of two pulleys, B and C, one being fixed on the axle, and the other loose, and the belt or band which conveys the motion may be shifted at pleasure, either upon one pulley or the other; by that means putting in or out of motion the axle A.

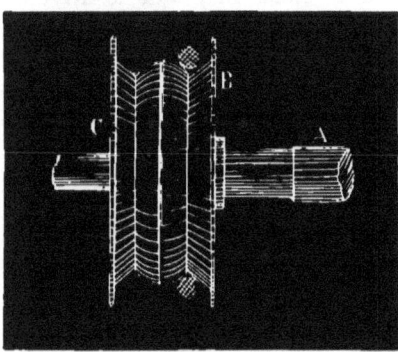

Fig. 2.

It may be proper here to mention that, in order to make a belt run properly on a pulley, it is necessary to have the rim a little rounded or swelled in the middle. The belt always inclines to that part of the pulley which is of greatest diameter.

"This curious property is of great practical use, and, until it was known, it was found very troublesome to prevent belts slipping off the pulleys."

From "The Science of Modern Cotton-Spinning," by Evan Leigh, C. E., Manchester, England, 1875.

4. Mr. Leigh advises us to "seize upon truth where'er 't is found," and we therefore transfer a few of the fine specimens he has given us in his excellent work on "Cotton-Spinning."

"Simplicity, which in all mechanism is desirable, is more especially so in mill gearing. Heavy, cumbrous, and rumbling gearing should be avoided as much as possible. It is always disagreeable and dangerous, because the breaking of a cog, from any hard substance getting in the wheels, often causes a fearful crash. The constant greasing required is also expensive, and produces much filth and unpleasant smell."

"Looking around, one must accord to America the honor of many useful inventions. Give a man a certain thing to do with limited means, and his ingenuity suggests a way of doing it; so in America heavy gearing is almost entirely discarded, and broad belts or straps substituted.

"Much may be said *pro* and *con* on this subject. The wisdom or folly thereof depends upon the mode of application.

"When *properly applied,* there is no question that the noiseless and practical way in which belts do their work is preferable to gearing.

"If belting be *improperly* applied it makes all the difference. A main driving-belt, to be rightly applied, should go through 3000 or 4000 feet of space per minute, and be sufficiently wide to drive all the machinery and shafting it has to turn quite easily, when running in a slack state. A wide belt moving with that velocity, on drums of large diameter, possesses enormous power. After a new belt has been tightened up once, it should work many years without again requiring tightening, and will do so if properly applied and made of good material, saving, in the meantime, all the grease and labor of putting it on which gearing requires, to say nothing of the horrid noise which heavy gearing makes. In America the main driving-belts are open straps, worked in this manner and neatly boxed up, so that nothing is seen, nothing heard; whilst in this country the disagreeable rumbling noise of the heavy gearing of some mills can be heard, in country places, a mile off."

Ask "John Bull" whether he would prefer driving his machinery by gearing or belting, and he will shake his head and tell you he never minds the noise; he likes to be *sure*. "John Bull" is generally a shrewd fellow, but, as a rule, he does not at present understand what a belt is capable of when it runs through 3000 or 4000 feet of space per minute.

"Driving by belt or band has ultimately to be resorted to in all cotton-spinning machinery; therefore the question of certainty goes for nothing when properly done, as one way is just as certain as the other. To apply belting to slow-running shafts would be simply ridiculous. As speed has finally to be attained, it should be gained, as much as possible, at once from the periphery of the fly-wheel of quick-running engines. The speed of engines and diameter of fly-wheel should be so adapted to each other that the rim of the latter will give off a speed of 3000 to 4000 Fpm at least; it being borne in mind that the power of a belt is exactly as the speed or space it runs through per minute. For example, a belt or strap of 6 inches wide, running through 4000 feet of space per minute, will turn as much machinery or give off as much power as a belt of 24 inches will do that moves only at the rate of 1000 Fpm. Therefore the quicker the speed the less is the expense of the belt.

"What has been said about slack straps applies to all heavy running machinery throughout. It will be found also a great saving of power to have larger pulleys than is usual both on the shafting and frames, so that the straps can do their work easily. This saves wear and tear also to a great extent. When a strap is obliged to be tight in order to do its work, it pulls down at the shafting and up at the pedestals of the frame it is driving, thereby wearing out the steps, consuming more oil, and absorbing power, besides pulling itself to pieces, in addition to which it slips and loses time.

"After a mill is settled to work there ought to be scarcely any piecing or tightening of straps, and if the precautions above enumerated be taken at the commencement, production will go on with greater regularity, and a very large saving in the aggregate will be effected.

"As an example of what may be done with belts, the first which the author saw in an American factory was one driving 140 horse-power from a drum of 9 feet diameter, and going at the speed of 130 Rpm, and driving a shaft which had a drum of 7 feet diameter upon it. The strap was 24 inches wide, of double leather sewn together. It was asserted that this strap had run for seven years without piecing or tightening, having been tightened only once since it was newly put on. Being surprised at this statement, further inquiry was made in different mills, which fully confirmed what had been said as to the durability and ease with which these large belts do their work. If reflected upon, what an impressive lesson this teaches!

"How delightful it would be in a mill if all the straps would run so long without piecing or giving trouble! Yet so it would be were the same conditions observed. How much we vary from those conditions will be seen upon examination. For instance, we often see carding-engines with pulleys on the main cylinders 12 inches diameter, running at a speed of 140 Rpm, which is equal to barely 440 Fpm of space through which the strap moves, whilst the big driving-strap, above alluded to, goes more than eight times faster, being at the rate of over 3673 Fpm. The straps which drive frames and other machinery are not much quicker than those which drive the cards.

"Therefore the lesson taught by the big belt is imperative, namely, that there should be very light shafting run at a very quick speed, with larger drums and pulleys; then very little would be heard of strap piecing, or wear and tear of belts, working with less power and steadier production all the while. 'Our American cousins' have taught some good things, and this is one of them.

"In new countries men have new ways, and do not fix their principles by inheritance, as they do in old countries.

"In some American factories, one long belt is made to run the whole round, from bottom to top of mill, turning every main shaft, passing, where necessary, over carrier-pulleys, and working its way to and fro. This is not a good plan, as the belt is required to be of enormous length, and, having all the stress upon it, is required to be sufficiently wide to take off all the power. It is likewise more costly than necessary, besides having other disadvantages. The simplest

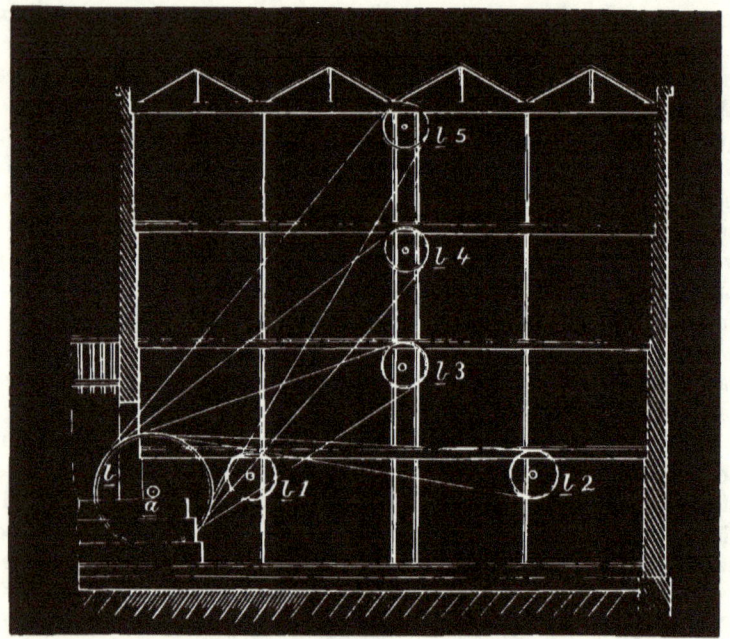

Fig. 3.

and best method of driving by belt, also the cheapest and most durable, is to convey the power from the main driving shaft direct to each room by a separate strap; and if more than one shaft is wanted in any one of the rooms, to drive it from the other direct by a separate strap, apportioning the width of each strap to the power it is required to drive, and where a belt is necessarily short, allowing a little extra width.

"The example below shows the best method of driving a mill of four stories, in which two shafts are required in the bottom room, which may be driven direct from the first strap, as will be seen in Fig. 3, in which a represents the main driving shaft, running 80

Rpm, driven direct from the steam-engine or other motor; b is a strong, well balanced drum, of 15 feet diameter, and about 3 feet wide, keyed on the shaft a, which, in making 80 Rpm, gives off a speed on its periphery of 3768 Fpm; $b\,1, b\,2, b\,3, b\,4, b\,5$, are strong pulleys about 6 feet in diameter, and 6 inches wide, keyed on the respective shafts they have to drive, which will in this case make 200 Rpm, but may, of course, be varied to run faster or slower by putting on smaller or larger pulleys; but whatever else is done, the speed of the straps must be kept up, for in that lies the whole secret of success in belt driving.

"The shaft a, if the power be steam, will be the crank shaft, and the drum upon it will act as fly-wheel, and have great centrifugal force, without being heavy, by reason of its speed. The pulley must be turned a little convex at the top, where every strap comes upon it, having a little flat space of 3 or 4 inches between every hump, to admit of boxing up each belt separately, and insuring them running in their proper places.

"Should any belt break, as it runs in a separate box all the way up, it cannot in any way interfere with the others. When a belt wants piecing or tightening (a very rare occurrence) the ends are fixed in cramps, which are drawn together by screws.

"As the straps or belts in the above example are supposed to be 6 inches wide, each belt is capable of driving horse-power, as the following rule for calculating the power of belts will show.

Rule to find the Horse-power that any given Width of Double Belt is easily Capable of Driving.

"Multiply the number of square inches covered by the belt on the driven pulley by *one-half* the speed in feet per minute through which the belt moves, and divide the product by 33,000, the quotient will be the horse-power.

Rule to find the Proper Width of Belt for any given Horse-power.

"Multiply 33,000 by the horse-power required, and divide the product first by the length in inches covered by the belt on the driven pulley, and again by *half* the speed of the belt.

"If these rules, which the author has devised after very careful study of the subject, be compared with the single straps as at present used in cotton-mills, it will be found that they considerably overshoot the mark; yet, theoretically, single belts, being so much weaker and more liable to stretch than double ones, ought to have less strain upon them. The secret of the wide double driving belts running so mysteriously long without attention, will at once be seen, when it is considered that *single* belts are, as generally used, made to drive three or four times more than they ought to do for their width and speed.

"For existing establishments, where it is not convenient to alter the speed of shafting or size of drums, in driving machines with single straps, the following will come nearer to actual practice.

Rule to find the width of Belt for any given Horse-power.

"Multiply 33,000 by the horse-power required, and divide the product, first by the length in inches covered by the belt, and again by its speed.

"This, and more than this, is what *single* straps are made to do when driving machinery. Comparatively, then, the strong double belts, working as per first rule, have exceedingly light work, which can be done with great ease while running in a slack state. Hence their durability; and the nearer a user of belts can approach the rule given for double belts, the longer his straps will last."

Duration of Belts.

5. "The power is taken from the jack-shaft from pulleys 12 feet diameter, 30 inches face, and communicated direct to main lines in each room by belts 24 inches wide, double leather.

"These run 3780 Fpm, and are six in number, driving from highest to lowest power 175 horse each; required tightening three or four times in the first three months, and never since. With proper care, will last 20 years; with English leather, would last much longer."—*Harmony Mill, Cohoes, N. Y., E. Leigh.*

"In Pittsburgh, a 20-inch gum belt has been in constant use $10\frac{1}{2}$ years, and notice is given of three 18-inch belts which have been running for 9 years."—*English paper.*

Example.

6. A 12-inch belt, driving a $5\frac{1}{4}$ feet pulley, turning 45 Rpm, will carry away 12 horse-power.

This is the equivalent of 64.8 square feet of belt per minute per horse-power.

Rule for Horse-power of Belt.

7.
$$W = \frac{350\ HP}{D \times Rpm.}$$

In which $W =$ width of belt in inches.

$D =$ diameter of pulley in feet.

This gives a strain of 30 lbs. per inch of width of belt, and 91.63 square feet of belt transmitted per horse-power per minute.

Rule.

8. "An empirical rule for ascertaining the width of belts that we know to be in use by some good practical men is as follows:

$$B = \frac{314\ N}{n\ d}$$

In which $B =$ width of belt in inches, thickness taken at $\frac{3}{16}$ inch.

$N =$ number of horse-power.

$n =$ number of revolutions per minute.

$d =$ diameter of pulley in feet.

Which equals 82.163 square feet of belt in motion per minute per horse-power."—*Lond. Mech. Mag., March, 1863.*

Example.

9. "A leather belt, $19\frac{1}{2}$ inches wide, is driven by a drum 11 feet in diameter, having iron arms and wooden lagging, and making 92 Rpm; consequently, the belt moves at the rate of 3179 Fpm. The amount of power transmitted by this belt is estimated at 175 horse-power, corresponding to a tension of the tight side of the belt of not less than

$$\frac{175 \times 33{,}000}{3179} = 1817\ lbs.$$

The pulley driven by the belt is 6 feet in diameter, and is entirely of iron; the peripheries of both drum and pulley are covered with leather. The belt is made of two thicknesses of leather, cemented together, and is about $\frac{3}{8}$ inch thick; it was slightly greased on the inside with a mixture of tallow and neat's-foot oil. The slack side running upwards nearly vertically."

The data above give 29.5 square feet of belt per minute, per horse-power, and a tension of 93.18 lbs. per inch wide.

Example.

10. "At a speed of 1800 Fpm on pulleys over 36 inches diameter, every one inch wide will give 2 horse-power."

This equals 75 square feet per minute per horse-power.

Example.

11. A certain 6-inch × 12-inch cylinder horizontal engine, with plain slide valve, arranged to cut off at ⅝ths the stroke, works under 80 pounds of steam, has a 7-inch belt on a 4-feet pulley on engine shaft making 100 Rpm, and drives a 30-inch pulley on the "line" shaft about 4 feet above the cylinder.

A 24-inch pulley on the other end of this "line," carrying a 7-inch belt, with a "half-twist," drove a 10-inch pulley on a shaft about 18 feet beneath the former.

The 10-inch pulley shaft, in its turn, drove a certain machine, which consumed more power than the engine was capable of giving.

The result was, the 7-inch belt from the line to the 10-inch pulley would continue to slip, even when very tight and well covered with rosin, while the 7-inch belt from the line to the pulley on the engine shaft would hold firmly to its pulleys, and stop the engine.

Crafts & Filbert's Loose Pulley.

12. "The loose pulley shown herewith was patented Feb. 29, 1876.

"Connected with machinery run at a high rate of speed, loose pulleys have been a source of continual annoyance. They not only require constant attention, but are hard on the belt and take much oil. Fig. 4 represents the loose pulley 2 inches smaller than the tight pulley and provided with a conical flange for the belt to run up on.

Fig. 4.

"The difference between the pulleys will slacken up the belt 3 inches, taking the strain off the belt and the friction from the pulley, and allowing the belt to contract when thrown off the tight pulley. The belt has a chance to give and take, as it is always in a slack condition when on the loose pulley, and should contract enough to keep it tight for a long period; and whatever will relieve the belt of strain will add to its durability. There is considerable wear and tear on a belt in shifting it with the ordinary pulley. In starting a heavy machine it is necessary to hold the belt on with the shifter until the machine is under full headway. During that time the edge of the belt is rubbing against the shifter, tearing up the corners of the laps and wearing away the edge of the belt. But the flanged pulley should require very little aid from the shifter. When the belt touches the flange it immediately climbs to the tight pulley, and remains there — starting the machine quickly." — *The Polytechnic Review, Philadelphia.*

Example.

13. The width of a certain belt is 18 inches, speed of same 1500 Fpm, angle of belt with horizon 45°, distance between centres of drums 25 feet, diameter of driving drum 8 feet, of driven drum 4 feet.

When this belt transmitted 20 horse-power it worked quite freely and well; when the power was increased to 28 horse, a tightener had to be applied.

From the above data we deduce the following formula:

$$\frac{HP \times 3\frac{3}{5}}{Diam. \; small \; pulley \; in \; feet} = width \; of \; belt \; in \; inches.$$

If we consider this belt as transmitting $22\frac{1}{2}$ horse-power, we shall have a constant travel of 100 square feet of belt per minute per horse-power, assuming the above conditions.—*Appleton's Dict. of Mech.*

Example.

14. "A 4-horse engine transmits its power through a leather belt over a cast-iron pulley 4 feet diameter, running 100 Rpm, and embracing .4 of its circumference.

"In this example the thickness of belt is taken at .15 inch, and the strain at 210 lbs., which gives 4.67 inches for width of belt, and 122.26 square feet of belt per minute per horse-power, or 101.88 square feet, when .5 of the circumference of pulley is embraced, using Morin's ratio of 2 : 2.4. If thickness of belt be taken at $\frac{3}{16}$ inch, and half the circumference be embraced by belt, then we have 81.375 square feet per minute per horse-power.

"An 11-inch belt on a 4-feet pulley running from 1200 to 2100 Fpm will transmit the power of a double steam-engine with 6-inch

× 11-inch cylinders, 125 Rpm, under a steam pressure of 60 lbs. per square inch."—*Haswell*, 1867.

Example.

15. A single leather belt in good driving condition has been frequently used in testing car-wheel forcing-presses. Under the following conditions this belt was repeatedly found to just do the work: Belt 6 inches wide, pulley 24 inches diameter, of iron, smooth turned, crank 2 inches radius, plunger $1\frac{8}{16}$ inch diameter in a $1\frac{1}{4}$ inch barrel, hydraulic pressure 7000 lbs. per square inch; hence we have:

$$\frac{(area\ of\ 1\tfrac{1}{4}\ inches - area\ of\ \tfrac{15}{16})\ 7000}{6 \times 6} = 104.36.$$

From this we may safely conclude the maximum driving force of a single leather belt to be near 100 lbs. per inch of width.—*A. B. Couch.*

16. Molesworth's Pocket-Book of Engineering Formulæ. 17th Edition, London, 1875.

Leather Belting.

$V =$ velocity of belt in feet per minute.
$HP =$ horse-power (actual) transmitted by belt.
$S =$ strain on belt in lbs.
$W =$ width of single belting ($\frac{3}{16}$ thick) in inches.
$S = x + k\,x.$
$W = .02\ s.$
$x = \dfrac{33{,}000\ HP}{V}.$

$k = 1.1,\ .77,$ and $.62$ when portion of driven pulley embraced by belt $= .40, .50,$ and $.60$ of the circumference, respectively.

For double belting the width $= W \times .6.$
Approximate rule for single belting $\frac{3}{16}$ thick:

$$W = \frac{1100\ HP}{V}. \quad \ldots \quad (a).$$

"The formulæ above apply to ordinary cases, but are inapplicable to cases in which very small pulleys are driven at very high velocities, as in some wood-cutting machines, fans, etc. The acting area of the belt on the circumference of the driven pulley being so small that either great tension or a greater breadth than that determined by the formula is required to prevent the belt from slipping.

"In such extreme cases of high-speed belts, find the breadth of the first-motion belt, by the formula for ordinary belting above (a), then if—

$A =$ acting area of first-motion belt.
$v =$ velocity of first-motion belt.
$a =$ acting area of high-speed belt.
$V =$ velocity of high-speed belt.

$$a = \frac{Av}{V}.$$

"The acting area of either belt $= l \times b$.

Where $l =$ length of circumference of driven pulley embraced by the belt,

$b =$ breadth of the belt.

$\therefore b = \dfrac{a}{l}$ in the case of the high speed belt.

"If there is no first-motion belt exclusively for the machine, it will be easy to suppose a hypothetical case from which the breadth of the high-speed belt may be calculated."

Rule (a) is the equivalent of $91\frac{2}{3}$ square feet of belt per minute per horse-power.

Rule.

17. Several years of satisfactory use of a table which purports to give reliable data about belts, would seem to clinch the truth of size for power, and to show its value, we take from it a 6-inch belt running 2200 Fpm, and find it capable of doing $12\frac{1}{4}$ horse-power, which is the equivalent of 89.8 square feet per minute per horse-power.

Rule.

18. A rule to find the power of a belt may be stated thus: Divide 1070 into the product of the belt's width in inches, by its velocity in Fpm, which proves it the equivalent of 89.17 square feet of surface velocity.

Rule.

19. Another takes this shape:

$$W = \frac{HP\ 5400}{V\ d}.$$

In which W = width of belt in inches.
" HP = horse-power.
" V = velocity of belt in feet per minute.
" d = diameter of smaller pulley in feet.

This rule is handed down to us by a good engineer, who used it with success for many years; it has also the advantage of giving a margin of 25 per cent. of adhesion before slippage will take place.

Example.

20. An horizontal non-condensing engine, with a cylinder 12 inches diameter, 30 inches stroke, running 66 Rpm, under 72 pounds of steam in boiler, and an average pressure of 19.7 pounds of steam on piston, has a $13\frac{1}{2}$-inch belt on an 8-feet fly-wheel pulley, which runs over a 4-feet pulley on a shaft 18 feet vertically above. Speed of belt 1658.58 Fpm. On one occasion the indicator showed 21 horse-power transmitted; this would give 88.85 square feet of belt per horse-power per minute.

Example.

21. An engine similar to that above, with 10 inches × 24 inches cylinder, making 80 Rpm, under 100 pounds of steam in the boiler, and showing by the indicator a constant work of 33 horse-power, has a $11\frac{1}{2}$-inch belt on a 5-feet pulley on engine shaft. This belt is crossed and runs over a 34-inch pulley on the "line" shaft 8 feet above and 18 feet distant. Number of square feet of belt transmitted per horse-power per minute 36.5. Speed of belt 1256 feet per minute.

Example.

22. An engine with steam cylinder 15 inches diameter, 36 inches stroke; fly-wheel pulley 12 feet diameter, carrying a $17\frac{1}{2}$-inch belt, which passes over a 5-feet pulley, 6 feet above and 18 feet 6 inches distant; top fold of belt slack.

Engine makes 48 Rpm under 65 pounds of steam, and by the indicator shows a work done of 40.7 horse-power. Speed of belt 1809.12 per minute, and square feet of belt transmitted 64.61 per horse-power per minute.

Example.

23. An horizontal non-condensing engine, with 11 inches × 30 inches cylinder, arranged with two steam and two exhaust valves of the double beat balanced Cornish style, each operated by a cam, the two former under the control of the governor. A 10-feet fly-wheel pulley carries a 20-inch single leather belt. On the line shaft 6 feet above and 15 feet distant is a 5-feet pulley, the belt passes over this pulley, the top fold running slack. Under 80 pounds of steam in the boiler a pressure of $60\frac{1}{2}$ pounds per square inch on the piston is maintained to the point of cut-off, which was one-third the stroke on one occasion when the indicator showed a work of 29.27 horse-power; speed of engine 56 Rpm. The load of this engine is very variable; the average of a number of cards taken shows 25 horse-power. Speed of belt 1758.4 Fpm, and number of square feet of belt per horse-power per minute transmitted 117.2.

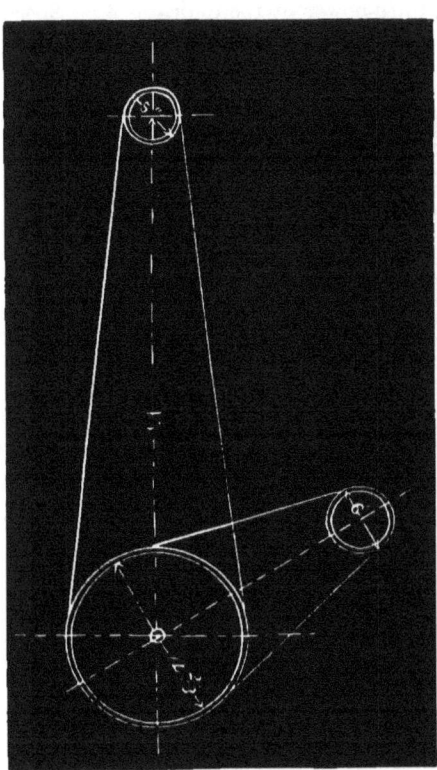

Fig. 5.

Example.

24. A 20-inch × 48-inch cylinder horizontal non-condensing engine has a 16-feet fly-wheel pulley carrying a 24-inch belt, and runs about 50 Rpm. This belt drives a 5-feet 4-inch pulley, 32 feet distant in an angle of about 40°, the top fold slack. By the indicator the engine is working up to 150 horse-power. Speed of belt 2513 Fpm, and square feet transmitted per minute per horse-power equals 33.5.

Example.

25. An 18-inch × 36-inch engine, having a 14-feet 8-inch fly-wheel pulley, making 52 Rpm, carries two belts (see Fig. 5); a 15-inch running over a 5-feet pulley directly above, 45 feet distant, and a 16-inch, running over a 6-feet pulley 20 feet distant, at an angle of about 30°.

This engine, under a boiler pressure of 85 lbs., shows by the indicator, 114.6 horse-power. Speed of belts 2392 Fpm, and square feet of belt transmitted per minute per horse-power 54.

Example.

26. A 14-inch by 36-inch cylinder engine has a 12-feet fly-wheel pulley carrying an 18-inch belt, tight fold below, at an angle of about 30°. Pulley on line shaft 6 feet diameter, and about 25 feet distant. Speed of engine 56 Rpm. Horse-power by the indicator 49. Speed of belt 2111.2 Fpm, and square feet per horse-power per minute 64.63.

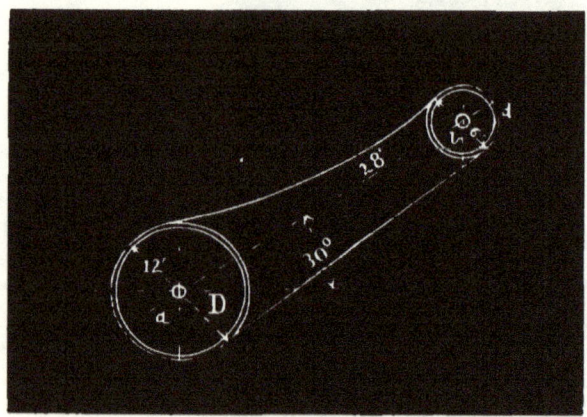

Fig. 6.

From J. W. Nystrom's "Elements of Mechanics." Philadelphia, 1875.

27. "The best and simplest mode of transmitting motion from one shaft to another, is by a belt and pulleys, which is very extensively used, and it gives the smoothest motion. The motion is transmitted by the frictional adhesion between the surfaces in contact of the belt and pulleys, for which reason that friction must be greater than the tension of the belt; otherwise the belt will slip and fail to transmit all the motion due from the driving pulley. There is always some slip in belt and pulleys, for which reason that mode of transmission is not positive or exact, and cannot be used where precise motions are required.

"Fig. 6 represents a belt transmitting motion between two parallel shafts, a and b. If the motion is transmitted from a to b, the pulley D is called the driving pulley, and d the driven pulley. The diameters of the pulleys can be of any desired proportions to suit the work of the machine.

D = diameter, and R = radius in inches of the driving pulley.
d = diameter, and r = radius of the driven pulley.
L = length, and B = breadth of the belt in inches.
F = force of tension in pounds of the pulling side of the belt.
f = force of tension on the slack side.
V = velocity of the belt in feet per second.
S = distance in inches between the centres of the two pulleys.
N and n = numbers of revolutions per minute of the respective pulleys, D and d.
ϕ = angle in degrees occupied by the belt on the small pulley.
HP = horse-power transmitted by the belt.

"Revolutions $N : n = d : D$, diameters. The revolutions are inverse as the diameters.

$$N = \frac{n\,d}{D}, \quad n = \frac{N\,D}{d}, \quad d = \frac{N\,D}{n}, \quad D = \frac{d\,n}{N}.$$

"The force bearing in the journals of each shaft is $F + f$, or the sum of the tensions of each side of the belt.

"The force which transmits the motion is $F - f$, or the difference between the two tensions.

"The effective power transmitted is equal to the product of the transmitting force and the velocity, and this power divided by 550 gives the horse-power.

"The length, L, of the belt will be found by the following formula:

$$L = \pi (R + r) + 2 \sqrt{S^2 + (R - r)^2}.$$

"When the diameters of the pulleys are alike, or $D = d$, the length of the belt will be

$$L = \pi D + 2 S.$$

"The slip of belt on equal pulleys has been found by experience to vary between 2 and 3 per cent. under ordinary circumstances.

$n =$ theoretical revolutions per minute of the driven pulley.
$n' =$ actual revolution after the slip is deducted.

$$n' = \frac{n\,\phi}{182}.$$

Formulas for Leather Belts on Cast-iron Pulleys.

Force and Power of Transmission.   **Breadth of Belts from Experiments.**

1. $F - f = \dfrac{126500\ HP}{d\,n}$

9. $B = \dfrac{777600\ HP}{n\,d\,\phi}$

2. $F - f = \dfrac{126500\ HP}{D\,N}$

10. $B = \dfrac{2.4\ F}{d}$

3. $F - f = \dfrac{550\ HP}{V}$

11. $B = \dfrac{432\ F}{d\,\phi}$

4. $HP = \dfrac{V(F-f)}{550}$

12. $B = \dfrac{4320\ HP}{n\,d}$

5. $HP = \dfrac{d\,n\,(F-f)}{126500}$

13. $B = \dfrac{7.8\ F\,V}{n\,d}$

6. $HP = \dfrac{D\,N(F-f)}{126500}$

14. $HP = \dfrac{B\,n\,d}{4320}$

7. $V = \dfrac{d\,n}{230} = \dfrac{D\,N}{230}$

15. $HP = \dfrac{B\,n\,d\,\phi}{777600}$

8. $n = \dfrac{126500\ HP}{d\,(F-f)}$

16. $F = \dfrac{B\,d}{2.4}$

Treatment and Condition of Leather for Belts.

28. "I stuff my belts with a composition of two pounds of tallow, one pound of bay-berry tallow, and one pound of beeswax, heated to the boiling-point, and applied directly to both sides by a brush, after which the belts are held close to a red-hot plate to soak the beeswax in, which does not enter the pores of the leather from the brush."

"Care must be taken to have the leather perfectly dry to prevent burning. I placed a kettle of the composition over a blacksmith's fire, and after melting it, I put in a coil of 2-inch belting about 16 feet long, and boiled it 45 minutes in the greatest degree of heat I could produce by blowing the fire continually, and the belt when taken out was not in the least injured by the heat of the composition. I then tried a piece of belting damped with water, and found it burnt and crisped in less than half a minute."

"The application of neat's-foot oil to belts, opens the pores of the leather, and destroys the adhesion of its parts, and in a very short time renders it flaccid and rotten, and a belt will not last half so long stuffed with oil as with the composition above named. Belts stuffed with the composition are impervious to water, and will run well for six months."—*I. H. B., Frank. Ins. Jour., June,* 1837.

"Fat should be applied to belts once every three months. They should be first washed with lukewarm water, and then have leather-grease well rubbed in. A good leather-grease may be made from fish-oil, 4 parts; lard or tallow, 1; colophonium, 1; wood-tar, 1."—*Workshop Receipts, by Ernst Spon, London,* 1875.

Leather belts may be kept in good working condition by the judicious use of fish-oil, mixed with spent grease of journal-boxes, and also by neat's-foot oil, which may be applied by a brush two or three times swept over after the belt has been soaked some ten minutes in water.

Whitaker's castor-oil dressing is one of the best modern adhesives for leather belts.

"In order that belting of cotton or linen should have both strength and flexibility, together with increased adhesive power, they should be thoroughly soaked in linseed-oil varnish. If the belting be new, the varnish may be applied with a brush, until no more will be taken up, whereupon it may immediately be used without any preparatory drying. After having been in use for some weeks, a second application of the varnish should be put on. Cotton or linen belting thus prepared will neither contract nor stretch, and will always be pliable and unaffected by change of temperature. The adhesion of the belt to the pulley is likewise increased by the varnish, while steam and acid fumes have no effect upon the belting at all."—*Polytechnic Review.*

Belts stuffed with tanners' dubbing on the flesh side, will become as smooth as the hair side, and will outlast six belts which are run on the hair side exclusively.

Frequent application of neat's-foot oil promotes regularity of speed, durability of leather, and economy of use.

Three times the adhesiveness is gained by softness and pliableness of belting leathers over those which are dry and husky.

A right good way to oil a belt is to unreel from one coil to another, allowing the loose fold to draw through a pot of oil, with rubbers at the outgoing part to wipe back the superfluous grease.

"Keep belts clean by washing them with warm water and soda, scrape them well and apply freely the spent grease of journal boxes."

Influence of the Thickness of Belts.

29. "When bent round the circumference of a wheel, the outer parts of the belt are distended, the inner parts relaxed; and supposing the section of the belt to be rectangular, the amount of force expended in making these changes is proportional directly to the breadth, to the square of the thickness, and inversely to the diameter of the wheel. Hence if two belts be of like strength, but the one broad and thin, the other narrow and thick, the amounts of force expended in bending them must be proportional directly to their thicknesses, and hence the advantage of using broad thin belts."

"The practice of strengthening belts by riveting on an additional layer must be exceedingly objectionable: indeed, it is difficult to see how any additional strength is gained, for the outer layer must be tight when on the wheel, and slack when free, so that in reality, the strength of only one layer can be available, the parts of the compound belt are puckered and opened alternately, as evinced by the crackling noise."

"The proper procedure is to increase the breadth of the belt."

"So far as we have yet seen, it is preferable to use heavy belts."— *Prac. Mech. Journal, November*, 1866, *p.* 240.

Variation of Speed.

30. From experiments made, it has been ascertained that about two revolutions per hundred are lost in the transmission of motion by a belt. In ordinary practice this would be a slight loss, and would in no wise interfere with the usual manufacturing processes, but where there is a long train of gear repeated from shaft to shaft by belts, the loss becomes serious.

It is clear if the co-efficient of loss by slippage be .98 for a single pair, which has been verified with great certainty by varying the tensions of the same belt, it will become equal to the successive powers: .98, .96, .94, .92, .90, and so on; so that after a succession of five speeds the loss amounts to $\frac{1}{10}$th of the calculated speed, and that at the end of thirty-four speeds the velocity will be reduced to half.

From these considerations it appears that where it is required to transmit speeds as near determinate as may be, by means of bands and pulleys it is necessary to increase the diameter of the driving pulley by its fiftieth part, or diminish the driven pulley in the same ratio.—*See Lond. Mech. Mag., March,* 1863.

Prof. L. G. Franck, in Jour. of Franklin Inst. (May, 1875), gives a theory of the tension of belts, from which we take the following.

31. "The tension of the belt is counteracted by the cross-section of the belt, that is $2.4S = wtK$, where w denotes the width of the belt, t the thickness, both given in inches, and K the number of pounds which one square inch of belt will fairly resist, found by experiment. From equation (5), in article referred to, we have

$$S = \frac{M}{R},$$

which, if introduced into the above equation and solved with respect to w, will give

$$w = \frac{2.4M}{tKR}. \qquad (1)$$

"In general M is not directly known, but the number of horse-powers the pulley shall transmit is given, and the number of Rpm of the pulley, or the number of feet that the belt travels per minute. From these data we are enabled to express S. Putting formula (1) in the form

$$w = \frac{2.4S}{tK}, \qquad (2)$$

the dynamical effect of S for n revolutions in one minute is expressed.

"Number of horse-powers $= N = \dfrac{2\pi R S n}{33000}$,

from which we get

$$S = \frac{33000 N}{2\pi R n},$$

which when introduced into equation (2) gives

$$w = \frac{2.4 \times 33000 N}{2\pi RntK}$$

where w denotes the width of the belt in inches, N the number of horse-powers, R the radius in feet, n the number of revolutions, t the thickness of the belt, and K the resistance expressed in pounds which one square inch of belt can fairly counteract. Taking $t = \frac{3}{16}$ inches, and for K, after Morin, 275 pounds. K depends on the quality of the leather, and ranges from 275 to 550 pounds per square inch. 275 pounds are recommended, however, by good authorities. Applying the latter we shall find, after reducing the above numerical values,

$$W = \frac{250 N}{nR}, \qquad (3)$$

where the numerical value is rounded off to an even number.

Example 1.

"A pulley of $1\frac{1}{2}$ feet radius makes 80 Rpm, having to transmit one horse-power. What should be the width of the belt?

Here $N = 1$; $n = 80$, and $R = 1\frac{1}{2}$.

Hence $W = \dfrac{250}{80 \times \frac{3}{2}} = \dfrac{25}{12} = 2\frac{1}{12}$ inches.

Example 2.

"A pulley of 3 inches radius makes 900 Rpm, and has to transmit 2 horse-powers. What should be the width of the belt?

$N = 2$; $n = 900$; $R = \frac{1}{4}$ ft. $\qquad W = \dfrac{250 \times 2}{900 \times \frac{1}{4}} = 2\frac{2}{9}$ inches.

"Solving equation (3) with respect to N, we find:

$$N = \frac{w\,n\,R}{250}. \qquad (4)$$

"Giving to w the exceptional width of 6 inches for a single belt, and assuming $n = 100$ Rpm, and further the radius of the pulley $R = 1$ foot, we find the number of horse-powers:

$$N = \frac{6 \times 100 \times 1}{250} = \frac{60}{25} = 2.4 \text{ horse-powers.}$$

"The number of horse-powers that are obtained is comparatively small, and it indicates that with pulleys and belts we cannot produce a very great effect unless we make the radius of the pulley very great, and apply an exceptionally great speed.

"I should mention that the above formulæ refer to pulleys with open belts only, and that it is of no consequence whether the radius and respective number of revolutions are taken from the greater or smaller pulley, as the numbers of revolutions are in an inverse ratio to the radii of the pulleys. That is:

$$\frac{R}{R_t} = \frac{n_t}{n}. \quad \text{Hence, } R\,n = R_t\,n_t.$$

"If the belt is made up of two layers or thicknesses, so that such a belt of the same width as a single one contains the double cross-section, we still may apply the upper formula, if we multiply it by $\frac{2}{3}$, owing to the greater stiffness of the belt.

Example.

"In order to transmit 4 horse-powers, we have a pulley 1 ft. 8 in. by 120 Rpm. What should be the width of the belt?

$$N = 4;\ R = 1\tfrac{2}{3} ft.;\ n = 120.$$

$$W = \tfrac{2}{3}\ \frac{250\,N}{n\,R} = \tfrac{2}{3}\ \frac{250 \times 4}{120 \times \tfrac{5}{3}} = 3\tfrac{1}{3}\ inches.$$

"For the single belt we should have received:

$$W = \tfrac{3}{2} \times 3\tfrac{1}{3} = 5\ inches."$$

NOTE 1.—"As great nicety is not required in these calculations, the co-efficient of friction may be taken in general as 0.25 and the arc covered by the belt as $\tfrac{4}{10}$ of the circumference of the smaller pulley."

NOTE 2.—The reader should observe that the figures of these examples show a velocity area of the belt of 130.83 square feet per minute per horse-power, which, with good single leather belts in fair working condition, is an allowance of nearly double the quantity needed under ordinary circumstances; for proof of which see other articles.

Running Conditions.

32. "The slack side on top, with large pulleys at high speed, is undoubtedly the true philosophy of transmitting power by belts."

Not speed alone but adhesive force must be gained to do work without destructive tightness or slippage of the belt, therefore, there should be a proper proportion of pulley diameter and belt contact.

Long belts are preferred to short ones, but care must be taken that the length be not too great.

We have a case in point where a 60-inch pulley at 45 Rpm drove a 15-inch pulley, about 50 feet distant, by an 11-inch belt, 109 feet long. The tops of the pulleys were nearly on the same level, and the belt was crossed.

This belt was continually flapping about, soon became crooked and irregular in width, and was frequently torn asunder at the lacings by excessive tension, and the whole arrangement proved very troublesome until changed to the following: The speed of the 60-inch and the diameter of the driven pulley were doubled, and the distance between their centres was reduced to 15 feet. The belt now drives with more power, gives greater regularity of speed, and works better every way.

Another case of excessive length, which has come under our notice, is that of an 11-inch open belt on a 4-feet pulley, running horizontally at a speed of 2261 Fpm over a 32-inch pulley, $30\frac{1}{2}$ feet distant. To prevent surging, this belt must be drawn and laced very tightly; too much so for economical running.

Some facts illustrating the evils of short belts were given to me by a friend.

A 30-inch pulley, running 127 Rpm, drove a 9-inch pulley by a 5-inch belt 14 feet long, the shafts were nearly in a horizontal plane, and the lower fold of the belt did the driving.

To do a certain work this belt frequently slipped, even when tightly drawn, so much so that it tore out at the lacings almost daily, and sometimes three times a day.

After much inconvenience it was changed for a belt 44 feet long; the 9-inch pulley shaft being removed horizontally to accommodate the increased length, while all the other parts remained the same.

It now performs most satisfactorily; it does not slip, holds at the lacings, and the slack fold above sometimes nearly touches the driving one beneath.

The 9-inch pulley shaft carries an 18-inch pulley, which, in its turn, drives a 7-inch pulley below on an Alden fan spindle.

"A belt adheres much better and is less liable to slip when at a quick speed than at a slow speed. Therefore it is better to gear a mill with small drums, and run them at a high velocity, than with large drums, and to run them slower; and a mill thus geared costs less and has a much neater appearance than with large, heavy drums; and in belting, if the power of a belt 18 inches wide were required, it would be better to put in two 9-inch belts than one so wide, owing to the greater inequalities of leather in such large pieces causing loss of adhesion."—*I. H. B., in Frank. Inst. Jour., June, 1837.*

Convexity of Pulleys.

33. Morin says. "The pulleys over which leather belts pass ought to have a convexity equal to about $\frac{1}{10}$ of their breadth."

London *Mech. Mag.* for March, 1863, says: "Belt pulleys should be made slightly convex, in a ratio of $\frac{1}{2}$ inch per foot of breadth." Molesworth says the same.

Another proportion is $\frac{1}{8}$ inch rise in 8 inches of width. Still another, $\frac{1}{8}$ inch to the foot.

"The rounding should be made as slight as is consistent with security, since every deviation from the cylindric form is accompanied by a loss of force."

"In their progress round the wheels, the different parts of the belt are stretched and relaxed alternately. Now, if the material were perfectly elastic, the force expended on the distension would be reproduced on the contraction of the belt. As the loss by this imperfect elasticity is not known, it will be enough to observe, for the present, that the loss of force will certainly be greater, the greater the disturbance of the particles — the higher the rounding of the pulleys."

Why a Belt Runs to the Higher Part of a Pulley.

34. Much disputation has been published in efforts to solve the question: Why does a belt run to the higher part of a pulley? There may be several causes, but the chief one is embodied in the following words: "That edge of the belt which is towards the larger end of the cone is more rapidly drawn than the other edge; in consequence of this the advancing part of the belt is thrown in the direction of the larger part of the cone, which obliquity of advance towards the cone must lead the belt on its higher part.

"It may here be observed that this very provision — the rounding of the face of the pulley — which keeps the belt in its place so long as the machinery is in proper action, tends to throw it off whenever the resistance becomes so great as to cause a slipping."

"To maintain a belt in position on a pulley, it is necessary to have the advancing part in the plane of the wheel's rotation."

Superiority of the Driving-Belt.

35. "There is no simpler or smoother means of communicating motion than that afforded by the noiseless agency of cords, bands, or straps. The very means by which the motion is maintained — namely, by the frictional adhesion between the surfaces of the belt and the pulley — is a safeguard to the whole mechanism, as, if any unusual or accidental obstruction should intervene, the belt merely slips, and breakage and accident are thus prevented." — *London Mech. Mag., March, 1863.*

"The facility with which this communication of rotary motion may be established or broken at any distance, and under almost every variety of circumstance, has brought the band so extensively into use in machinery, that it may be considered as one of the principal channels through which work is made to flow." — *Moseley.*

Covering for Pulleys.

36. Pulleys may be well covered in the following manner: — Take a piece of belt-leather of uniform thickness the width of pulley face, and of a length equal to circumference of pulley, plus the lap, but less $\frac{3}{4}$ inch for every foot of diameter of the pulley, then scarf and unite the lap so as not to increase the thickness when cemented together. When ready for use draw the covering on by means of iron hooks, observing to put hair side out and so that outer end of lap will not be raised when covering slips under the belt. Secure to the pulley rim by copper rivets, sinking heads beneath the driving surface.

The laps of all belts should be disposed in a similar way.

Effect of Disproportion of Connected Machinery.

37. Sometimes a belt works badly from causes outside of its own motion and proportions.

We have a case in practice which will forcibly illustrate this. A 46-inch pulley, on the "line" shaft, drives a 5-feet pulley on a 4-inch shaft, at the rate of 73 Rpm, by a 12-inch open belt. This shaft is 7 feet 8 inches below, and 2 feet aside of the "line" shaft, and carries an 8-feet fly-wheel of 3750 lbs. weight on its middle, and a crank, with a double pin, on its overhanging end, which latter is connected with and drives two marble saw-frames, one very heavy, the other of medium size. The belt runs slack and free, and has not been touched at the lacings during six months of very steady and satisfactory running.

Before the 8-feet fly-wheel was put on, a 6-feet fly-wheel, of about 1450 lbs. was used, which a long, troublesome experience proved altogether inefficient. The belt had to be run very tightly; it tore frequently at the lacings — even when the laced ends were doubled to make the stronger joining — and at all times while running the lack of momentum of the wheel caused unsteadiness of motion in the whole system of gearing in the mill.

Care of Belts.

38. In order to have belts run well, they should be perfectly straight and be of equal thickness throughout their length, have but one laced joint; but if circumstances, require any belts to be composed of several pieces, the ends should be evenly bevelled, and united by one or other of the permanent ways mentioned hereafter. The ends to be laced should be cut at right angles to the sides, the lace-holes formed by an oval punch, reducing the cross-section of belt the least, and the lacing put in evenly, of equal strength at the edges of the belt, and no crossing of laces on the inside. If copper or other rivets are used, the heads should be "let in" rather below the level of the inside surface of the belt to prevent contact with the pulley, and the washers placed on the outside surface. If the bevelled and lapped ends are sewed, the waxed ends should be "laid in" flush on the inside of the belt to prevent wear.

Belts and pulleys should be kept clean and free from accumulations of dust and grease, and particularly from contact of lubricating oils, some of which permanently injure the leather.

Quick motion belts should be made as straight and as uniform in section and density as possible, and endless if practicable, that is, with permanent joints.

Horizontal, inclined, and long belts give a much better effect than vertical and short ones and those which have the driving side below than otherwise.

Belts which run loose of course will last much longer than those which must be drawn tightly to drive, tightness being evidence of overwork and disproportion.

Tighteners should never be used; but when they must be, they should always be as large in diameter, and as free running as can be, and should be applied to the slack side of belts.

The most effective tightener is the weight of the belt on its slack side, which increases adhesion by increasing circumferential contact with the pulleys.

"Belts which run perpendicularly should be kept tightly strained, and should be of well-stretched leather, as their weight tends to decrease their close contact with the lower pulley." — *J. B. Hoyt & Co.*

"Belts of *coarse, loose* leather will do better service in dry, warm places; for wet or moist situations the *finest* and *firmest* leather should be used." — *J. B. Hoyt & Co.*

"Care should be taken that belts are kept soft and pliable. The question is often asked: 'What is best for this purpose?' We advise, when the belt is pliable, and only dry and husky, the application of blood-warm tallow; this applied, and dried in by heat of fire or sun, will tend to keep the leather in good working condition; the oil of the tallow passes into the fibre of the leather, serving to soften it, and the stearine is left on the outside to fill the pores and leave a smooth surface."

"The addition of resin to the tallow for belts used in wet or damp places, will be of service and help preserve their strength. Belts which have become hard and dry should have an application of neat's-foot or liver oil, mixed with a small quantity of resin; this prevents the oil from injuring the belt and helps to preserve it. There should not be so much resin as to leave the belt sticky." — *J. B. Hoyt & Co.*

Eel-skin Bands and Ropes.

39. "I have used eel-skin upwards of twenty years, for drilling holes for pearls and diamonds, by which means I have a knowledge of its utility. I have tried whip-cord, which will not last an hour; I have tried also cat-gut, and that indeed is very little better. An eel-skin cut in three or four pieces of the same size as the gut, or string, will last for three or four months certain, which shows the little wear to which it is subject. I have had them on the shelf for from six to twelve months, in the dusty shop, till they have been quite hard, yet they are as good as ever. . . . My business is that of a goldsmith and jeweller." — *Joseph Williams, England,* "*Journal Franklin Inst.," April, 1844.*

Driving Power of Belts.

40. "As regards the width of the belt, this will be found ample with respect to friction, if we calculate the cross-section of the same for the strain to be transmitted, in which case $\frac{1}{8}$ of an inch square is allowed for every 5 lbs. strain."—*C. D. Abel, in Weale's Series.*

"Morin concludes that we may, without any risk, and with the certainty that they will run a long time, make them support tensions of 355 lbs. per square inch of section."—*Frank. Inst. Jour., July, 1844, p. 27.*

"Good belting of, say, $\frac{3}{16}$ inch thick, should sustain a tensional strain of 50 lbs. per inch of width, and without serious wear, for a long time."—*Appleton's Mech. Mag.*

Haswell, in his Engineer's Pocket-book for 1867, says: "A leather belt will safely and continuously resist a strain of 350 lbs. per square inch of section."

We are indebted to Prof. R. H. Thurston, for the following:

$$w = \frac{7000 \times HP}{SV}.$$

In which $w =$ width of belt in inches.

$HP =$ indicated horse-power transmitted.

$S =$ portion of circumference of smaller pulley covered by belt, in feet.

$V =$ velocity of belt, in feet, per minute.

Prof. Thurston considers 100 lbs. per inch of width on ordinary belting of, say, $\frac{3}{16}$ inch thick, a fair working load. Then calling $t =$ tension, and inserting same in the formula above, we have:

$$w = \frac{700,000 \, HP}{SVt}.$$

Cone Pulleys. (From "Rankine's Rules and Tables.")

41. "To find the *ratio of the speed of turning* of two pulleys connected by a band. Measure the *effective radii* of the pulleys from the axis of each to the centre line of the band; then the speeds of turning will be inversely as the radii.

"To design a pair of *tapering speed-cones*, so that the belt may fit equally tight in all positions.

"When the belt is crossed, use a pair of equal and similar cones tapering opposite ways.

"When the belt is uncrossed, use a pair of equal and similar conoids tapering opposite ways, and *bulging* in the middle, according to the following formula: Let c denote the distance between the axes of the conoids; r_1 the radius of the larger end of each; r_2 the radius of the smaller end; then the *radius in the middle*, r_0, is found as follows:

$$r_0 = \frac{r_1 + r_2}{2} + \frac{(r_1 - r_2)^2}{6.28\,c}.$$

"Line upon Line, here a Little and there a Little."

42. Experience says, the grain side of a belt put next to the pulley will drive 34 per cent. more than the flesh side.

"Every one knows that the strength of belt leather is on the hair side." To be convinced of the contrary, see Article No. 49.

If variation of speed has resulted from changing belts, then the thickness of the belt is the cause. Some engineers add the thickness of the belt to the diameter of the pulley in their calculations for exact transmission of speed.

It is not adhesion alone we want to prove the better belt; beyond a certain amount it is rather an injury to the belt than an advantage in its use; for slippage is to be preferred to abrasion, when rapid destruction of the belt would result from the closeness of its sticking.

Put the horse-power above and the speed of the belt in hundreds of Fpm below, draw a line between and you have a fraction whose equivalent is the width of the belt in feet. To make this rule apparent, consider the following example, in which 36 horse-power is to be transmitted by a belt moving 1800 Fpm; how wide should the belt be?

$$\frac{36}{18} = 2\,ft.$$

Experiments show that a ¼ inch round belt is more than equal to a one inch flat, and a ½ inch round more than a 3 inch flat, *in so far as adhesive quality is concerned*. The rounds, of course, must be used in V-grooved wheels; comparative durability will depend much upon quality of material and circumstances of use.

Don't put a crossed belt on so that at the place of passing the laps will be torn up, and joints severed in a short time.

Don't cross the lacing of a belt joint on the inside, when there is a way to avoid crossing the lacing at all. See Article No. 144.

Of course belts are weakened by punching for laces or rivets just in proportion to the amount cut out; it is therefore necessary to narrow the punch, say to an oval form, to reduce the number of holes to the least, or to range them fore and aft, preserving the most substance in any straight line across the belt.

Experiments have shown that $\frac{5}{8}$ of the breaking strain of the solid part will start rupture at the lace holes in leather belts, and that they will endure $\frac{1}{3}$ of the breaking strain for a week without appearance of fracture.

We thank a correspondent kindly for having exercised his ingenuity upon methods of strengthening lace holes, especially in gum belts, and after much experimenting has generously presented to us this result: "With large oval eyelets, securely put in, the breaking strength of the joint was nearly up to that of the solid section."

A Rule.— $$HP = \frac{velocity\ in\ Fpm \times width}{1000}.$$

Driving value of double belts as compared with single leather, 10 to 7.

Leather belts should not be forced over $\frac{1}{16}$th, and rubber belts not over $\frac{1}{8}$th of their breaking strength.

"All belts riveted to run with the *grain* side next pulley. They will do *one-third* more work than with flesh side to pulley — will last longer, and will never crack."— *English Advertisement.*

Thoroughly stretched belting leather is more liable to tearing at the lace holes under undue strains, because less elastic; but it must not be condemned on that account. A belt poorly stretched will yield too readily to strains, and will require re-tightening often.

Rule for Piecing out Belts.

43. In order to calculate the changed length of belt when a different size pulley is put on in place of one removed, take out of the belt or put in $1\frac{1}{2}$ times the difference of the diameters of the pulleys.

Thus: If you take off a 24-inch pulley and put on a 30-inch one, you will want to add $30 - 24 \times 1\frac{1}{2} = 9$ inches of new belt to the existing one.

This rule provides simply for the approximate difference of semi-circumferences of the two pulleys thus:

$$\begin{array}{r} \textit{The cir. of 30} = 94.24 \\ \textit{The cir. of 24} = 75.39 \\ \hline 2)\overline{18.85} \\ \hline 9.425 \end{array}$$

Example.

44. In hoisting the materials for the towers of the Cincinnati bridge, Mr. John A. Roebling, C. E., used engines of 10 inches bore and 20 inches stroke, making 80 to 150 Rpm, and working under a steam pressure ranging from 60 to 80 pounds.

The power of these engines is transmitted by a 9-inch leather belt, from a 4-feet iron pulley on the engine shaft, to another 4-feet pulley on the pinion shaft. This pinion is $14\frac{1}{2}$ inches diameter, and drives a 6-feet spur-wheel: on the shaft of this latter is another $14\frac{1}{2}$ inch pinion, gearing into another 6-feet spur-wheel, on the shaft of which is secured a 3-feet drum. This drum carries a $1\frac{1}{2}$ inch diameter wire rope, connected directly to the loads to be lifted.

A block weighing 8400 pounds can be raised at the rate of 50 feet per minute by pressing the tightener down so that the belt laps on $\frac{7}{8}$ths of the circumference of the 4-feet pulleys.

With a load of 10,200 pounds the belt slips, and its splicings and safety are endangered by too severe an application of the tightener which is necessary to lift this weight. A load of 8000 pounds may therefore be considered a fair working condition of the belt, which indeed it has endured nearly three seasons without failing.

Blocks weighing 8000 pounds have been frequently raised 150 feet high in two and a half minutes without slippage of the belt. This speed is equal to 60 Fpm, and the duty performed is equivalent to $60 \times 8000 = 480,000$ lbs. $= 14.54$ horse-power, speed of belt being 1885 Fpm.

Quantity of belt running per minute, per horse-power $= 97.232$ square feet.

Combined Strap-Shifter and Stop-Motion.

45. In the use of sewing-machines it is as necessary to stop them instantly as to drive them rapidly, particularly for manufacturing purposes, where a speed of some 6000 stitches per minute is made, and where the machines are repeatedly started and stopped for changing the direction of the seam as well as the pieces to be sewn.

Fig. 7 illustrates a little device which answers this purpose so well that it deserves a record among the good things employed in the best use of belting.

The shifting-bar, A, guided by the staples, B and C, secured to the table, K, has a notch in its side for embracing the round belt, G, which drives the machine, and carries a brake, H, of leather, secured by an adjustable screw, to the bar, and in such position that when the

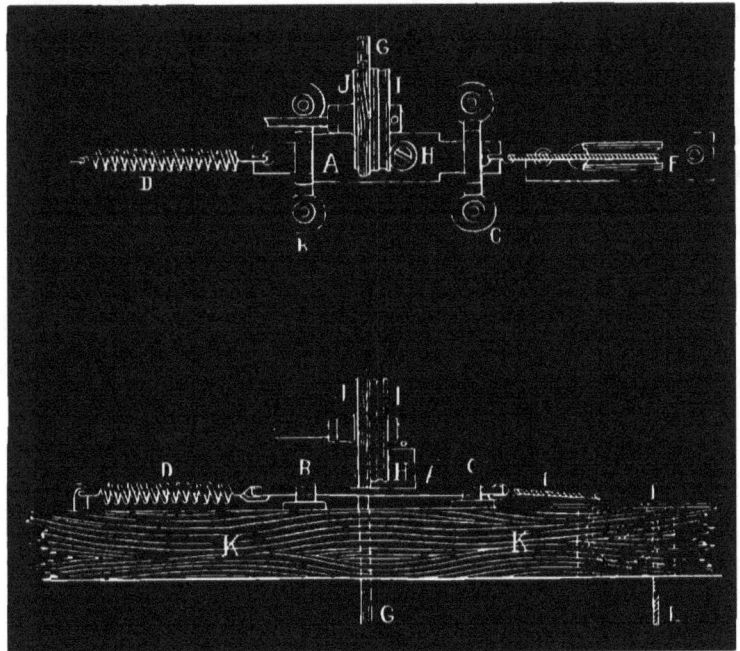

Fig. 7.

cord is on the loose pulley, J, as shown, the brake is held firmly against the tight pulley, I, by the spring, D, which is always in action.

To start the machine, the foot of the operator is pressed upon a hinged treadle, which is secured to the floor, and attached by a cord, E, to the shifting-bar, the direction of its motion being changed by a pulley, F, also fixed in the table, K. This action draws the brake, H, from, and puts the belt on, the tight pulley, I, at the same time distending the spring, D. When the foot is lifted, the recoil of the spring pulls the belt over on the loose pulley and holds the brake against the tight pulley, instantly stopping the machine and holding it still, without attention or effort of the operator, until motion is needed again, which the simple act of pressing the foot produces at will.

Atmospheric Influence on Adhesion.

46. The adhesion of belts to pulleys is frequently attributed to the pressure of the atmosphere, and in order to show how much the air influences belts in this particular, the following simple experiments are presented.

Take a circular disc of leather, say 3 or 4 inches diameter, with knotted string secured in its centre, and, when well water-soaked, press it upon any level wetted surface. The "boys" call this apparatus a "sucker," and it well illustrates the phenomenon of atmospheric pressure, or "suction," as it is usually called.

If an effort be made to draw it away from this surface by the string, it will be found resisting very forcibly, but the gentlest pressure will slide it on the wetted surface; it does not offer the slightest opposition to motion in the direction of its face, nor will it resist removal if raised first at the edge and then peeled off.

The atmosphere does not press two bodies together when it can get between them; it is only when excluded by a tight joint that the development of its pressure is possible, and it becomes sensible only when an effort is made to separate them by a force acting at right angles to the plane of their faces.

Another simple experiment shows that when two level, smooth, and clean surfaces come together by a motion like the closing of a book— which is similar to that of a belt coming in contact with its pulley — there will be retained between the two a thin film of air, and, while this remains, the contact of the two is imperfect, and the sliding of one over the other is easily performed. Take two iron "surface-plates" which have been scraped down to a practically perfect plane, and lay one of these on the other like a belt goes to a pulley; they will be found not in contact at all, but as if floating one on the other, and the top one will slide off by its own weight at the least inclination of the lower one.

Much of this interposed film of air can be displaced by a sliding of one plate on the other, starting, say at one corner, with the plates in close contact, and carefully pushing one over the other, holding it the while close to, as if to keep the air out. Then, indeed, an obstinate resistance to sliding will be felt, and the friction of nearer contact will be made thoroughly sensible.

But this way of bringing surfaces into contact has nothing to do with belt action, except to prove the need of a plastic surface on belt and pulley which will enable them to adhere, while in contact, with sufficient force to prevent sliding, and at the same time be uninfluenced by the intermedium of air.

And lastly, in order to put the matter to actual test, an apparatus was constructed, such that a leather belt was made to slide on the face of a smooth iron pulley, and also to drive the same iron pulley up to slipping of the belt. In both cases the adhesion or driving power of the belt was held by a spring-balance, so that the work of the belt could be observed.

Experiments were tried with this mechanism placed in a bell-glass jar on an air-pump plate, with and without air in the jar, and if any difference was observed in the adhesion of the belt to the pulley, it had more in vacuum than when the atmosphere was present.

Messrs. J. B. Hoyt & Co. kindly permit me to reproduce their valuable practical statements and experiments.

47. "Good leather belts can only be obtained by employing good materials. We stretch every piece by powerful machinery, joint it and secure it in such a manner as that both sides will present an even surface to the pulley, and run on it as though it were one strip. The fact that there is a great want of information in relation to the selection and use of belting must be apparent to all who have given thought to the subject. While the 'what, how, and why' of every other subject connected with machinery seems to have been extensively considered, that of belting has been almost entirely passed over.

"We have inserted a useful table, with calculations and deductions from it. These we believe to be correct, the experiments having been made at our factory.

"This table gives the relative driving power of leather belting, with both grain and flesh side to pulley; also, of rubber, gutta-percha, and canvas. The pulleys on which the experiments were made were the same in size, on one shaft, and their surfaces severally of leather, polished iron, rough-turned iron, and of polished mahogany. The bands were passed over the pulley, one end made fast and stationary, and on the other one pound weight was suspended to every square inch contact surface of the band and pulley.

"The number of pounds required to slip the band is given; also, number of pounds strain on the band at which it will cease to slip; and also, number of pounds required to make it continue to slide.

"The belts were in like condition, and had the same contact surface, the same strain; consequently it is easy to determine the relative value of each for driving machinery, also that of the pulleys.

	Leather. Grain side to Pulley.			Leather. Flesh side to Pulley.			Rubber.			Gutta-Percha.			Canvas.			
	Commence to slip.*	Cease to slip.†	Slide.‡	Commence to slip.	Cease to slip.	Slide.	Commence to slip.	Cease to slip.	Slide.	Commence to slip.	Cease to slip.	Slide.	Commence to slip.	Cease to slip.	Slide.	Relative value of different Pulleys.
Pulley with Leather surface.....	6	2½	10	3½	2¼	7	2½	1½	5	2½	1¼	3½	1¾	1	1¾	52
Polished Iron surface..........	1½	1	9	1¼	¾	6½	1¼	¾	4½	¾	½	2½	1	½	2	33¾
Rough Iron surface...............	1½	¾	3	1½	¾	2¼	1¼	¾	4	¾	½	1½	1	¾	1¼	21½
Smooth turned Mahogany......	3¾	2¼	4	3	1½	3¼	2¼	1½	4½	2½	1	2½	1¾	1¼	1¾	36¾
Relative value of each belt..	45¼			33½			29¾			19¾			15¾			

* "Commencing to slip" refers to that point when the resistance is sufficient to make the belt (almost, not quite) slip over the pulley.

† "Cease to slip" refers to that point when the belt has just slipped over the pulley and takes a new hold.

‡ "Slide" refers to that condition when the motion of either belt or pulley ceases while the other passes over it.

☞ Belts are liable to stretch more or less, decreasing their tension, so that they will slip—hence the necessity for the above distinction.

Deductions and Conclusions drawn from Table.

"Pulleys covered with leather, with grain side of band to pulley, will sustain *50 per cent. more resistance* than without the pulley being covered. The per cent. of resistance of the bands on the different pulleys is nearly as follows, and this per cent. will indicate the relative working value of each pulley respectively:

Iron pulley covered with leather..................	36	per cent.
" polished..................................	24	"
" rough turned..........................	15	"
Wood pulley, polished mahogany................	25	"
	100	

"Full 6 per cent. should be added to the polished iron pulley, to make allowance for the difference between commencing to slip and its sliding; thus making polished pulley 30 per cent., or next in value to leather.

"The relative or comparative working per cent. of the different bands, as indicated by the table, is nearly as follows:

Leather, grain side to pulley..................	31 per cent.
" flesh " "	23 "
Rubber. ....·........	21 "
Gutta-Percha.......................·............	14 "
Canvas..	11 "
	100

"Thus leather belts, grain side to pulley, will drive 34 per cent. more than flesh side to pulley; 48 per cent. more than rubber; 121 per cent. more than gutta-percha; 180 per cent. more than canvas; consequently, the very best arrangement for belting is to use it with grain side to pulley, and have the pulley covered with leather.. This is best in all cases. The next best pulley is polished iron, especially for quick motions; polished wood next, and rough iron least in value.

"Leather, used with grain side to pulley, will not only do more work, but last longer than if used with flesh to same. The fibre of the grain side is more compact and fixed than that of the flesh, and more of its surface is constantly brought in contact, or impinges on the particles of the pulley. The two surfaces — that of the band and that of the pulley — should be made as smooth as possible, the more so the greater the contact surfaces, and the more the particles of each impinge on the other. The smoother the two surfaces, the less air will pass under the band and between it and the pulley — the air preventing the contact of band with pulley — the greater this contact, the more machinery will the band drive. The more uneven the surface of band and pulley, the more strain will be necessary to prevent bands from slipping. What is lost by want of contact must be made up by extra strain on the band, in order to make it drive the machinery required — oftentimes, if the band is laced, causing the lacings to break, the holes to tear out, or fastenings of whatever kinds to give way.

"'This want of contact is noticeable on most of new bands used with flesh side to pulley, and is distinctly marked by dark impressions on the band where it comes in contact with the pulley. Oftentimes not half of the surface will be found to have come in contact, and, until it is worn smooth, or filled in with other substances, the full extent of the power of the band is not obtained. . . .

"Bands used with grain side to the pulley will never crack; as the strain, in passing it, is thrown on the flesh side, which is not liable to crack or break, the grain not being strained any more than other portions of the band.

Rule for determining the Width of Belts.

48.
$$W = \frac{5334}{V} \frac{HP}{C}.$$

In which W = width of belt in inches.
" HP = horse-power transmitted.
" V = velocity of belt per minute in feet.
" C = part of circumference of smaller pulley in contact with belt in feet.

We take pleasure in presenting the following facts contributed by Alexander Brothers, manufacturers of Oak-Tanned Leather Belting, Philadelphia.

49. "It is conceded by all practical men, that oak-tanned slaughter leather is the best material for belting, and as inferior stock such as *chemical* tanned and hemlock leather, colored in *imitation* of oak, by means of quercitron, sumac, etc., is largely used in the manufacture of belting, purchasers should be cautious, in buying, not to get any but pure oak-tanned leather.

"The strongest part of belt leather is near the flesh side, about one-third the way through from that side. It is, therefore, desirable to run the *grain* side on the pulley, in order that the strongest part of the belt may be subject to the least wear.

"In order to prove the above assertion, we split, in our machine, a strip of ordinary belt leather exactly in the middle of its thickness, and then subjected each half to a breaking tension, which gave the following results:— Grain side half broke under a direct strain of $468\frac{1}{2}$ lbs. Flesh side half sustained $740\frac{1}{2}$ lbs.

"A part of the grain side half is of the same kind of fibre as that of the flesh side, and as the grain extends but about one-fourth the way through a hide, much of the strength of the grain half in this experiment is due to the flesh part adhering to it.

"The flesh side is not liable to crack, as the grain sometimes will do when the belt is old, hence it is better to crimp the grain than to stretch it.

"Another important reason for running belting with the grain side on the pulley, is to get greater driving power; that being the smoothest side, it will hug closer, is less liable to slip, and will drive 30 to 35 per cent. more than if run the other way; and if the pulley is covered with leather, grain side out, there will be still greater friction. Therefore a belt will do more work and wear longer on a leather-covered pulley than on any other.

"Belts should *not* be *soaked* in water before oiling, and penetrating oils should but seldom be used, except occasionally when a belt gets very dry and husky from neglect, it may be moistened a little and then have straits or neat's-foot oil applied. Frequent applications of such oils to a new belt renders the leather soft and flabby, thus causing it to stretch and making it liable to run out of line. A composition of tallow and oil, with a little resin or beeswax, is better to use. Whitaker's castor-oil dressing is good, and may be applied with a brush or rag while the belt is running."

"We find the average permanent stretch of oak-tanned leather for belting to be about .725 in. per lineal foot, or as follows:

"Average stretch of back pieces per foot, .562 inch.
" " middle cuts " .75 "
" " lower edge " .875 " "

Cones of Pulleys.

50. When 2 cones of pulleys, or "stepped cones," are made alike, *i. e.*, with equal steps, such that the sum of the diameters of each belted pair is the same, a *crossed belt* will run with equal tension on any pair so made, when the shafts are parallel and the cones reversed; but an *uncrossed belt* will not so run on such cones.

To show why the open belt will not have equal tension on all the pulleys, let A B, Fig. 8, be two equal stepped cones on parallel axes A E and B D. Now, if the sum of the diameters of the extreme pulleys E and F be equal to the sum of the diameters of C and D, the connecting strips E F and C D of the belt will be equal, because the enrolled parts are equal by the construction of the cones; but E F and C D cannot be equal, because they are not parallel, and hence it plainly appears that C D, being at right angles to the shafts, is shorter than E F; therefore, to preserve a certain tension of the belt when on the extreme pulleys, the middle pulleys must be larger than the size given by equal steps in order to take up this difference.

To find the proper diameters of the intermediate pulleys for open belt cones, first get, by Rankine's rule, Art. 41, the radius N O from the given radii I J, K L, and distance between shafts and through the points J O L describe a circular arc, upon which draw the faces of all the pulleys in the series as shown. Make both cones by this rule, and an open belt will run upon them with equal tension.

Unequal cones may be made in like manner by drawing 2 similar cones, and then using, say S T V of the large end of one, and X Y Z of the small end of the other, as needed to serve the purpose.

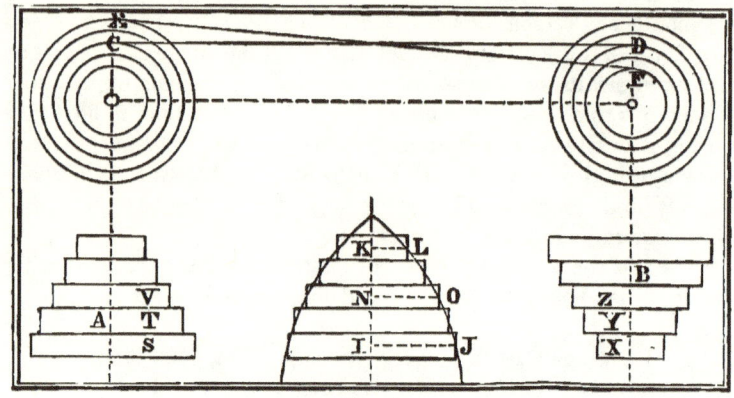

Fig. 8.

For wheels driven by round bands in V grooves the same rules apply, observing that the *effective* diameters must be taken at the line of *band contact* in the grooves, which are the acting circles of adhesion. All the grooves must be alike, and should have *concave* instead of straight sides, like a Gothic arch, and the band must not touch the bottom of the groove.

Belts.

51. "The effective radius of a pulley is equal to the radius of the pulley added to one-half the thickness of the belt.

"In ordinary cases a belt will last about three years.

"Leather belts must be well protected against water, and even moisture.

"India rubber is the proper substance for belts exposed to the weather, as it does not absorb moisture and stretch and decay.

"It is quite probable that these belts will increase in popularity among manufacturers.

"In joining the ends of a belt, it is generally considered best to cut small holes through the ends and lace them with a leather strap.

"It is found, in practice, that belts should not be subjected to a strain of over 300 pounds per square inch of section.

"Good belting of $\frac{3}{16}$ inch thick will sustain a strain of 50 pounds per inch width without risk, and without serious wear, for a considerable time."—*Amer. Artisan, July 1, '68, p. 394.*

Belts and Pulleys.

52. "We would not venture to prophesy that belts will continue for ever in general use, especially if made of leather, in the old way. Nor is our faith in the perpetuity of small pulleys very resolute. Robertson's frictional gearing seems to work more steadily than belts; but if pulleys are large, the friction of their faces against each other is sufficient for most work, and they will run with less power than belts or Robertson's gearing. To make pulleys work well against each other, they must be turned with more truth than is usual; and the shafting must be better lined; and there must be something elastic in the journal-boxes, such as a little rubber, to press the pulleys together.

"The vibration or chattering of lathes is probably due to the spring of the belts, more than to the spring of the metal of the lathes. So far as this is the case, the frictional gearing, of either kind, will get rid of it.

"In driving fans there is frequent trouble with belts, and there is more expense than need be; and the noise and vibration are evidences of serious imperfection, which, perhaps, is unavoidable when a soft material is used that cannot have an exact shape.

"A friend of ours proposes to use steel belts for certain cases, such as fans. The steel must be welded perfectly, and must be of mild quality, and the pulleys must be large, to make them work well. Steel belts would run with less power than leather, because the power exerted in bending them would be restored when they straightened themselves; but leather does not return much of the power required to bend it around the pulley. . . ."—*Amer. Artisan, Aug. 2, '65, p. 201.*

Leather Belts.

53. "A friend of ours who has had experience in tanning leather for belts, and has conversed about belts with many machinists, and is familiar with their experience, assures us that the adhesion of belts is much better when the hair side is against the pulleys, and that the belt used in this way is more durable. The belt will not crack on the flesh side so readily as on the hair side, and seldom cracks at all, if the hair side is against the pulley. He has known cases in which belts slipped and would not do the work required of them when the flesh side was against the pulley; but when they were turned hair side to the pulleys they ceased to slip, and did much more work than they could do with the flesh side to the pulleys. His theory is, that the belt, as well as the pulley, adheres best when smooth, and the hair side adheres best because it is smoothest."—*Amer. Artisan, Aug. 16, '65, p. 234.*

From "Power in Motion," by J. Armour, C. E. Lockwood & Co., London, 1871.

54. "Good new belt leather has been found to break with an average tension of 5000 lbs. applied quietly per square inch of sectional area.

"The working tension for continuous service ought not to be more than about $\frac{1}{14}$ of this, or about 350 lbs. per square inch.

"A thickness of $\frac{3}{16}$ of an inch, which is the ordinary thickness, equals .186 inch : therefore, for an inch of breadth, we have .186 × 5000 = 930 lbs. breaking strain, and .186 × 350 = 65.1 lbs. continuous service strain.

"With the same working tension, when we double the breadth, we reduce the strain per square inch of section to one-half the strain for the single breadth, and thereby save the belt. The axle pressure is the same, however, because the belt of double breadth is simply doing the same amount of work upon the rim of the pulley as the single breadth had to perform.

"When we double the diameter, the revolutions of pulley per minute being as before, we may reduce the tension to one-half; because we have the speed at the circumference equal to 2, and this multiplied by .5 tension = 1 power; the same as 1 speed × 1 tension = 1 power.

"When two pulleys at rest are connected by a belt, the tension on each connecting part is nearly equal; when motion begins, the driving pulley has to stretch the pulling parts to the tension required to overcome the resistance before the driven or loaded pulley can move; and, in doing so, the driver is passing a corresponding amount of slack into the returning part.

"Should the resistance of the load grow less from any cause, less tension will be required to balance it, and the driven pulley will be moved by the excess of the pulling tension a fractional quantity faster than the driver, thereby throwing part of the slack of the returning part into the pulling part, until the reduced load resistance and the pulling tension come to a balance; this diminishes the amount of slack on the returning part.

"On the other hand, should the load increase from any cause, greater tension is required; the driver must move a fractional quantity more than the loaded pulley, to put the greater strain upon the belt, and the amount of slack is increased correspondingly.

"Hence, in a narrow belt the returning part will be slacker than when a broader belt is employed, because it will stretch more with a given tension.

"Short belts require to be tighter than long ones. A long belt, working horizontally, increases the tension by its own weight, acting in the curve formed between the pulleys.

"One of the properties of this curve is to make the tension greater than is due to the simple weight of the belt; that is greater than when the belt is hanging vertically; besides, it never loses contact.

"In vertical belts so little stretch is needed to make them lose contact with the lower pulley, that the tension for the state of rest requires to be greater than is found necessary for a horizontal belt, if the breadth be not increased to reduce the stretching stress per sectional square inch.

"In ordinary leather belts, on large pulleys, the bending resistance is so small that it may be disregarded.

"Ropes of hemp or wire are often employed for driving bands. Their resistance to bending is greater than that of flat leather belts, and as the surface in contact with the pulling is less, the pressure per square inch of actual contact must be greater, and therefore more severe upon the material.

"This, however, does not affect the amount of tension required for work, because, as friction is independent of the extent of surface, we get the same driving power from 10 lbs. pressure or tension on the narrow line of contact with the pulley, in the case of a circular rope, that we would get from the same pressure supposing the rope flattened out so as to have a surface of contact many times greater.

"When we know the weight per foot of a long belt or rope working horizontally, we find the tension in the curve of the belt between the pulleys by multiplying the whole weight of the part between the pulleys by the distance between the same, and dividing by eight times the deflection. This rule, however, applies only to curves in which the deflection is small compared with the span; so that the flatter the angle of suspension the closer the approximation."

I. H. Beard on Belts.

55. "Throughout New England, until within a few years, it was generally thought, by engineers and millwrights, that cotton-mills and woollen-mills, and all others requiring a very considerable power, could not be run effectively without large and ponderous lines of upright and horizontal shafts of either cast or wrought iron, and heavy trains of cog-wheels of cast iron, or partly of iron and partly of wood.

"And when large leather belts began to be introduced for the main gear of mills, as a substitute for gear-wheels, it was thought by some of our best engineers to be an experiment, at the least, of very doubtful result, if not altogether impracticable; and, indeed, at the present time (1837), notwithstanding all the evident advantages of belts over gear-wheels, many still adhere to the old mode of gearing; and this, doubtless, not so much from a want of discernment and sound judgment, as from a lack of opportunity of comparing and testing the advantages of belts and the disadvantages of gear-wheels; or, perhaps, they may have formed an erroneous opinion of the utility of using belts from the inspection of some mills that have been belted on a bad principle, or from belts injudiciously managed."

"Having, from a constant practical experience of both modes of gearing mills, for more than ten years, at Lowell, Saco, and other places, become fully satisfied of the utility of belting mills, instead of running them with gear-wheels, and that they run much lighter, stiller, and with far less friction, and a proportionately less motive power, with belts than with gear-wheels."

"A cotton-mill of dimensions adapted to the convenient operation of 4000 spindles of cotton machinery, including all the preparation for making yarn and weaving cloth, ordinarily required four trains of upright shafts, extending through the height of four stories, the trains usually commencing in the basement story."

"To each train of uprights were attached from two to four pairs of heavy gear-wheels, and, in addition to these, in many mills all the counter lines of shafts were geared off at right angles with the horizontal main shaft, which required a very large number of gears and shafts. A mill thus geared is a full load for the power of a moderate sized water-wheel without any machinery, and a great proportion of this unnecessary weight and friction may be saved by the judicious use of belts instead of gears. And besides the disadvantages before named, the trains of gear-wheels require the constant extra expense of careful attendance and of oil or some unctuous matter to lubricate and keep them from heating, friction, and abrasion. And again, all the gears must be closely boxed in, and supplied with tight dripping-pans, or the mill-grease will be liable to drop into the work, and greatly damage if not entirely ruin it.

"These are serious inconveniences and evils that may be avoided by substituting belts in place of gear-wheels. The first expense and the constant repairs will be as little with belts as with gear-wheels, and the risk and hindrance that may be caused by belts are far less; for if a main belt breaks, it is the work of a few minutes only to repair it or replace it with a new one, whereas the breakage of a single gear-wheel may cause the hindrance of a week, and the almost entire loss of the wheel broken, together with a hundred times the labor and expense in exchanging the broken wheel for the new one that would be caused in repairing or exchanging the belts. And, again, if it should be found desirable at any time to change the velocity of any part of the mill gear, it is much more easily done, and with far less expense, by varying the size of the pulleys and drums than by changing gear-wheels."

"But to gear a mill wholly with belts, and to do it judiciously and to the best advantage, doubtless requires more nice calculation, careful judgment, and practical experience than to do it with gear-wheels; for many mills have been so belted as to cause more friction, trouble, and expense than would be caused or required in the use of gear-wheels."

Therefore, to enable those who may wish to calculate mill-gear, and who may not have had the means in forming a correct judgment by practical experience, to judge correctly of the advantages, as well as of the disadvantages, of several modes of gearing, I shall first introduce such modes as I consider objectionable, and then bring forward a mode that I consider the least objectionable and the best now in use. To know how to avoid an evil is frequently as beneficial as to know how to remedy it.

"In Fig. 9 A represents the main driving-pulley, geared from and driven by the water-wheel, and is made from 8 to 12 feet in diameter; B the water-wheel; C the basement; D the carding-room; E the spinning-room; F the weaving-room; G the dressing-room.

"a, b, c, d, e, f, g, h, i, and j, represent the lines of drums in the carding and weaving rooms. These lines of drums extend very nearly the whole length of the mill inside, and for a mill of 4000 spindles are driven by two belts operating in the same manner. 1, 2, 3, and 4, represent the belt-binders to lead, or bind, the belt in the required directions.

"The belt here represented must be about 320 feet long, and from 12 to 15 inches wide, and will require from 600 to 700 lbs. of stout belt leather to make it. These belts are bulky, ponderous, and unmanageable; and whenever a lacing breaks, to which accident they are frequently liable, they are likely to run nearly, or quite off, of the drums, and it would cause the hindrance of the whole work of the machinery, and the work of some half a dozen men half a day to put one of them on again."

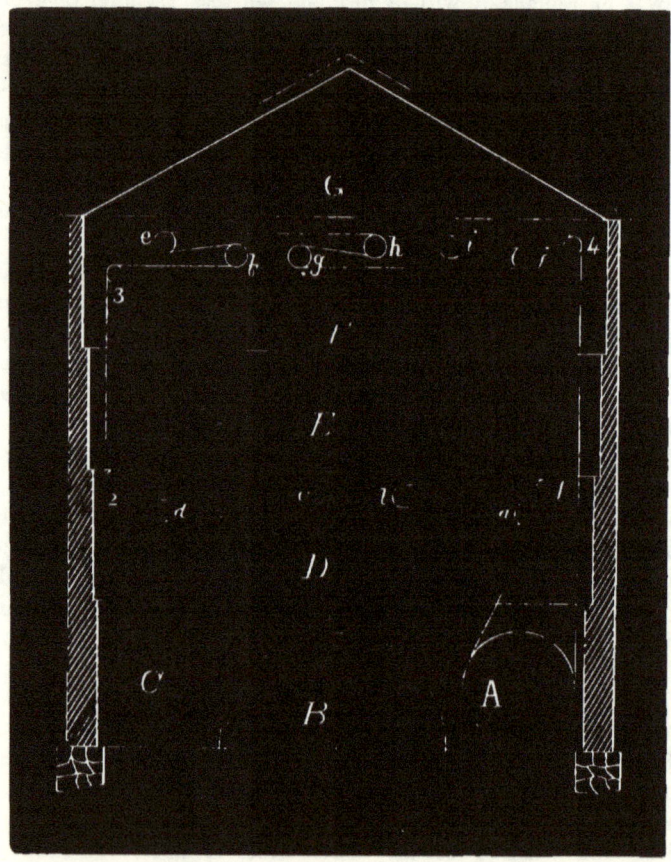

Fig. 9.

"In laying out the gear of a mill, it is worth much time and pains to arrange the drums and belts in such a manner that, so far as may be practicable, the stress of one belt upon the journals shall be counteracted by that of another belt in an opposite direction, referring to the stress upon the line of main drums, the counter drums being of minor consequence; but where the main power is to be exerted to throw the stress upon one belt into that of another, is economy in the wear of the whole mill gear, as well as in power, both of which are points of great importance to the manufacturer. This point has not always been observed; for it is sometimes more convenient in arranging the gear and machinery of a mill to place the line of main drums upon one side of the mill instead of in the centre. And the effect of this arrangement is to throw the whole stress of the belts upon one side of the journals of the main drum shafts, which ought ever to be avoided."

. . . "It is further of great importance that each belt should be of such a length that it will adhere to the drum so much as to prevent it from slipping, and that without the necessity of pulling on the belt so tight as to cramp the drums and wear the bearings. Every belt, to run easy and well, should be so slack, when running, that the slack side should run with a waving, undulating motion, without any tension except on the leading side, and when belts will so run without slipping upon the drums or pulleys, they will wear for a great length of time; for although a belt may be heavily loaded, yet, if, at every revolution, it can have an opportunity for relief from its tension so as to contract to its natural texture, it will prevent it from breaking by the stress upon it. But if, otherwise, it be kept strained so tensely as to be constantly strained to its greatest extent on both sides of the drums, it will wear but a short time without cracking at the edges, and will shortly be destroyed."

"Sufficient care is seldom taken to have belts to run free and easy, and it has been one of the greatest errors, more or less prevalent in all cotton or woollen mills, to run the belts so tense as greatly to injure the belts and rapidly increase the wear of the bearings."

"It has been customary in almost all belted mills to affix heavy cast-iron or wooden binders (tighteners) weighted, to the belts which drive the main mill gear, to prevent them from slipping, and it has been generally thought impracticable to keep them from slipping on the pulleys and drums without binders; but this opinion is wholly erroneous, and without any true foundation, if the belts are properly prepared and pulleys and shafts arranged so that the belts pull against one another, and thus relieve the shaft bearings of much of the strain as when they pull the same way."— *Journal Franklin Inst., 1837.*

Wrapping Connectors.—From "Fairbairn's Mills and Mill Work," London, 1865, p. 1, Part II.

56. "Considerable difference of opinion exists as to the best and most effective principle of conveying motion from the source of power to the machinery of a mill. The Americans prefer leather straps, and large pulleys or riggers.

"In this country, and especially in the manufacturing districts, toothed wheels are almost universally employed.

"In some parts of the South, and in London, straps are extensively used; but in Lancashire and in Yorkshire, where millwork is carried out on a far larger scale, gearing and light shafts at high velocities have the preference.

"Naturally, I am of the opinion that the North is right in this matter, and that consistently, as I was to a great extent the first to introduce that new system of gearing which is now general throughout the country, and to which I have never heard any serious objection, I have been convinced by a long experience that there is less loss of power through the friction of the journals, in the case of geared wheel work, than when straps are employed for the transmission of motive power. Carefully conducted experiments confirm this view, and it is therefore evident which mode of transmission is, as a general rule, to be preferred."

"There are certain cases in which it is more convenient to use straps instead of gearing. With small engines driving saw-mills, and some other machinery where the action is irregular, the strap is superior to wheel work, because it lessens the shocks incidental to these descriptions of work. So, also, when the motive power has been conveyed by wheel work and shafting to the various floors of a mill, it is best distributed to the machines by means of straps."

"In some of the American cotton factories, however, there is an immense drum on the first motion, with belts or straps from 2 to 3 feet wide, transmitting the power to various lines of shafting, and these in turn, through other pulleys and straps, giving motion to the machinery.

"From this description it will be seen that the whole of the mill is driven by straps alone, without the intervention of gearing.

"The advantages of straps are the smoothness and noiselessness of the motion. Their disadvantages are cumbrousness, the expense of their renewal, and the necessity for frequent repairs. They are inapplicable in cases where the motion must be transmitted in a constant ratio, because, as the straps wear slack, they tend to slip over the pulleys, and thus lose time. In other cases, as has been observed, this slipping becomes an advantage, as it reduces the shock of sudden strains, and lessens the danger of breaking the machinery.

"Very various materials are employed for straps, the most serviceable of all being leather spliced with thongs of hide or by cement. Gutta-percha has been employed with the advantage of dispensing with joints, but it is affected by changes of temperature, and it stretches under great strains. Flat straps are almost universally employed, in consequence of the property they possess of maintaining their position on pulleys, the faces of which are slightly convex.

"Round belts of cat-gut or hemp are sometimes used, running in grooves, which are better made of a triangular than a circular section — so that the belt touches the pulley in two lines only, tangential to the sides of the groove; in this case the friction of the belt is increased in proportion to the decrease of the angle of the groove."

"The strength of straps must be determined by the work they have to perform. Let a strap transmit a force of n horse-power at a velocity of v feet per minute, then the tension on the driving side of the belt is $\dfrac{33,000\, n}{v}$ lbs., independent of the initial tension producing adhesion between the belt and pulley. For example, let v be 314.16 Fpm, or the velocity of a 24-inch pulley at 50 Rpm, and let 3 horse-power be transmitted, then $\dfrac{33,000 \times 3}{314.16} = 312$ lbs., the strain on the pulley due to the force transmitted."

"The following table has been given for determining the least width of straps for transmitting various amounts of work over different pulleys. The velocity of the belt is assumed to be between 25 and 30 feet per second, and the widths of the belts are given in inches. With greater velocities the breadth may be proportionably decreased."

The following formula will meet every requirement of the table:

$$W = \frac{5940\, HP}{1650\, d}.$$

In which W = width of belt in inches.
    " HP = horse-power.
    " d = diameter of smaller pulley in feet.
    "   1650 = average speed in feet per minute,

from statement above, and which might be changed for $v =$ velocity of belt in Fpm to make the rule more general, which would seem to be allowed by the closing paragraph of the quotation.

The use of d in this formula forbids the naming of any definite area of belt running per horse-power per minute. If we take, however, the cases of belts from 12 inches to 24 inches wide on a 6-feet pulley, transmitting from 20 to 40 horse-power, and giving $82\frac{1}{2}$ square feet of belt per minute per horse-power, we would not be stepping outside the line of usual good practice in selecting average examples from the table. But if we select from one extreme of the table a 1.4 inch belt running on a 10-feet pulley, transmitting 4 horse-power, or 48.125 square feet of belt per minute per horse-power, we will find it, if not out of the limits of possibility, certainly not within the ordinary economy of practice as to width of belt and diameter of the pulley. On the other extreme, the proportion of a 43.2 inch belt, running on a 12-inch pulley, and transmitting 12 horse-power, or, measuring off 495 square feet of belt per minute per horse-power, is such an excessive one that perhaps it never has, and certainly never should be used in practice.

From J. Richard's "Treatise on Wood Machinery." E. & F. N. Spon, London, 1872.

57. "Belting, gearing, and unbalanced pulleys or wheels represent transverse strain, which must be a matter of judgment rather than estimates. It would be folly to predicate the transverse strain upon a shaft as being simply the tension of belts, or the strain of gear-wheels working under ordinary conditions. A rule in the author's practice has been in the case of belts to provide sufficient strength in shafts and supports to tear them asunder, without damage to the machinery. This is the only safe rule, for there is no means of always guarding against winding belts.

"Calling the distance between the hangers or bearings b, the diameter of the shaft d, and width of belt w, a rule for ordinary cases would be $w = d^2$ and $d \times 25 = b$. This is, of course, arbitrary; and presuming the pulleys to be in the centre between the bearings, and not more than five faces in diameter. In proportioning shafts for belting, much must be left to judgment, and be dictated by that peculiar sense of realizing what is wanted from previous experience.

"There are, in fact, so many obscure conditions that have to do with the matter, that any rule must be an arbitrary one, if given for general application. The above is, however, safe, so far as strength is concerned, for gearing, shafts must, as a rule, be stronger than for belts. The motion is positive, and lacks the elasticity that exists in belt connections. Shafts are in general made strong enough to crush cast-iron gearing; practice has given larger proportions to shafts that receive gearing, no doubt for the reasons stated, that of positive motion; yet the proper plan in the construction of wood machines would in all cases be to drive the first movers with belting so proportioned and arranged that it would be sure to yield before breaking the gearing. Ordinary belting, with its surfaces dry, as they must be when operated on wood-working machines, has much less driving power than the belting on metal-working machines, when the surfaces become covered with oil or gum, and the leather soft and pliable. Assuming the belts to be dry, a good rule for belting and gearing for feeding wood machines would be as follows: Let V be the velocity of the belt, and v that of the pinion or first mover, the width of the belt to be the same as that of the gearing $\frac{V}{6} = v$; or, in other words, the diameters of the pulleys to be to the pinion as 6 to 1, with equal faces: variations as to relative width should be directly as the proportion between pinion and pulley; if, for instance, the face of the pinion was reduced to 2 inches, and the belt remain 3 inches wide, their velocities would require to be $v \times 4 = V$, or diameter as 1 to 4, the diameter of the shaft being equal to the square root of the face of the pulley. There would with these proportions be no danger of breaking either shafts or gearing, it being understood, of course, that in a train of gearing such as is used in planing-machines, the force and pitch of each wheel and shaft should be inversely as their velocity.

"The belting for circular saws is, as a rule, too narrow, or upon pulleys of too small diameter. To drive a saw well and without injurious strain upon the bearings, belts should be one-third the diameter of the saw in width, and the pulley equal in diameter to the width of the belt, which is a very simple rule, and does not give any more than the needed driving force, under fair conditions.

"One-fourth the diameter of the saw for the diameter of pulleys on cross-cutting spindles. Their faces can be one and a half diameter in length.

"That speed should be an element in estimating belt contact is apparent in looking at the spindle pulleys in wood-cutting machines. The degree in which belts are affected by centrifugal force in running at high speed is dependent upon the tension, weight, and flexibility of the belt and the diameter of the pulley. At 5000 feet a minute, with belts of ordinary harness leather, running on pulleys six inches or less diameter, the amount of contact is not more than three-fifths of what would be shown in a diagram, and is often much less. Coupled with this, however, is the strange fact that the tractive force does not seem to be as constant as the amount of contact. That the pressure on so much of the surface as has contact is increased by the belt lifting, is unquestionably the case, but it hardly accounts for the want of proportion between the power transmitted and the amount of contact. This matter is mentioned as an experimental fact, and merely to stand as a reason for saying that the width of the belts need not be predicated directly upon the pulley contact for high speed spindles.

"For spindles having unusually high speeds, the writer has found belts of cotton webbing to be preferable. Such belts, if closely woven and of the best material, will, when waxed, be found to have a high tractile power and wear well, while their comparatively light weight avoids their lifting from centrifugal force.

"The convexity of pulleys to keep belts central should be sufficient for the purpose, and no more. It is difficult to account for the practice of many builders of wood machines, especially in England, who give a degree of convexity to pulleys that interferes with the contact and tends to the destruction of the belt, unless both pulleys have their faces the same, a thing impossible in the case of shifting belts. Without entering into an examination of the laws and conditions that govern the matter, the following rule is given:

"For pulleys from 2 to 24 inches face, the convexity should be from $\frac{1}{8}$ to $\frac{1}{18}$th of an inch to a foot, graduated inversely as the width of the faces; for pulleys of narrower face, the convexity can be slightly increased.

"This is quite sufficient to govern the running of belts, and a necessity for more can safely be construed as a fault in the position of the shafting.

Mr. James Christie, of Pittsburgh, gives the following:

58. "I can give an example of engine and connected belting, during the performance of which I had frequent opportunities of observation.

"The engine has a cylinder, 8-inch diameter by 12-inch stroke, and made 150 Rpm under a boiler pressure of 75 lbs. to the square inch, and piston pressure of 40 lbs., doing an effective service of 18 horse-power. It drove a brick machine by a first-class double leather belt, 6-inch wide and nearly ⅜-inch thick, laced in the usual way. The smooth turned iron driving pulley or engine shaft is 22-inch diameter; the driven pulley, of like material and finish, on brick machine, 48-inch diameter; distance, horizontally, between shafts, 12 feet; top fold of the belt slack, no tightener applied, belt was well stretched, no resin or unguent used.

"Both engine and belt were insufficient for the purpose, but I considered each about the equal of the other. When first used the engine would drive ahead and slip the belt; as belt and pulley became more *attached*, and belt was tightly laced, the belt would 'stall' the engine."

The above gives 24.27 square feet of belt per horse-power per minute, and 114.8 pounds strain to one inch width of belt, and may be considered as a very reliable example of a hard-worked belt.

"In driving 8-inch trains of rolls, the practice here is, when driving direct, to use engines of from 12 to 16-inch diameter of cylinder, similar in construction to the usual slide-valve engines the country over, and working under a boiler pressure of say 90 lbs. to the square inch, and average piston pressure of 50 lbs. The 12-inch cylinders have proved inadequate, the 14-inch have ample power, and the 16-inch have an overplus.

"The piston speed of these engines is invariably high, seldom less than 400 Fpm, frequently 600 and 700, and sometimes as high as 800.

"14-inch belts have proved insufficient in driving such a train, 16-inch do very well, and 18-inch are wider than needed. The 16 and 18-inch are, however, commonly used, and are always double leather, or two or three-ply gum belts.

"When driven from line shaft, which is the usual plan, 6-feet pulleys are used, both on line and on rolls, say 25 feet between centres, and the angle of belts with a vertical line about 20°. Sometimes the pulley on rolls is made heavy, but commonly separate fly-wheels, of about 8 feet diameter and 4 tons weight, are used. Tighteners are seldom employed, speed of rolls from 150 to 250 Rpm.

"In a steel mill here, a 20-feet fly-wheel pulley on engine shaft, running at 60 Rpm, drives a 6-feet pulley, 4 tons weight, on the train at 20 feet distant, between shafts, horizontally. The train is a 9-inch one, belt 17 inches wide, of three-ply gum, and no tightener used.

"The belt slips while rolling long lengths, which would indicate insufficient momentum in and lack of belt contact on the fly-wheel pulley on the train.

"Greater distance apart of shafts and less difference of diameters of pulleys would give, therefore, a better belt result.

"I have never heard of any objection to the use of belts in rolling-mills on the score of durability.

"In order to obtain the full value of a belt, it is first necessary that it should be in thorough contact with the pulleys. A new belt will be found touching in spots, and will not pull well for want of entire contact with pulley. Any unguent put on the belt will be found of immediate benefit, as it softens its surface, and brings it in complete contact with the face of pulley. The hair side of belt, on account of its smoothness and closeness of texture, seems to conform to this necessary condition much sooner than the flesh side, which is open in texture and rough on the surface. But after the belt is once worn to the proper condition, I doubt if there is any appreciable difference in the two sides in value. In fact, with *well-worn* belts, which have been used alternately with each side to pulley, it is often difficult to distinguish the hair side from the flesh side. By *well worn* I do not mean *injured by use*, but simply that condition of belt in which the color of the sides is rendered uniform by absorption of oil, and in which the surface gloss and texture are made nearly uniform by contact with pulleys. Intimate contact between belts and pulleys is undoubtedly necessary. The utility of smooth faces to pulleys is also well established."

The following Notes on Belting are taken from Mr. B. F. Sturtevant's Illustrated Catalogue of Pressure-Blowers, Boston, Mass.

59. "Double belts are calculated to be $\frac{1}{3}$ of an inch thick, single belts $\frac{1}{5}$ of an inch thick. The holding power of the belt is governed wholly by its tension or tightness and the condition of its surface, the diameter of the pulleys having nothing to do with it, and is capable of being increased, by tightening, to 350 lbs. to one square inch of cross section.

"As a general rule, belts running from a large to a small pulley slip on the large and not on the small one, as is commonly supposed. This is explained as follows: The pressure per square inch with which the belt hugs the small pulley is just as much greater as the small pulley is less in diameter than the large one, and the friction of the belt more than follows the increased pressure. For example: If you cover the small pulley on the counter with leather, to prevent slipping, you must also cover the driving pulley on the main line, or you get no benefit from covering the small pulley. These remarks apply to horizontal belts.

"A table is given for finding the width in inches of double main belts for driving his patent pressure-blowers, which is based on the statement that a belt one inch wide, having a velocity of 1000 Fpm, will give one horse-power. This is also equivalent to a surface velocity of $83\frac{1}{3}$ square feet of belt per minute per horse-power, and a working strain of 33 lbs. per inch width of double belt, or 99 lbs. per square inch of cross section of leather.

"New belting, such as is generally used for the driving-belts on polished iron pulleys, will only transmit from one-third to one-fifth the power, without slipping, that the same belt will after it has been in use from one to two months.

"The belts for driving the blowers are not exposed to the wear and tear of shippers, as lathe belts are; all of them run over small pulleys in proportion to the width of the belts. Hence I have recommended a scale of thicknesses of belts in proportion to the widths.. The lighter the belt the tighter it hugs the pulley, for the reason that all unnecessary weight of leather tends to lift itself off the pulley when going around it at a greater velocity, and some of these belts travel at a velocity of 5000 Fpm.

"Belting, as it is generally manufactured as an article of merchandise, is intended for all purposes for which belting is used; but when parties are desirous of having everything about the blower just as it should be, they will see that their belts are made exactly to suit their work.

No. of Fan.	Diam. of Pulley.	Face of Pulley.	Breadth of Belt.	Thickness of Belt.	Revs. of Fan per min.	Diam. Driving Pulley.	HP required.
00	$1\frac{7}{8}$	$1\frac{1}{2}$	$1\frac{1}{4}$	.08	......	21	......
0	$2\frac{1}{4}$	$1\frac{3}{4}$	$1\frac{1}{2}$	.09	......	21	......
1	$2\frac{5}{8}$	2	$1\frac{3}{4}$	.11	4135	21	.5
2	3	$2\frac{3}{8}$	2	.12	3756	24	1
3	$3\frac{1}{2}$	$2\frac{3}{4}$	$2\frac{3}{8}$	.13	3250	28	1.8
4	$4\frac{1}{4}$	$3\frac{1}{4}$	$2\frac{3}{4}$	.15	3100	32	3
5	$4\frac{7}{8}$	$3\frac{7}{8}$	$3\frac{1}{4}$	.16	2900	36	5.5
6	$5\frac{3}{4}$	$4\frac{1}{2}$	$3\frac{7}{8}$	.18	2820	42	9.7
7	$6\frac{7}{8}$	$5\frac{1}{8}$	$4\frac{1}{2}$	.20	2600	48	16
8	$7\frac{7}{8}$	6	$5\frac{1}{4}$	.22	2270	54	22
9	$9\frac{1}{8}$	$6\frac{7}{8}$	6	.24	2100	48	35
10	$10\frac{1}{4}$	8	7	.25	1815	54	48

"In the above table all the dimensions are in inches. To give some idea of the proper length of belt, the centre of the driving pulley is placed 3½ times its diameter from the centre of the fan and 2¼ times above it. The driven pulley on the driving pulley-shaft is 18 inches diameter, and that, in turn, is driven by a 54-inch diameter pulley and belt of double thickness. The centres of the 54-inch and 18-inch pulleys are placed about 3 times the diameter of the larger pulley apart and nearly at same height from floor."

On the Creep of Belts.

60. " It is generally considered by those who have had experience in the matter, that there is a slight slip in all belts, however large they may be in proportion to the power transmitted, and however tightly they may be stretched. Perhaps in the case in which a belt is much larger than is required, it would be better to say that there is a slight *creeping*, instead of a slip, this creeping being caused by the change in tension of the belt as it moves from the tight to the slack side in passing over the pulley.

"Where belts are driven at a high velocity it is found that the centrifugal force still further changes the ratio of speeds of the driving and driven pulleys, in some cases decreasing the tension of the belt, and in others acting in the contrary way. This may be explained in the following manner: If the pulleys over which the belt is stretched are quite close together, the two parts of the belt will be nearly straight, so that there will be little change of tension, whether the belt is at rest or in motion, and the centrifugal force diminishes the tension. If the pulleys are a considerable distance apart, the two portions of the belt will be curved, and part of the centrifugal force acts in increasing the length of the belt, and thus increases the tension, instead of diminishing it, as in the former case. In ordinary cases, where the velocity of a belt is not very great, it is probably not necessary to consider the action of centrifugal force.

"The writer is frequently engaged in making tests of machinery, and finds it convenient to obtain a record of the speed of shafts and engines. At a recent test of power in a large factory some interesting data were obtained in relation to the creeping of belts, and it is believed that these results may be of interest and value to other engineers."

The case was one in which the size of the belts was largely in excess of that actually required for the transmission of the power, and the velocity with which they moved was quite moderate. Under these circumstances it was to be expected that the difference between the actual ratio of speeds, and that which should have been obtained if there had been no slip or creep, would be less at an increased than at a diminished rate of speed, if the changes in speed were so slight as not sensibly to vary the centrifugal force. For the purposes of the test, counters were attached to the driving engine and to the shaft in a distant part of the building, the power being transmitted to that shaft by five belts. Simultaneous readings of the two counters were taken every minute for the space of an hour. The result of several of these observations for intervals of five minutes each is given below:

REVOLUTIONS OF ENGINE.	REVOLUTIONS OF SHAFT.	RATIO OF SPEED.
159	417	2.623
153	397	2.595
154	403	2.617
148	383	2.588
Total, 1848	4816	2.612

"It will be observed that, as the speed was decreased, there was a perceptibly greater creep in the belts. It having been found advisable to increase the speed of the engine and shafting, similar tests were made, when this change was effected, and a few of the results are given below:

REVOLUTIONS OF ENGINE.	REVOLUTIONS OF SHAFT.	RATIO OF SPEED.
205	538	2.624
204	535	2.623
203	532	2.621
206	541	2.626
208	544	2.615
Total, 2457	6432	2.618

"It is generally known that there is a best speed for a belt under given conditions, and that, if this speed is either increased or diminished, the tension of the belt will be diminished. It appears, from these latter results, that this speed was about reached by the change, so that the belts were transmitting under the most favorable conditions as regards speed.

"A careful measurement of all the pulleys transmitting the power from the engine to the shaft in question, increasing their diameters by the mean thickness of the belts passing over them, shows that the actual ratio of speeds of engine and shaft should have been as 1 to 2.625, so that there was an average slip or creep in the first case of 0.495 per cent., and in the second 0.267 per cent.—*R. H. Buel, M. E., 80 Broadway, N. Y., The Engineering and Mining Journal, February 28, 1874.*

Leather Pulley.

61. As there is "nothing like leather" for many purposes, especially in places where there is much rubbing and at high velocities, it has occurred to many, perhaps, and the transition of thought seems natural indeed, that if leather will endure so much wear and tear at the circumference of a wheel, it must resist a like action at its axle, and therefore it would make a good bush for a loose pulley, or, where small, the whole pulley might be made of leather. Accordingly, this has been done with good effect; leather discs were bolted together, bored out, turned off, soaked in oil, and put in place as if made of iron in the usual way.

Driving Power of Belts Compared with that of Friction Gear.

62. Newton's Journal for 1857, Vol. VI., New Series, p. 163, presents Mr. James Robertson's paper on *grooved surface frictional gearing*, from which we take the following:

"The object of this paper is to describe a system of frictional gearing recently introduced by the writer, intended chiefly for high speeds, and to give such information regarding its action and driving capabilities as the several applications of it in use will afford.

"The grooved surface frictional gearing consists of wheels or pulleys geared together by frictional contact, communicating motion independently of teeth or cogs; the driving surfaces are grooved or serrated annularly, the ridges of one surface entering the grooves of the other. The extent of contact is thus increased in the direction of the breadth of the rim, and a lateral wedging action is obtained, which augments the effect of the pressure holding the wheels in gear, the necessary amount of which is felt to be so injurious to the bearings of the shafts when the power is communicated by plain driving surfaces.

"The grooves are made V-shaped, and are found to suit best when formed at an angle of about 50°. The pitch of the grooves is varied to the velocities of the wheels and the power to be transmitted; the smallest pitch employed is $\frac{1}{2}$-inch, and that required for the heaviest operations about $\frac{3}{4}$-inch. The ordinary pitch is about $\frac{5}{8}$-inch. The wheels are turned up truly, and the grooves equally pitched and made exactly alike on each face; so that, on applying the surfaces to each other, a well-fitted contact throughout the faces is obtained. In order to increase and sustain the wedging action, the points of the ridges are left blunt, to prevent them from reaching the bottom of the grooves.

"Cast-iron has as yet been the only material used in the construction of grooved wheels, and its action has been found so satisfactory that there is no necessity for trying any other. The surfaces, after working a short time together, assume a smooth, polished appearance, taking a greater hold in proportion to the smoothness they acquire; and when a sufficient breadth for the speed and power to be transmitted gets into contact, there is afterwards no perceptible tendency to wear.

. . . "The points that have to be attended to, so far as the power or driving contact is concerned, are the angle of the grooves and the pressure holding them in contact; the extent of surface in contact being determined so as to prevent abrasion and withstand the wearing action. . . . Wheels of large diameter show a decided superiority of action.

"In order to obtain a high-speed from a driving belt, without the usual arrangement of counter-shafts and belt-pulleys between the main driving shaft and the machine to be driven, and without the disadvantage of passing the belt over a small pulley, a small grooved pulley is keyed on the shaft to which the high velocity is to be communicated, and upon it is placed a loose, inflexible ring, of two or three times the diameter of the pulley, grooved internally to fit it, and turned up smoothly on the outside to receive the driving belt. The belt gives motion to the speed-ring, the inner grooved surface of which communicates a higher speed to the pulley. The speed-ring is held in effective driving contact simply by the tension of the belt. For obtaining increased lateral steadiness at high speeds a double speed-ring may be used if required. By these arrangements a belt may be passed over a speed-ring of 16 inches diameter, and yet communicate the same speed to the shaft as if it were passed over a pulley of only 4 inches diameter.

"In Fig. 10 the driving pulley, D, carries a belt, B, which passes over and drives the speed-ring, R, of large diameter as compared with the small grooved wheel, P, increasing circumferential contact thereby.

"Enlarging the diameter of the speed-ring will permit the driving pulley to be placed nearer to, without altering the speed of the driven wheel, P, and will also increase the adhesion and driving power of the belt.

"The upper 'cut' shows an enlarged section of the speed-ring exhibiting its grooves, and those of the grooved wheel, P, with which it is engaged.

"Clutches may also be arranged for engaging and disengaging by means of grooved surfaces.

"In applying this system of driving, where a reverse motion is

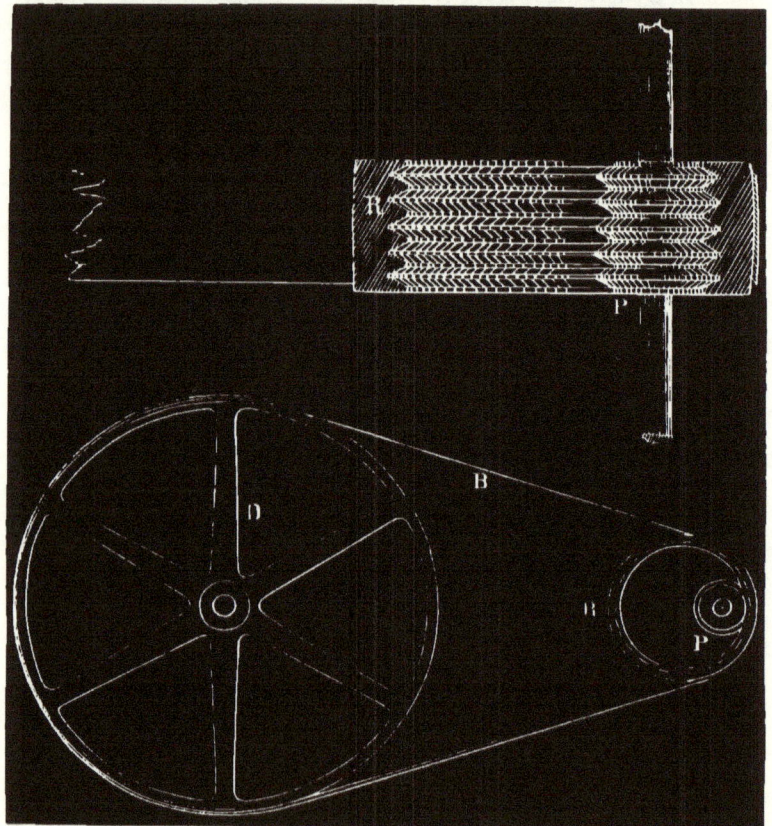

Fig. 10.

required, a disc, having an outer and an inner rim, is keyed to the main driving shaft — the outer rim grooved on the inside, and the inner rim on the outside, with corresponding grooves. The shaft that is to be driven carries a small grooved pulley, the diameter of which is slightly less than the distance between the two grooved rims. The motion of this pulley will be reversed by moving it slightly nearer to or farther from the main driving shaft, so as to throw it into gear with the inner or outer rim respectively. . . .

"For comparing the pressure required to hold the grooved surfaces in gear and the power transmitted, various opportunities have occurred in the actual use of the frictional gearing, and arrangements have been made for purposes of experiment. One method of comparing its driving capabilities with those of belts is directly obtained by the simple speed-ring movement already described for raising high speeds. One of these speed-rings has been working satisfactorily on a large foundry fan for some time; and, from the circumstances that the fan was previously driven by a belt of the same size, over a plain pulley of the same diameter as the small grooved pulley now used, this case affords a certain practical means of comparing the efficiency of these two methods of communicating motion. Before the application of the ring the belt was passed over a pulley 6 feet diameter, keyed on the driving shaft, and over a pulley $7\frac{1}{2}$ inches diameter on the fan spindle; but the continual bending of a large heavy belt over a pulley of so small diameter made it difficult to keep up the proper driving tension, and the belt was speedily cut up. The ring now interposed between the belt and pulley is $13\frac{1}{2}$ inches diameter, and saves the belt from injury by the greater diameter over which it bends. The ring works steadily, and drives the fan at the same speed as when the belt was passed directly over the small pulley; thereby showing that the grooved metal surface does not strain the bearings more than the ordinary arrangement of driving by belts.

"Another method has also been employed for comparing the driving capabilities of the grooved-surface gearing with those of belts, by means of a testing apparatus, having the same pressure on the bearings of the axis as is produced by belts. The testing apparatus is made by gearing together two spur grooved wheels, each 21 inches diameter and $3\frac{1}{2}$ inches face: the grooves being cut $\frac{3}{8}$-inch pitch, and at an angle of 50°. Motion was communicated to the driving wheel by a 7-inch belt, over a pulley 30 inches diameter, so disposed that there was no pressure to hold the two wheels in gear but the pull or strain of the belt. A plain friction strap wheel was keyed on the spindle of the driven wheel, by a strap and break handle attached, so that it could be either retarded or stopped. On applying the break, it either caused the belt to slip or the driving engine to stop, without the grooved wheels showing any tendency to slip.

"There is a slight slip in the rolling action of the grooved wheels which does not occur in the action of plain surfaces, which arises from the difference of the diameters of the points of the ridges and bottoms of the grooves; but this slipping is little felt in practice, and, when measured, is inconsiderable in amount. In a pair of grooved wheels, 8 feet diameter and 1 foot broad, with 24 grooves working together, there is a slip of only 10 square inches in an entire revolution; whereas, in toothed wheels of the same breadth and diameter, with cogs of 3-inch pitch, of the ordinary proportions, there is a surface to slip over on each cog of about 24 square inches, or nearly the entire area of one side of the cog; making a total slip of about 16 square feet in every revolution.

"Lengthened experience is necessary to ascertain the smallest breadth of face that will be sufficient for transmitting a given amount of power without abrasion or wearing action; and it is therefore preferred at present to make the grooved wheels broader in every position than seems to be absolutely necessary. The general proportion of toothed wheels, as regards both breadth of face and other dimensions, are sufficiently strong for transmitting the same power of grooved surfaces; but the writer is of opinion that less breadth of face and lighter proportions of arms and rims can be used with safety. If the grooved wheels are employed in every position in a factory where wheel gearing is required, no shocks or jolting action can take place; and therefore all the wheels themselves, and also the shaftings and supports, may be made much lighter than can be used with ordinary gearing.

"One of the principal advantages of these grooved wheels is their smoothness of action, in positions and at speeds when ordinary toothed gearing produces a disagreeable jarring noise, their action is scarcely audible."

From a Circular issued by P. V. H. Van Riper, of Paterson, N. J., we select the following:

63. "Having been engaged in the manufacture of oak leather belting for the past fifteen years, I would respectfully call attention to the essential points necessary to the manufacture of good belting, the first of which is the selection of the leather, which should be *oak tanned*, it being more pliable than any other, and as durability is required, it should be thoroughly tanned and made from young hides, they having more strength than the hides from old animals.

"Suitable leather having been chosen, though it may be ever so good, may be spoiled in currying, and as this is an important part, it is conducted under my own supervision, where I have the *shoulders* cut from the hides, and nothing but 4 *feet in length of the choice butts*, curried for belting purposes, as the *shoulder naturally stretching in a different direction* from the butts, causes that great annoyance in factories of belts running crooked.

"The putting on of belts should be done by a person acquainted with the use of belting, and too much judgment cannot be exercised in this respect, as the wear of the belt depends considerably on the manner in which it is put on, therefore, the following suggestions, if practised, will be of much service to persons employed in this capacity. The butts to be joined together, should be cut perfectly square with the belt, in order that one side of the band may not be drawn tighter than the other. For the joining of belts, good lace leather, if properly used, being soft and pliable, will always give better satisfaction than any patent fastening or hooks which have yet been invented.

"Where belts run vertically, they should always be drawn moderately tight, or the weight of the belt will not allow it to adhere closely to the lower pulley, but in all other cases they should be slack.

"In many instances, the tearing out of lace holes is often unjustly attributed to poor belting, when, in reality, the fault lies in having a belt too short, and trying to force it together by lacing, and the more the leather has been stretched while being manufactured, the more liable it is to be complained of.

"All leather belting should occasionally be greased with the following mixture, or it will become dry and will not adhere to the pulleys: one gallon neat's-foot or tanners' oil, one gallon tallow, twelve ounces resin, dissolved by heat, and well mixed together, to be used cold, the belt having been previously dampened with warm water, except where it is spliced together. During the winter season, an extra quantity of oil should be added to the mixture.

"To obtain the greatest amount of power from belts, *the pulleys should be covered with leather;* this will allow the belts to be run very *slack,* and give 25 per cent. more wear. I drive a large circular saw, requiring 15 horse-power, with a very slack belt, the pulleys being *covered with leather.*

"For heavy counter belts not intended to be used on cone pulleys, or at half cross, I recommend double belts, made from shoulders only, which I furnish at the price of single belting; and as the *stretch is taken out from the shoulders after they are cut from the side,* they are guaranteed to give better satisfaction as a counter belt than a single belt will.

"More power can be obtained from using the grain side of a belt to the pulley than from the flesh side, as the belt adheres more closely to the pulley; but there is this about it, the belt will not last half so long, for when the grain, which is very thin, is worn off, the substance of the belt is gone, and it then quickly gives out; so that I would advise the more saving plan of obtaining power by driving with wider belts, and covering the pulleys with leather.

"Where belts are to run in very damp places, or exposed to the weather, I would recommend the use of rubber belting; but for ordinary use it will not give the satisfaction which is so generally obtained from using oak leather belting, as it cannot be run on cone pulleys through forks, or at half cross, and with fair usage would be worn out, while a leather belt was regularly performing the work allotted to it; for when the edge becomes worn, the belt soon gives out."

We formularize the following rules from the text: —

$$HP = \frac{3\,Wcv}{16000}$$

$$W = \frac{16000\,HP}{3\,cv}$$

In which $HP =$ number of horse-power transmitted.
" $W =$ width of belt in inches.
" $c =$ belt contact with smaller pulley in lineal feet.
" $v =$ velocity of belt in feet per minute.

The following examples are given: "A $13\frac{1}{3}$-inch belt, running at the rate of 1600 Fpm, over a 4-feet pulley, and touching 5 feet of its circumference, gives 20 horse-power." This is equal to 88.888 square feet of belt per minute, per horse-power.

"A 20-inch belt at 2000 Fpm, on 6 feet of the circumference of a 4-feet pulley, gives 45 horse-power." This is equal to 74.074 square feet of belt per minute per horse-power.

To Measure Belts in Coil.

64. In order to calculate the length of a belt in coil, we may consider the coil as composed of a series of concentric circles, and then calculate the sum of their circumferences for the total length of the coil; but this would be tedious, and a simple method should be devised.

To obtain this, first, find the mean diameter of the extreme coil diameters by taking half the sum of the diameters of the outer and inner coils, multiply this number by 3.1416, and then by the number of coils, this will give the total length of the coil in inches if the diameters of the coils were taken in inches.

A formula expressing this rule will be:

$$L = 3.1416 \, n \left(\frac{D+d}{2} \right)$$

and may be simplified thus:

$$L = 1.5708 \, n \, (D + d).$$

In both of which L D and d must represent like units of measurement.

To simplify calculations still further by getting the length L in feet, and the diameters D and d in inches, the rule may be put into this form, which is probably the best for use:

$$L = .1309 \, n \, (D + d).$$

From Appleton's "American Cyclopædia" we take the following:

65. "Belts are used instead of gearing when the shafts to be connected are far apart.

"Belts in general are used between parallel shafts, and when the direction of motion is desired to be reversed the belt is crossed.

"The diameters of belt pulleys are in the inverse ratio of their speeds.

"To modify the velocity while in motion, conical drums are employed on parallel shafts, with the cones reversed in position, and a shifter used to move the belt, and hold it in any desired position.

"When shafts are not parallel, but are in the same plane, the only way to connect them by belts is to use a third shaft placed across both, and at equal angles, or nearly so, with each, and to which each is connected by a belt.

"When shafts are neither parallel, nor in the same plane, they can be connected by a belt, but there is only one place on the shafts for the pulleys. These must be at the ends of a straight line, perpendicular at the same time to both axes. There is only one such line. This theoretical place has to be corrected in each particular case according to the diameters of the pulleys, by taking care that the belt arrives square on each pulley, no matter how obliquely it leaves the other.

"As a consequence of this unavoidable connection, the motion of the shafts cannot be reversed without securing the pulleys in other places.

"A careful attendant will make a belt last five years, which through neglect might not last one.

"It has been found in practice that belts must not be run faster than 30 feet per second, nor have a tension of above 300 lbs. per square inch of section.

"The friction of a belt is double on wood what it is on cast iron.

"If a belt is passed over a horizontal cylinder with a known weight suspended at one end, and a spring balance attached to the other, and gradually let go until the belt begins to slide, the suspended weight, *minus* the weight indicated by the balance, is the friction.

"It has been found that by taking a turn and a half around a rough cylindrical post, 1 lb. will hold 110 lbs. in check, and that by taking $2\frac{1}{2}$ turns 1 lb. will hold 2500 lbs.

"A 12-inch belt over a 4-feet pulley, at 30 feet per second, will transmit the power of a 6-inch cylinder engine having 12-inch stroke, running 125 Rpm, under 60 lbs. of steam."

If we allow 30 lbs. average pressure per square inch on piston, this engine will give 6·4 horse-power, and the data above, 281.25 square feet of belt per horse-power per minute, which is a liberal allowance.

Pulleys with Leather Covering.

66. "The sliding or slipping of belts on the pulleys is an evil experienced by almost every one whose business depends on machine power. Various means have been devised to avoid it. One of them is to strew powdered rosin or pitch on the inside of the belt. Another is to cover the pulleys with wood. A third is to give the rim of the pulley a curved surface. These means are only palliatives, and lack a thorough, steady, and continued action. Rosin and pitch are soon pressed into the leather, when they not only lose their efficacy, but contribute to the rotting and destruction of the belts. A wood covering on the pulley gets polished in a short time, and is then as slippery as iron. It is therefore necessary to frequently roughen its surface, by which operation the diameter of the pulley is diminished, and the proportions of the transmission are altered. A convexity of the rim of the pulley is very effective to prevent the dropping off of the belt, especially when the pulley has a horizontal position; but it counteracts the slipping of the belt to but a small extent.

"We therefore take pleasure in communicating to the public a mechanical contrivance, which completely prevents the sliding of belts, and all the great disadvantages resulting from it. It consists in covering with leather the working surface of the pulleys. As the friction of leather on leather is equal to five times that of leather on iron, and as leather can be roughened and be easily kept in that condition, it is evident that a sliding of the belts cannot take place on pulleys covered with leather, not even when the belts have to transmit the very highest amount of power. We have seen such pulleys working in sugar factories, breweries, in manufactories of German silver, in paper mills, machine shops, sawing mills, and in many other mechanical establishments, in all of which they have proved of eminent usefulness and great practical value. With pulleys which have to run at a great velocity, as, for instance, those which drive blowers and saw-frames, as well as with pulleys of small diameter, which transmit powerful strains, the advantages of a leather covering are especially great. But besides these evident advantages resulting from the avoidance of the slipping, a leather covering on the pulley preserves the belt, in the first place, because the belt does not require tightening so hard, the friction being considerably increased; and in the second place, because there is no occasion for a rapid rotting of the belt. For this rapid rotting is generally caused by the fact that, under the influence of the heat produced by friction, the tannic and sebacic acids contained in the leather of the belts combine chemically with some of the iron of the pulleys, forming a hard compound in the belts, which produces what is called rottenness, and frequently causes breakages. This evil is, of course, avoided by covering the pulleys with leather. . . . . The coverings are fixed to the pulleys by a kind of paste or glue, which hardens in a very short time, and sticks so well to iron and leather, that the greatest forces can be transmitted by the pulleys without loosening the leather. The operation of covering is very simple, and can be done and renewed by every intelligent workman. The price is $1\frac{5}{8}$ Prussian thalers per square foot of leather, including the glue."—*Translated from Polyt Centralblatt. From Van Nostrand's Eclectic Engineering Magazine, July, 1869, p. 604.*

From a "Practical Treatise on the Manufacture of Worsteds and Yarns," M. Leroux, H. C. Baird, Philadelphia, 1869, we extract the following:

67. "It is rare that a force of over 10 horse-power is transmitted by means of belts.

"For a force of 8 or 10 horse-power the belts should be double, which prevents their stretching; that is to say, two belts are superposed and sewed together at their edges. Thus, for a 9 horse-power two belts are sewed together, one of which is 1 millimetre thicker than the other: 5.5 below and 4.5 above, making 10 millimetres, the thickness of a belt which will resist the action of this power and even a greater one. For low powers the thickness is always from 4 to 5 millimetres.

Table for ascertaining the Width of Belts.

Velocity per Minute in Metres.	Width of Tanned Leather Belts in Millimetres. Force in Horse-Power.						
	$\tfrac{1}{10}$	$\tfrac{2}{10}$	$\tfrac{5}{10}$	$\tfrac{9}{10}$	1	2	3
20	68	132	328	......	......	......	......
30	44	88	220	394	......	......	......
40	34	66	164	296	......	......	......
50	26	53	132	237	......	......	......
60	22	44	110	197	220	440	......
70	19	38	94	170	188	377	565
80	17	33	82	148	165	329	494
90	15	29	73	132	147	293	440
100	13	26	66	119	132	264	396
120	11	22	55	99	120	220	330
140	9	19	47	85	94	188	283
160	8	17	41	74	82	165	247
180	......	15	37	66	73	147	220
200	......	13	33	55	66	132	198
240	......	11	28	47	55	110	165
280	......	9	24	41	47	94	141
300	......	8	22	39	44	88	132
360	......	......	18	33	37	73	110
400	......	......	16	28	33	66	99
500	......	......	13	24	26	53	79
600	......	......	......	......	22	44	66
700	......	......	......	......	......	38	56
800	......	......	......	......	......	......	50
900	......	......	......	......	......	......	44
1000	......	......	......	......	......	......	40
1200	......	......	......	......	......	......	33
1500	......	......	......	......	......	......	26
2000	......	......	......	......	......	......	20

"The transmission of motion from one shaft to another, by means of belts, depends entirely upon the friction produced by their tension upon the pulleys or drums around which they are made to move. If the force transmitted by them is augmented, the friction is in like manner increased; and, if in that case the tension of the belts remains the same, their friction surface, or, what amounts to the same, their breadth must be increased: *the powers to be transmitted are to each other as the product of the width of the belts multiplied by the velocity.*

"M. Morin has found that belts of tanned leather will resist a tension computed at 2 kilogrammes for every square millimetre of their section.

"When it is desired to determine the width of a certain belt, *multiply the number of revolutions of the pulley or drum made in one minute by its circumference, and the product will express in metres the desired velocity.* The width in millimetres will then be found opposite this number and in the column of the foregoing table of the given power. If pulleys, however, are not in the relation of identical diameters, but are in the relations about to be mentioned, *then multiply the width given in the foregoing table by the coefficient of transformation.*

Coefficient of Transformation of the Width of Belts According to the Relations of the Diameters of Pulleys.

"For pulleys, the diameters of which are to each other as 1 : 2, the width indicated in the table will be multiplied by 0.75. For the ratio of 1 : 3, the multiplier will be 0.65; and for the ratio of 1 : 4, 0.58."

Experience shows that belts ought never to be less than 20 millimetres wide, as they are subject to stretching and breakage. Their width should also exceed that ascertained from the table by at least one-sixth. Machines working different materials, with varying quantities, undergo more or less strain. Thus, a spinning-mule, after having worked ten hours, absorbs $\frac{1}{8}$ more power than at the outset. Wet weather occasions the same effect, while obstructions, want of oiling, materials more or less difficult to spin, etc., are so many causes which have to be neutralized by developing the friction surface of the belts.

Loss of Velocity Suffered by Belts while in Motion.

The variable length of the belts has an influence upon their slipping. When they are crossed they are less liable to slip.

"The loss of velocity suffered by belts when mounted depends upon their friction surface.

"Long belts are less liable to slip than short ones, for the latter are always stretched in a manner injurious to the journals and brasses, and, notwithstanding this amount of tension, they are still subject to a considerable loss in velocity.

"I have undertaken some experiments in regard to losses of this nature to which belts are liable relatively to their lengths, and I have thought it well to prepare a table for calculating the amount of motion transmitted by belts which no operator can well do without.

Table Showing the Slip of Leather Belts Relatively to their Lengths.

Parallel Belts.		Crossed Belts.	
Length in Metres.	Percentage of Velocity Lost by Slipping.	Length in Metres.	Percentage of Velocity Lost by Slipping.
2	4.2	2	3.5
4	3.9	4	3.2
6	3.6	6	2.9
8	3.3	8	2.6
10	3.0	10	2.3
12	2.7	12	2.0
14	2.5	14	1.8
16	2.3	16	1.6
18	2.1	18	1.4
20	1.9	20	1.2

"Belts, after having served for a certain length of time, and having withstood more or less tension, become greatly impaired by stretching and narrowing.

"The width of belts diminishes in proportion to the strain upon them. Experience shows that on the first day a belt is used it suffers an elongation of one per cent. This action continues to diminish till the third day, after which the belt works on without much change in its dimensions.

"The causes producing loss of velocity in belts are, imperfect lubrication of machinery, obstructions in the journal boxes, wheel gearing out of line, inferior quality of leather, couplings and sewing, and oil on the pulleys.

"When a belt slips, the difficulty is remedied by sprinkling the rubbing surface with a mixture of Spanish white and resin. If the belt is smeared with oil, Fuller's earth is employed, which has the property of absorbing greasy substances, and the rubbing side of the belt is then scraped with a wooden blade.

"Very often a badly-made knot in a coupling joint will cause the belt to lose 1 or 2 per cent. of velocity.

"To transmit and secure the motion to be imparted, the belts are sewed in such a manner as best to insure against their slipping; but as they always tend to elongate, in order to obviate this difficulty, the ends are bound together by a leather thong. These are generally of Hungarian leather, cut into thin and narrow strips, so as to be readily handled, as well as to avoid the necessity of punching large holes in the belt, which greatly lessens its strength.

"The flaxen or hempen thread, intended for sewing belts, ought to be of superior quality, and smeared with some pitchy substance to prevent ravelling.

"M. Hunebelle, of Amiens, manufactures very durable belts, the couplings of which are made of Hungarian leather prepared in some peculiar manner. I have substituted animal substances for thread. I have had good results from eel-skins, and have also tried small cat-gut. My experience has been that the belt of a spinning-frame, sewed with this material, may last two years without suffering any deterioration; and the cost of this article is not so great as to oblige us to reject its employment.

"Pulleys should have a rise of $\frac{1}{12}$ of their face width.

"Sometimes, to impart motion to a machine situated at a distance from the transmission, we resort to what is called a binder or carrying pulley, which consists of two small wooden drums, having a face convexity of $\frac{1}{20}$ of their width, secured to iron axles running in brasses, the whole made adjustable. These drums should never be less than 20 centimetres ($0^m.20$) in diameter; a larger diameter never does harm. This contrivance, which was first introduced by a foreman named Buignet, stretches the belt in every direction."

Mr. C. R. Rossman, in "Technologist" for Oct., 1871, suggests:

68. As a safe working tension for single leather belts the allowance of 45 lbs. to the inch of width, and adds the following:

A 125-horse engine drives two 18-inch belts over 8-feet pulleys, making 75 Rpm. This gives a velocity of 1875 Fpm, and a tension of 61 lbs. to the inch of belt.

"This is in excess of the safe limit of tension generally recommended; but we may here remark that belts of the width here mentioned are generally thicker and stronger than the average belts used, and from which the ordinary data were taken. But, from a careful examination of a great number of cases of belts of ordinary width and strength, we find that a safe and judicious limit lies between 40 and 50 lbs. In order to increase this, however, it is not unusual for engineers to double the thickness of the belt by cementing or riveting two thicknesses of leather together. But this plan, though advisable in some cases, is not so economical of power and material as the equally efficient plan of increasing the width of the belt.

"The tensile strength of good ox-hide, well tanned, has been carefully examined, with the following results:

The solid leather will sustain, per inch of width,			675 lbs.
At the rivet holes of the splices,	"	"	382 "
At the lacing,	"	"	210 "
Safe working tension,	"	"	45 "

"The belts are assumed to be $\frac{1}{4}$ of an inch thick."

From C. F. Scholl's "Mechanic's Guide," pages 483–5.

69. "Pulleys must be true and concentric and their shafts parallel, otherwise belts which run upon them must be guided, and the guiding device will wear their edges rapidly. To prevent belts from running off, pulleys should be made convex on their faces, much convexity, however, is destructive to thick and to double belts. Pulleys for shifting belts should be parallel-faced, except where the shafts are far apart, when they may be convex. Flanges to pulleys and belt-guides should be avoided, except to pulleys on upright shafts which have a slower motion, and where two belts run closely together on the same pulley, or on two pulleys of like size; but for high speed they may be discarded.

"The softer woods are better for pulleys than the harder kinds, but pear-wood and nut-tree are best for cord-wheels. Grease must not be put on wooden wheels on which belts run.

"Tighteners must be applied to the slack side of belts.

"Good oak-tanned wild leather is the best for belts, and not that prepared with alum.

"The belts should be cut from the centre of the skin, stretched, and of even thickness throughout. The ends are joined by leather laces, rove through uniformly-punched holes.

"New belts are liable to stretch, and should be unlaced, shortened, new holes punched, and then laced again.

"The most practical method of fastening the ends of belts is by the screws shown in Fig. 11. The belt must travel in the direction of the arrow, and never allowed to run against the joint.

"The method shown in Fig. 12 is also recommended. The plate is of brass, rather narrower than the belt, curved to the pulley, lapping the joint, and receives countersunk head-screws from each end of the belt.

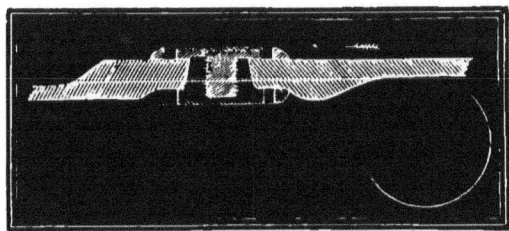

Fig. 11.

"In Fig. 13 another plan is shown in which incurved teeth of malleable iron connected to a plate, are driven through each end of the belt when butted, and are then clinched on the inside.

Fig. 12.   Fig. 13.

"Thickness of belt does not always give strength. Small pulleys injure the structure of the belt by too great flexure. Single belts are relatively more durable than compound belts; the latter should be used only on large diameter of pulleys. Gum belts and belts of which gum forms a part, are preferable in damp localities, but they should not be shifted much.

"Horizontal running belts should be long; their own weight doing the required work without excessive tension, but is limited to a shaft distance of about 28 feet. At a greater distance than this, especially at high speeds, belts sway injuriously from side to side.

"Powdered Colophonium may be applied to a slipping belt.

"Belts driving mills at high speed work better than toothed wheels, and all machines subject to intermittent motions, or liable to sudden stoppage, should be driven by belts.

"Fat should be applied to belts, say every three months of their use: they should first be washed with lukewarm water, and then have leather grease well rubbed in.

"A good composition, which should be applied warm, consists of:

Lard or Tallow...	1 part.
Fish Oil...	4 "
Colophonium...	1 "
Wood Tar...	1 "

An Engineer of considerable Practical Experience with Machinery replies:

70. Rule for horse-power of a belt —

$$\frac{HP \times 26000}{V \times C \times 6} = W.$$

In which $HP =$ horse-power.
" $V =$ velocity of belt in feet.
" $C =$ circumferential contact with smaller pulley in feet.
" $W =$ width of belt in inches.

"Single belts have given us the best satisfaction, taking less power to drive them, and adhering to the pulleys much better, and do not crack in bending.

"The area of belt contact determines its driving power.

"For fastening the ends we consider hooks the best for small belts. For large belts we use hooks, with the addition of a piece of leather riveted over the lap.

"The best composition for preserving belts and giving them adhesion, is oil with a small quantity of rosin in it.

"Thin belts are better than thick ones.

"The convexity of pulleys should be the least possible, not over $\frac{1}{8}$ inch to the foot of breadth."

Another Engineer gives the following:

71. "To find the number of horse-power which a belt will transmit, multiply the number of square inches of belt contact with the pulley by the velocity of the belt in feet per minute, and divide the product by 64000."

A Skilful Machinist, of much Experience with Experimental Machinery, says:

72. "The best method of joining belts that I know of is by the ordinary lacing. The holes therefor should be punched about $\frac{3}{4}$ of an inch apart, or as near that as the equal division across the belt will allow, and for wide belts 2 rows of holes should be punched, in a zigzag direction, commencing with the lace in the middle of the belt, and lacing singly to each edge, returning to the centre, and securing the ends, crossing the lace on the outside of belt.

"I have found this plan preferable to the one generally adopted, that is, of lacing double at once across the belt finishing at one edge, the objection to which is in the loosening of the ends of the lace and yielding of the joint at one edge, which will crook the belt and render it liable to lateral running, off the pulleys, against flanges, into gear perhaps, and may permanently stretch the belt so unevenly as to make straight running again matter of impossibility."

Belts for Rolling-Mills.

73. "Nearly all rolling-mills in Pittsburgh are driven with belts, from 20 inches in breadth upwards; something like the proportions may be guessed at from the following: One mill has a pulley 26 feet diameter, 66-inch face, with two 32-inch belts running to a pulley overhead, about 8 feet diameter. Another is 25 feet diameter, 45-inch face. Engines with such wheels and belts are being built every day, giving great satisfaction.

"The question of driving rolling-mills with belts might be dwelt upon, but its advantages must be apparent to any practical man."—*The Engineer, March 4, 1871.*

A Large Leather Belt.

74. "At the New Jersey Zinc Co.'s works at Newark, N. J., is a quadruple leather belt of unusually large dimensions. It is 102 feet long, 4 feet wide, and weighs 2200 lbs. The outside layer consists of 2 widths, the second and fourth layers of 3 widths, and the third layer of 4 widths; all the layers being riveted and glued together, and the end joints of the pieces forming the several layers are lapped to give the greatest tensional strength to the whole.

"It runs on an engine band wheel 24 feet diameter, with straight face 4 feet wide, of smooth turned iron, and over a driven pulley on the line shaft of 7 feet diameter having similar face, the centre of which lies 5 feet above the engine shaft.

"It has been in use three years, is doing well and giving no trouble, even when doing its heaviest work.

"The engine has a cylinder of 28 inches diameter and 5 feet stroke, makes 35 Rpm under 70 lbs. of steam and 14 lbs. of vacuum; estimated capacity of belt 250 horse-power, velocity 3080.7 Fpm, and

49.3 square feet of belt travelling per minute per horse-power."—
J. M. Hartman.

Mr. Benjamin Clement, M. M., of the Calico Print Works, Dover, N. H., communicates the following:

75. "For estimating the horse-power of belts, I use the formulæ of Hoyt Bros., of New York, which are:

$$W = \frac{36,000 \times HP}{V \times F \times 6} \text{ and } HP = \frac{\frac{1}{2} S \times V}{36,000}$$

In which $W=$ width of belt in inches.
" $V=$ velocity of belt in feet per minute.
" $HP=$ horse-power.
" $S=$ square inches of belt in contact with smaller pulley.
" $F=$ length of belt in feet in contact with smaller pulley.

"The above supposes each square inch of belt in contact to raise $\frac{1}{2}$ lb. one foot high per minute.

"All other things being equal, the area of contact will govern the driving power.

"Castor-oil drainings are the best for leather belts.

"The grain side in all cases should go next the pulley.

"My practice is to give $\frac{5}{16}$" per foot of breadth for the convexity of pulleys.

"For covering pulleys I use a mixture of three pints of glue made up with vinegar, to which I add a common teacup full of Venice turpentine.

"We have at these works an 18" single belt driven by a pair of bevel gears. The belt pulleys about 30 feet between centres, 6 feet in diameter, and make 100 Rpm. The gears are $2\frac{1}{2}$" pitch, 7" face, and have 48 teeth; the driven one wooden cogs. According to the relative wear I consider them well balanced.

"We have a double leather belt 18" wide, 35 feet between centres of pulleys; the driven pulley 5 feet 10" diameter, running 116 Rpm, which has given off 130 horse-power by actual tests with indicator applied to the engine.

"This is a belt velocity of 2,117 Fpm, and a surface velocity of 24.42 square feet of belt per minute per horse-power."

The Largest Belt in the World.

76. "Messrs. J. B. Hoyt & Co., of New York, exhibited in Machinery Hall a double belt of oak tanned leather, 186½ feet long, 60 inches wide, and weighing 2212 lbs.

"This is believed to be capable of transmitting 600 horse-power, and is made for the 'Augustine' Mill of Jessup & Moore, paper manufacturers, Wilmington, Del."— *Polytechnic Review, Philada.*

A Heavy Belt.

77. "Messrs. P. Jewell & Sons, Hartford, show a double belt 147½ feet long and 36 inches wide, and weighing 1130 lbs., or over 2.57 lbs. per square foot. This is claimed to be the heaviest belt, per surface, in the Exhibition."— *Polytechnic Review, Philada.*

Example.

78. "At Druid Hill, Baltimore, Md., a Babcock and Wilcox engine turns a 28-feet pulley fly-wheel, lagged with wood, 52 Rpm, carrying three single leather belts, each 16 inches wide, and developing 500 horse-power, which equals a belt transmission of 36.6 square feet per minute per horse-power; velocity of belt, 4579 Fpm."— *G. H. Babcock.*

Strength of Band-Saw Blades. From "Polytechnic Review," Phila.

79. "Test of the strength of eight specimens of Perrin's Band-Saw Blades, with brazed joints, by Richards, London & Kelly, made on Riehlé Bros. Testing Machine, July 19, 1876:

No.	Thickness.	Width.	Width nearest $\frac{1}{16}$ inch.	Breaking weight.	Strength per square inch.
1	.0346	1.05	$1\frac{7}{16}$	7600	209,193 lbs.
2	.0353	.62	$\frac{10}{16}$	4000	182,765 "
3	.0365	.745	$\frac{12}{16}$	6000	220,649 "
4	.0337	1.062	$1\frac{1}{16}$	3000	83,823* "
5	.0310	.625	$\frac{10}{16}$	2230	115,090† "
6	.0310	.490	$\frac{8}{16}$	2000	131,060† "
7	.0335	.280	$\frac{9}{32}$	2000	213,210 "
8	.0310	.094	$\frac{3}{32}$	485	16,430 "

* Broke at end of joint.   † Broke across centre of joint.

"The average strength of the unjoined pieces was 446 lbs. for each $\frac{1}{16}$ inch in width, and the strength of the weakest (which were the narrowest also), 323 lbs.; while the average strength through the joints for each $\tfrac{1}{16}$ inch in width was 206 lbs. per $\tfrac{1}{16}$ inch; in the weakest, 176 lbs. All the blades for the ordinary saws are made of No. 19 B. W. G. steel, and vary only by the inequalities caused by grinding or filing the joints. The knowledge that when a band-saw is being strained to the amount of 175 lbs. for each $\tfrac{1}{16}$ inch in width is strained to nearly its limit of endurance, may be of some value to the makers and users of band-saws."— *John E. Sweet.*

From Dr. H. M. Howe, Franklin Sugar Refinery, Philadelphia:

80. "A 24-inch diameter iron cylinder, nearly horizontal, has a stream of bone-black running constantly through it at a temperature of 350° Fah. This is caused to revolve by a 7-inch wide gum belt, applied directly to the naked cylinder, and lasted just three months. The cylinder was used 20 to 22 hours in the 24. Before applying the next belt the cylinder was lagged with wood 1 inch thick; two months of use showed no appreciable deterioration of the belt.

"In our refinery gum belts do better than leather in hot and in damp situations, and in places where they are subjected to the combined influence of heat and moisture.

"Single leather belts seem to be more economical than double ones in places where the atmosphere is charged with sugar and bone-black dust. Double belts stiffen and crack, then the pores become filled, and dubbing does not relieve them, but forms with the dust a glazing on the surface.

"Single belts are not likely to become so unpliable, and therefore hug the pulleys better, and work *more than half* as long as double belts. Gum belts make better elevators than double leather belts.

"Our belts are generally put *flesh side* next the pulley; have had some put *hair side* next the pulley, which we prefer, believing the adhesion is greater, and have not observed any less liability to crack when so used.

"In our 'Mill Room,' where there is much sugar dust, but no great heat, if we keep the belts dry, they work and last well; while in the 'Black' rooms, where there is high heat and dust combined, belts are short lived."

The Value of Rubber, Gutta-percha, and Canvas Belts as compared with Leather.

81. "Under the same circumstances, and on the same machines, these bands will not last or wear one-fourth as long as leather. When once they begin to give out, it is next to impossible to repair them.

"Wide bands cannot be used for or cut up into narrow ones, as leather can be.

"Leather belts may be used over and over again, and, when of no further value for belts, can be sold for other purposes.

"A rubber band, costing hundreds of dollars, may be spoiled in a few moments, by the lacing giving out and the band being run off into the gearing, or by being caught in any manner so as to damage the edge, or by stoppage of either the driving or driven pulley. A few moments of quick motion or friction will roll off the gum from the canvas in such quantities as to spoil the band.

"Leather belts may be torn or damaged, yet are easily repaired.

"Should a rubber or gum belt begin to tear by being caught in the machinery, if the rent strikes the seam, it is most certain to follow it, even the entire length, if the machinery is not stopped. It would be impossible to tear leather in like manner.

"Oil, in contact with rubber belting, will soften the gum.

"Rubber, gutta-percha, and canvas belts will continue to stretch as long as in use, rendering it necessary to shorten them continually.

"During freezing weather, if moisture or water finds its way into the seams, or between the different layers of canvas composing these bands, and becomes frozen, the layers are torn apart, and the band is spoiled; or, if a pulley becomes frosty, the parts of band in contact with it will be torn off from the canvas and left on the pulley.

"Gum belts will not answer for 'cross' or 'half-cross' belts, for 'shifting' belts, 'cone pulleys,' or for any place where belts are liable to slip, as friction destroys them. . . .

"A well-made leather band, if properly looked after — the width and pulley surface proportional to the amount of work to be done — will last 12, 15, or 20 years, and yet be of value to work over into narrow belts."—*J. B. Hoyt & Co.*

From "Rankine's Manual of Machinery and Mill Work" we take the following:

82. "The flexible pieces used in machinery may be classed under three heads: *Cords*, which approximate to a round form in section; *Belts*, which are flat; and *Chains*, which consist of a series of rigid links, so connected together that the chain, as a whole, is flexible. Mr. Willis gives them all the common name of *wrapping connectors*, and, for the sake of brevity in stating principles that apply to them all, they may conveniently be called *bands*.

"The *effective radius* of a pulley is equal to the radius of the pulley added to half the thickness of the band.

"Smooth bands, such as belts and cords, are not suited to communicate a velocity-ratio *with precision*, as teeth are, because of their being free to slip on the pulleys; but the freedom to slip is advantageous in swift and powerful machinery, because of its preventing the shocks which take place when mechanism which is at rest is suddenly *thrown into gear*, or put in connection with the prime mover. A band at a certain tension is not capable of exerting more than a certain definite force upon a pulley over which it passes, and therefore occupies, in communicating its own speed to the rim of that pulley, a certain definite time, depending on the masses that are set in motion along with the pulley and the speed to be impressed upon them, and until that time has elapsed the band has a slipping motion on the pulley; thus avoiding shocks, which consist in the too rapid communication of changes of speed.

"The swell usually allowed in the rims of pulleys is *one twenty-fourth part of the breadth.*"

In quarter twist belts, "in order that the belt may remain on the pulleys, *the central plane of each pulley must pass through the point of delivery of the other pulley*. It is easy to see that this arrangement does not admit of reversed motion.

"The safe working tension of leather belts, according to Morin, is 285 lbs. on the square inch. The ordinary thickness of belting leather is about .16-inch.

"The inside of the leather is rougher than the outside, and is placed next the pulleys, crossed belts being twisted so as to bring the same side of the leather in contact with both pulleys.

"*Leather belts*, when new, are not quite of the heaviness of water, say 60 lbs. per cubic foot; but, after having been for some time in use, they become thinner and denser by compression, and are then about as heavy as water. The weight of single belting is approximately .068 lbs. per one inch breadth and one foot length.

"*Raw-hide belts* have a tenacity about one and a half that of tanned leather. When raw hide is used for belts or for ropes it is soaked with grease, to keep it pliable and protect it against the action of air and moisture.

"*Gutta-percha* is sometimes used for flat belts. They are made of the same dimensions with leather belts for transmitting the same force, and are nearly of the same weight.

"*Woven belts* are made of a flaxen or cotton fabric, a sufficient number of plies being used to give a thickness equal to that of leather belts, and cemented together with india-rubber. When made of flax, they are said to be about three times more tenacious than tanned leather belts of the same transverse dimensions.

"Ultimate tenacity of leather rope, 10,000 feet, or 3360 lbs. on the circular inch; of raw hide, 15,000 feet and 5040 lbs. on the circular inch; safe working tension, one-sixth of these dimensions.

"The ordinary speed of wire ropes in Mr. C. F. Hirn's 'Telodynamic Transmission' of power is from 50 to 80 feet per second, and with wrought-iron pulleys it is considered that it might be increased to 100 feet per second.

"In order that the rope may not be overstrained by the bending of the wires of which it consists, in passing round the driving and following pulleys, the diameter of each of those pulleys should not be less than 140 times the diameter of the rope, and is sometimes as much as 260 times.

"The distance between the driving and following pulleys is not made less than about 100 feet, for at less distances shafting is more efficient; nor is it made more than 500 feet in one span, because of the great length of the catenary curves in which the rope hangs. When the distance between the driving and following pulleys exceeds 500 feet, the rope is supported at intermediate points by pairs of bearing pulleys, so as to divide the whole distance into intervals of 500 feet or less.

"The bearing pulleys have half the diameter, and are of similar construction with the driving pulleys.

"The loss of work due to the stiffness of the rope may be regarded as insensible; because, when the diameters of the pulleys are sufficient, the wires of which the rope is made straighten themselves by their own elasticity, after having been bent.

"Experience shows the loss of power by the axle friction of driving and following pulleys to be about $\frac{1}{40}$, and of the axle friction of each pair of bearing pulleys about $\frac{1}{500}$ of the whole power transmitted."

Belts.

"The driving pulleys fixed upon the shaft should be well centred, so that there may be no inequality of motion which would destroy the belts.

"To transmit motion to the apparatus without noise or loss of power, tanned leather belts of first quality are preferably used. They wear one and a half times as long as those of inferior qualities, which, although their low price is an inducement to purchasers, are more expensive in the end, by the stretching and rapid deterioration they undergo.

"The greater or less thickness of belts often contributes to their stretching, and the continual variations to which they are subject while extended over the circumference of pulleys or drums.

"For high powers, well tanned leather of sufficient thickness should be preferred. I have prepared the following table, which gives the thicknesses of belts calculated from the variable power of machinery, and the diameters of pulleys:

No. of Horse-power.	Thickness in Millimetres. Pulley Diam. at least = 0m .30.	Thickness in Millimetres. Pulley Diam. at least = 0m .20.
$\frac{1}{2}$	$5\frac{1}{2}$	5
1	6	$5\frac{1}{2}$
2	$6\frac{1}{2}$	6
3	7	$6\frac{1}{2}$
4	$7\frac{1}{2}$	7
5	8	$7\frac{1}{2}$
6	$8\frac{1}{2}$	8
7	9	$8\frac{1}{2}$
8	$9\frac{1}{2}$ doubled belt.	9 doubled belt.
9	10 " "	$9\frac{1}{2}$ " "
0	11 " "	10 " "

Extracts from paper "On the Centrifugal Force of Bands in Machinery," by W. J. M. Rankine, in "Engineer," March 5, 1869.

83. "It is well known, through practical experience, that a belt for communicating motion between two pulleys requires a greater tension to prevent it from slipping when it runs at a high than at a low speed.

"Various suppositions have been made to account for this, such as that of the adhesion to the belt of a layer of air, which at a very high speed has not time to escape from between the belt and the pulley. But the real cause is simply the centrifugal force of the belt, which acts against its tension, and therefore slackens its grip of the pulleys.

. . . "It can be proved from the elementary laws of dynamics, that if an endless band, of any figure whatsoever, runs at a given speed, the centrifugal force produces an uniform tension at each cross section of the band, equal to the weight of a piece of the band, whose length is twice the height from which a heavy body must fall, in order to acquire the velocity of the band.

"In symbols, let w be the weight of a unit of length of the band;

v the speed at which it runs, and g the velocity produced by gravity in a second ($= 32.2$ feet); then the *centrifugal tension* (as it may be called) has the following value: $\dfrac{w v^2}{g}$.

"The effect on the band when in motion is, that at any given point, the tension which produces pressure and friction on the pulleys, or *available tension* (as it is called), is less than the total tension by an amount equal to the centrifugal tension; for this amount is employed in compelling the particles of the band to circulate in a closed or endless path. It is, of course, to the total tension that the strength of the band is to be adapted, therefore the transverse dimensions of a band for transmitting a given force must be greater for a high than for a low speed.

"One of the most convenient ways of expressing the size of a band is by stating its weight per unit of length; for example, in pounds per running foot or in kilogrammes per metre. When the size is expressed thus, the corresponding way of expressing the intensity of any stress on the band is in lineal units of itself, such as feet or metres. Let b denote the greatest safe working tension on a band of a given kind, in units of its own length; w, as before, the weight of a unit of length; so that $w l$ is the amount of the safe working tension in units of weight. Let T be the amount of the available tension required at the driving side of the band for the transmission of power, being usually from two to two and a half times the force to be transmitted. Then the total tension is

$$T + \frac{w v^2}{g} = w l.$$

"Whence it is obvious that the required weight per unit of length is given by the following formula:

$$w = \frac{T}{l - \dfrac{v^2}{g}}.$$

"For example, suppose that the band is a wire rope, that the greatest working tension is to be equivalent to the weight of 2900 feet of the rope, and that it is to run at 100 feet per second: then we have

$$l = 2900 \text{ feet};$$
$$\frac{v^2}{g} = 310 \text{ feet};$$

"And consequently the weight per running foot of the rope required is:

$$w = \frac{T}{2900-310} = \frac{T}{2590};$$

"Or about one-eighth part heavier than the rope required, for a speed so moderate as to make the centrifugal tension unimportant.

"In fixing the value of the greatest working tension on a wire rope, a proper deduction must of course be made for the stress produced by the bending of the wires round the pulleys.

"That stress is given *in equivalent length of rope* by the expression $\frac{L\,d}{D}$, where D is the diameter of the smallest pulley round which the rope passes, d the diameter of the wire of which the rope is made, and L the modulus of elasticity of the wire, *in length of itself*, viz.: about 8,000,000 feet, or 2,400,000 metres. That is to say, let l_i be length of the rope equivalent to the greatest safe working tension on a straight rope; l as before, the length equivalent to the actual greatest working tension, then

$$l = l_i - \frac{L\,d}{D}.$$

"In the case of leathern belts, b may be estimated at about 660 feet, or 200 metres.

"In the case of a leather belt running at the rate of 100 feet per second, the weight per unit of length required, in order to exert a given available tension, is increased in the ratio of $\frac{660}{660-310} = \frac{660}{350}$, or to nearly double, as compared with that of a belt whose centrifugal force is unimportant.

"The *sectional area of a leathern belt* may be calculated approximately in square inches by multiplying the weight per running foot by 2.3; or in square millimetres, by multiplying the weight in kilogrammes to the running metre by 1000.

"The ordinary thickness of a single belt being about 0.16 inch, or 4 millimetres, the breadth may be deduced from the sectional area by dividing by that thickness.

"The length (L) equivalent to the modulus of elasticity of a leathern belt, as calculated from Bevan's experiments, is about 23,000 feet, or 7000 metres."

The Sliding of Belts and its Prevention.

84. "The *Praktische Maschinen-Constructeur* affords us an excellent paper upon the above topic, the translation of which reads as follows:

"The most convenient and least expensive mode of transmitting power consists, undoubtedly, in the use of belts. Nevertheless, this system, owing to the sliding of the belts, is connected with a great loss of power that is seldom observed. Let us suppose two corresponding pulleys, of the same diameter, in motion at a moderate velocity. In case the tension of the belt is sufficiently great, and the pulleys not too small, it may be difficult at first to perceive the sliding of the belt; but if the rotations of both the pulley and the belt are counted, a difference in their number will, in nearly all cases, be discovered, the belt performing a less number of rotations than the pulley. We thus become aware that the belt slides upon the smooth surface of the pulley, owing to the fact that the power sought to be transmitted by the belt is greater than the friction of the leather upon the iron, and that this power overcomes the friction, more or less slowly, according to the circumstances of the case. In regard to this, it is self-evident that the velocity and diameter of the pulleys must be taken into account, for the friction is more easily surmounted at a rapid velocity than at a slow one, and pulleys of large diameter offer to the strap more surface, and this increases the friction. The consequent sliding of the straps takes place the more easily the greater the peripherical velocity and the power to be transmitted, on the one hand, and the smaller the diameter and the breadth of the driver or pulley on the other.

"The sliding of the belt, however, represents a loss of power and fuel. Supposing that the two corresponding pulleys, of equal diameter, the turning one makes 100 rotations in a minute, and the turned one 95, there will be a loss of nearly 5 per cent. of the amount of force generated by the motor.

"It will become clear why the loss amounts to not quite 5 per cent. when it is taken into consideration that in sliding there will be a surplus of the transmitted force over the frictional resistance of the belt, which will be expended in the removal of the friction. This may be most easily recognized when we have a very considerable transmission of power, and when a perceptible sliding occurs, as, for instance, in driving a hydraulic press. If the pressure upon the piston of the pump is much greater than the friction of the belt upon the drum, the former is stopped in its motion, while the belt runs around the driving pulley, the motor itself thereby attains an accelerated velocity, from which it follows that the force required to overcome the friction is less than the one necessary for the working of the hydraulic press.

"In order to ascertain this loss of power for all cases, numerous and extensive experiments had to be undertaken. It may be stated here that this loss may amount to as much as 20 per cent., according to the circumstances. It may appear, at first, that the best means of guarding against this loss would be to construct the pulleys, in regard to diameter and width, so that the resistance of friction could not be overcome by the power transmitted. Nevertheless, upon a closer examination, it will become evident that this can, in most instances, only be accomplished at great expense. The outlay for leather belts is already considerable in an establishment of medium size, and would be still greater if all the belts were to be taken of such a width as is necessary for the prevention of sliding. Other means have, therefore, been proposed and applied. For instance, an endeavor was made to increase the friction of the leather upon the iron by spreading pulverized rosin or asphaltum upon the belt. In instances where the latter ceases to draw, the effect shows itself at once: but, nevertheless, only for a short time. On account of the pressure, the resinous powder penetrates the belt, so that its surface soon becomes as smooth as before, while the leather soon gets brittle and is gradually destroyed.

"The covering of the pulleys with wood is less objectionable, but it can find only a limited application. Only a wide pulley can be lined in this manner; but as the wood soon gets as smooth as the iron, it is necessary to roughen its surfaces repeatedly. This brings about a change in the diameter of the pulley, as well as in the amount of force transmitted.

"A third preventive, which seems to have found extended application, consists in giving to the pulley a convex surface. But whether this prevents the belt from running off is yet to be proven. The writer has ascertained that in consequence of the convexity of the drums, the belts will be stretched more in the centre than at the edges, and that as the frictional surface thus becomes smaller, the danger of sliding is increased.

"The covering of the pulleys with leather is undoubtedly the more advantageous, as thereby the frictional resistance is increased; the co-efficient of friction of which, with the leather, is considerably greater than that of the leather upon iron. The latter amounts to 0.25, the former to 1.25, which is just five times more.

"The resistance of friction may, nevertheless, be greatly increased by roughening the leather lining, and keeping it thus by means of an alum or salt solution. The great benefit to be derived from this system has been demonstrated by practical tests.

"We give some of the results of the new system. A spinning-machine was made to produce, continually, a uniform thread, while by the sliding of the belt it produced a thread containing knots and unequal spots. Ventilators which made only 1100 rotations per minute, in consequence of sliding, made 1400 after sliding was prevented. In a steam-mill, with five run of mill-stones, each set ground 27 bushels per day after the pulleys were covered with leather, while before the amount ground per day was only from 23 to 24 bushels. Moreover, the troublesome falling off of belts, which previously occurred very often, ceased altogether. In a paper-mill, a rag engine did 15 per cent. more work per day after its pulleys were covered with leather. In sugar-mills, for the beet crushers the centrifugal and other apparatus, this system has been fully approved, and it cannot be doubted that it will be of advantage wherever introduced. It may be remembered that leather may also be used in establishments where the power is transmitted by wire-ropes. It is in such cases preferable to wood, cork, asphaltum, and gum.

"The reason why the belts last longer is to be attributed to the fact that the increased friction allows a lesser degree of tension of the belts than would be the case if they were to run on a smooth iron surface. It is well known to every machinist, that for great transmissions of power the belts, if running on an iron surface, must be stretched to the utmost limit. This may be regarded as one of the causes of the rapid destruction of the leather.

"The same result is brought about sooner from the circumstance that, on account of the friction, fine particles of iron are detached, which, by combining with the tannic-acid and the fatty acids in the leather, form compounds, which, by penetrating the leather, cause the same to become brittle. By covering the pulleys with leather, this evil is prevented. But the chief cause of the rapid destruction of the leather is to be attributed to the sliding itself, which, as before mentioned, represents a useless loss of power. By the friction of the leather upon the iron heat is generated, which causes what is called the 'burning' of the leather.

"Therefore, by running belts upon smooth iron pulleys, not only the power to be transmitted acts destructively upon the leather, but other causes also, which is not the case when leather-band wheels are employed. The application of leather to the wheel is very simple and easy, and may be done by means of glue by any intelligent workman."—*Technologist, Nov., 1870.*

Mr. F. W. Bacon sends us the following:

85. "A 12" cylinder, 36" stroke engine, running 66 Rpm, with 10-feet fly-wheel pulley, drives a 6' 6" pulley, 19 feet horizontally away, and 8 feet above the engine shaft; both pulleys of iron, smoothly turned. Over these passes a 14" single leather belt, with strips 2" wide secured to each edge of the outer face; top fold of belt sags 20" from the straight line between pulley surfaces. The belt is about 80 feet long, and was dressed with castor oil when started; has been running 4 months, does not slip under heaviest load, adheres closely to pulleys, even at the edges, owing to the increased tension put upon the belt by the strips. The engine indicates 35 to 40 horse-power. At 40 it gives 60.475 square feet of belt per horse-power per minute.

"Another case which may be of interest is that of a belt 12" wide, 68 feet long, stripped at the edges like the above; is driven by a 53" diameter pulley, unturned, around two pulleys on a vertical shaft, thence to a 40" diameter smooth-turned pulley at right angles to the driver, which makes 101 Rpm. The work done is from 25 to 30 horse-power. The belt is dressed with castor oil, and is ample; at 30 horse-power it is 35.256 square feet of belt per minute per horse-power."

I have the case of an 18-inch diameter by 36-inch stroke horizontal steam-engine, running 60 Rpm, and indicating 77 horse-power. It has a 12-feet pulley fly-wheel, over which runs a $14\frac{1}{2}$-inch, much used, double leather belt. This belt drives a 5-feet pulley, the centre of which is $24\frac{1}{4}$ feet from and $10\frac{1}{3}$ feet above the centre of fly-wheel; the lower fold draws, and the belt runs quite freely without slipping. These figures give a velocity of surface equivalent to 35.5 square feet per horse-power per minute.

A certain 13-inch pulley on a shaft running 203 Rpm carries a belt 2.25 inches wide, and drives a 20-inch pulley on a shaft 20 feet vertically above. The pulleys are smooth turned iron, and the belt of single leather, with grain side to pulleys.

This belt had been running a year or more, under a tension which was limited only by the strength of the lacing. It was used to convey two horse-power to the upper shaft, but was considered by the lessee to be unequal to the task, even when tightly drawn, and its adhesion increased by free application of rosin.

It was admitted by both parties that the belt was worked to its fullest capacity.

In order to ascertain the exact amount of work done by the belt, the following experiment was made:

On the driven shaft above was a smooth turned iron pulley, 6.25 feet in circumference and 4-inch face; over this was thrown a 3-inch leather belt, with grain side to pulley, and to its ends were attached unequal weights, such that the 2.25-inch belt was subjected to its maximum working-power. These weights were 203 lbs. and 2.25 lbs. Speed of friction pulley was taken at 132 Rpm.

Then we have $132 \times 6.25 = 825 =$ velocity of 3-inch belt in Fpm, and $\dfrac{825 \times 200.75}{33,000} = 5.018 =$ horse-power of 3-inch belt.

This is equivalent to a driving-power of 41.1 square feet of belt per minute per horse-power.

A single leather riveted belt of ordinary make connects a 60-inch to a 30-inch pulley; both of smooth turned cast iron: the centre of the latter being 8 feet horizontally distant 5 feet above that of the former. The 60-inch pulley drives and makes 80 Rpm, the top fold of belt sagging 13 inches from the straight line. This belt does not slip, runs with the hair side to pulleys, has been in use more than six years, was originally 9 inches wide, is now 8 inches, runs one inch crooked, and during the first two years of its existence was exposed to the weather, frequently saturated with water, but is now soft and adhesive by the application of prepared castor oil.

Horse-power transmitted, 14.48 indicated, which is equivalent to 57.826 square feet of belt travelling per minute per horse-power.

Rule for Ascertaining the Horse-power of Belts, by Mr. Bacon.

We convert the text into the following:

$$\frac{HP\ 6000}{v\ c} = w.$$

In which $HP =$ horse-power transmitted.

$v =$ velocity of belt in Fpm.

$c =$ contact of belt with smaller pulley in lineal feet.

$w =$ width of belt in inches.

Page's Patent Tanned Leather Belting.

86. This excellent belting leather is made by Page Brothers, in Concord, N. H. It possesses greater pliability, strength, and durability than the ordinary tannage, will endure moisture better, is lighter, adheres to glue and cement as well as any belting, and, being softer, will answer better for round belts. It has been thoroughly and successfully tried, and costs no more than well-made oak-tanned belting.

A trial showed 25 per cent. more adhesive power than hard oak or hemlock-tanned leathers, and a test of strength proved that, while 1050 lbs. broke a $1\frac{1}{4}$-inch wide oak-tanned belt, it required 1850 lbs. to break the same size Page belt.

The single belts are $\frac{1}{4}$-inch, the light double belts $\frac{5}{16}$-inch, and the heavy double $\frac{3}{8}$-inch thick.

The light double belts, which are about the same price as best single, work well on cone and flange pulleys, and, of course, very well where running free and where much shifted.

Shafts and Pulleys.

"In the location of shafts that are to be connected with each other by belts, care should be taken to secure a proper distance one from the other. It is not easy to give a definite rule as to what this distance should be. Some have this rule: Let the distance between the shafts be 10 times the diameter of the smaller pulley. But while this is correct for some cases, there are many other cases in which it is not correct. Circumstances generally have much to do with the arrangement, and the engineer or machinist must use his judgment, making all things conform, as far as may be, to general principles. This distance should be such as to allow of a gentle sag to the belt when in motion.

"A general rule may be stated thus: Where narrow belts are to be run over small pulleys, 15 feet is a good average. For larger belts, working on larger pulleys, a distance of 20 to 25 feet does well. Shafts on which very large pulleys are to be placed for main or driving belts should be 25 to 30 feet apart. We know of shafting located as above stated, where the belts work in a very satisfactory manner, the slack side, while in motion, having a sag of $1\frac{1}{2}$ to 2 inches on the short distances above mentioned, while the larger belts show a similar sag of $2\frac{1}{2}$ to 4 inches, the main belts working well with a sag of 4 to 5 inches.

"If too great a distance is attempted, the weight of the belt will produce a very heavy sag, drawing so hard on the shaft as to produce great friction in the bearings, while at the same time the belt will have an unsteady, flapping motion, which will destroy both the belt and the machinery.

"The connected shafts should never, if possible, be placed one directly over the other, as in such case the belt must be kept very tight to do the work.

"It is desirable that the angle of the belt with the floor should not exceed 45°. It is also desirable to locate the shafting and machinery so that belts shall run off from each shaft in opposite directions, as this arrangement will relieve the bearings from the friction that would result where the belts all pull one way on the shaft.

"If possible, the machinery should be so planned that the direction of the belt motion shall be from the top of the driving to the top of the driven pulley.

"All pulleys should be carefully centred and balanced on the shafting. Driving pulleys on which are to be run shifting belts should have a perfectly flat surface. All other pulleys should have a convexity in the proportion of about $\frac{3}{16}$ of an inch to one foot in width. The diameter of the pulleys should be as large as can be admitted, provided they will not produce a speed of more than 3000 feet of belt motion per minute. When this speed has been obtained, the possible size of the pulley may be reduced in proportion to the speed greater than 3000 Fpm, as this speed is considered the limit of economy. The pulley should be a little wider than the belt required for the work. It is also well to consider the possibility of adding, at some future time, more machinery than at first contemplated, and to make all needed provision for such possible increase. Such a course often proves a large saving of expense, or, what amounts to the same thing in the end, guarantees the machinery against overwork. Every pulley not placed in a damp room should be covered with a good leather lagging, put on by an experienced workman. A pulley so covered is capable of much greater and better results, as the belt is not so likely to slip. Repeated experiments prove that the advantage of leather-covered pulleys over all others is fully $33\frac{1}{3}$ per cent.

"The whole arrangement of shafting and pulleys should be under the direction of a mechanical engineer, or a machinist thoroughly competent for such work. Destruction of machinery and belts, together with unsatisfactory results in the business, is a common experience which may, in most cases, be traced to want of knowledge and care in the arrangements of the machinery, and in the width and style of the belts bought, and in the manner of their use, while manufacturers of the machines and belts (especially the latter) are often blamed for bad results which are caused by the faulty management of the mill owner himself."

Purchasing Belts.

"Having properly arranged the machinery for the reception of the belts, the next thing to be determined is the length and width of the belts.

"When it is not convenient to measure with the tape-line the length required, the following rule will be found of service: Add the diameters of the two pulleys together, divide the result by 2, and multiply the quotient by $3\frac{1}{4}$. Add the product to twice the distance between the centres of the shafts, and you have the length required.

"The width of belt needed depends on three conditions. 1st, the tension of the belt; 2d, the size of the smaller pulley and the proportion of the surface touched by the belt; 3d, the speed of the belt.

"The average strain under which leather will break has been found, by many experiments with various good tannages, to be 3200 lbs. per square inch of cross section. A very nice quality of leather will sustain a somewhat greater strain. In use on the pulleys, belts should not be subjected to a greater strain than $\frac{1}{11}$ their tensile strength, or about 290 lbs. to the square inch of cross section. This will be 55 lbs. average strain for every inch in width of single belt $\frac{3}{16}$ inch thick. The strain allowed for all widths of belting — single, light double, and heavy double — is in direct proportion to the thickness of the belt. This is the safe limit; for if a greater strain is attempted, the belt is liable to be overworked, in which case the result will be an undue amount of stretching, tearing out at the lace or hook holes, and damage to the joints. When the belt is in motion the strain on the working side will be greater than on the slack side, and the average strain will be one-half the aggregate of both sides.

"The working adhesion of a belt to the pulley will be in proportion both to the number of square inches of belt contact with the surface of the pulley, and also to the arc of the circumference of the pulley touched by the belt. This adhesion forms the basis of all right calculation in ascertaining the width of belt necessary to transmit a given horse-power. A single belt $\frac{3}{16}$ inch thick subjected to

Care of Belts.

"Belts should never be oiled except when they become dry and hard, and then oil should be used very sparingly. Oil not only rots the leather, but it causes the belt to stretch. In oiling or greasing a belt avoid everything of a pasty nature. The belt should be made pliable, not covered with a sticky substance.

"In 'taking up' belts observe the same rules as in putting on new ones.

"Never add to the work of a belt so much as to overload it."

Friction of Belts.

87. "The friction of belts upon pulleys depends upon the extent to which they are tightened, the extent of circumference with which they are in contact, and their breadth. It is commonly believed that the greater the diameter of pulley, the more surely does the belt cause it to revolve without slipping. Theoretically, however, and we believe practically, it will be found that, with equal degrees of tightness, equal breadth of belt, and equal circumstances as to perfection of contact, the friction of a belt on the circumference of a pulley is the same, whatever be its diameter. The only circumstance that can affect the constancy of the result, is that belts not being perfectly flexible, lie more closely to surfaces curved to a large radius than to those of smaller radius. When a certain amount of power has to be communicated through a belt, the speed at which the belt moves has to be taken into account, because power being pressure multiplied by velocity, the greater the velocity with which the power is transmitted, the less the pressure that has to be communicated at that speed. In this sense, then, it appears that the larger the pulley the less is the slip of the belt, because the greater the circumference of the pulley, revolving at a given angular velocity, the greater is its absolute velocity through space, and therefore the less the pressure required to communicate a given power.

"It is found, practically, that a leather belt 8 inches wide, embracing half the circumference of a smoothly turned iron pulley, and travelling at the rate of 100 Fpm, can communicate one horse-power.

"When less than half the circumference of the pulley is embraced, the strap must be proportionally wider; and when more than half the circumference is embraced, its width may be less.

"The law according to which the friction of a belt increases with an increased arc of contact, is of a peculiar character; but may be readily understood by comparing the friction on arcs of different lengths. If a pulley (of any diameter whatever) were prevented from revolving, and a belt passing over part of its circumference were stretched by a certain weight at each end, additions might be made to the weight at one end until the belt began to slip over the pulley. The ratio which the weight so increased might bear to the weight at the other end, would measure the amount of friction.

"For example, in experiments made to test a theoretical investigation on this subject, a belt passing over a pulley in contact with 60° of its circumference, was stretched by a weight of 10 pounds at each end. One of the weights was increased until it amounted to 16 pounds, when the belt began to slip. The ratio of 16 to 10, or $\frac{16}{10} = 1.6$ was then the measure of the friction. When 20 pounds at each end were used to stretch the belt, the one weight was increased to 32 pounds, giving the ratio of $\frac{32}{20} = 1.6$, the same as before; and likewise, when 5 pounds were used for stretching, the weight at one end was increased to 8 pounds, giving still the same ratio, $\frac{8}{5} = 1.6$. So far, then, the friction was precisely proportional to the stretching weight, as might have been expected from the ordinarily received doctrine on the subject of friction. On extending the arc of contact to 120°, the ratio was found to be 2.56, or 1.6^2. And again, on embracing 180° the ratio was found to be 4.1, or very nearly 1.6^3.

"The theoretical investigation brought out this result independently, and the following law may therefore be taken as established:

"If, for any given arc of contact, the one weight bears to the other, at the point of slipping, a certain ratio — for double the arc, the ratio will be squared; for triple the arc, it will be cubed; for four times the arc it will be raised to the fourth power; and so on.

"In all cases, however, much depends on the tightness of the belt, the limits to the force with which it is strained being, first, the tensile strength of the belt itself, and, secondly, the amount of pressure that it may be convenient to throw upon the shaft and its bearings. New belts become extended by use, and it is therefore frequently necessary to shorten them. Before use, they should be strained for some time by weights suspended from them, so as to leave less room for extension while in use. Wherever belts are employed, they should be of the greatest breadth, and travel at the greatest speed consistent with convenience, as it is most important to have the requisite strength in the form best suited to flexure, and the least possible strain on the shafts and bearings.

"When ropes or chains are employed, as in cranes, capstans, windlasses, or the like, for raising heavy weights or resisting great strains, the requisite amount of friction is obtained by coiling them more than once round the barrel of the apparatus. It is found that one complete coil of a rope produces a friction equivalent to nine times the tension on the rope, the barrel being fixed. Two complete coils of the rope produces a friction equivalent to 9×9 times the tension, and so on. The diameter of the barrel does not affect the result.

"Having regard to these facts, we may readily understand the force with which a knot on a cord or rope resists the slip of the coils of which it consists, for the several parts of the cord act as small barrels, round which the other parts are coiled; and the yielding nature of the material of which the barrels are composed, permits the coils to become impressed into their substance on the application of force, and prevents them from slipping more effectually than if they were coiled on a hard and resisting barrel."—*From Wylde's Circle of the Sciences, London.*

Mr. A. K. Rider, of De Lamatre Iron Works, N. Y., has favored us with the following:

88. "Our rule, which appears to work well and gives very satisfactory results, is based on the assumption that a belt one inch wide, when properly surfaced and sufficiently tight, and bearing on not less than ¼ the circumference of smaller pulley, will transmit a force of $19\frac{1}{4}$ pounds *at any velocity*. The power of a belt, in foot pounds, is thus readily obtained by multiplying its velocity in Fpm by its width in inches, and again by $19\frac{1}{4}$. The generally received rule is, 144 square feet of surface passing per minute equals one horse-power, and the cohesive strength of good belting is taken at 4000 pounds per square inch of section as its breaking strength."

From Spon's "Dictionary of Engineering," p. 312.

89. "Belts and drums form very effective friction-couplings. If a machine driven by a belt becomes accidentally overloaded, the belt slips upon the drum, and a break-down is generally prevented. By the introduction of fast and loose pulleys the driven shaft can be set in motion or stopped with perfect safety, whilst the driving shaft is running at full speed. The motion of belts and drums is much smoother than that of gearing, and they can be readily applied to machines which require a high velocity, where ordinary gearing would be quite inadmissible.

"The best description of leather for belts is English ox-hide tanned with oak-bark by the slow, old-fashioned process, and dressed in such a way as to retain firmness and toughness, without harshness and rigidity. The prime part of the hide only, called the *butt*, should be used; these are cut out of hides in the preliminary preparing process, and tanned by themselves, afterwards stretched by machinery and allowed to dry while extended. Strap-butts of best leather can be permanently elongated 4 to 5 inches.

"For light work, belts of single substance are sufficient, the strips of leather being joined together by feather-edged splices, first cemented and then sewn. Single belting varies in thickness from $\frac{3}{16}$ to $\frac{1}{4}$ inch. For heavy work, double and sometimes treble layers of leather are required, cemented and sewn through their entire length. The material used for sewing is either strong, well-waxed hemp, or thin strips of hide prepared with alum. The latter is generally used in the North of England; but its advantages over good waxed hemp is doubtful. The thickness of double belting is from $\frac{5}{16}$ to $\frac{7}{16}$ inch.

"An improvement in the ordinary double belt has been introduced by Messrs. Hepburn & Sons, of Southwark, who have given much attention to this branch of leather manufacture. It consists in the use of a corrugated strip of prepared untanned hide for the outer layer of the belt, and the usual tanned leather for the inner layer, riveted together by machinery. The rivets are made of copper or malleable iron, and have their ends spread, bent, and driven in flush with the surfaces of the layers. Metallic sewing of this kind is also applied to double belting made entirely of leather, and has been found to work well, and is more durable than ordinary hand-sewing.

"The drum should be $\frac{3}{16}$-inch per foot of width, rounding except in the case of small high-speed pulleys, which should be $\frac{3}{8}$ to $\frac{1}{2}$-inch.

"In order that the natural tension of the belts shall remain constant, and not exceed, though equalling the value calculated, it is requisite to use *tension rollers*. The weight, W, of these rollers is found by the approximate expression, $W = \dfrac{2\,T\,\cos a}{\cos b}$; wherein a is half the obtuse angle A D B, formed by the belt upon which the weight rests, and may be assumed *a priori*; b the angle between the line A B and the horizontal line A C: that is, the angle B A C = b, and T = tension on tight side.

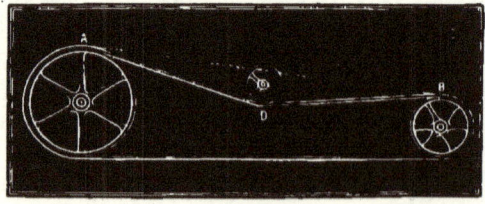

Fig. 14.

"In fixing the belt, care must be taken to give it such a length that, when at repose, it shall only have a minimum flexure."

From Leonard's "Mechanical Principia" we make the following extracts:

90. "If the power to be transmitted exceeds 20-horse, and circumstances will not allow the centre of the drums to be over 15 feet apart, the power should be transmitted by gearing."

A table is given which is based upon the following data: One horse-power is transmitted by belts, 1.8, 1.2, 0.9, 0.72, 0.6, 0.514, 0.45, 0.4, 0.36 inches wide, if carried over pulleys 2, 3, 4, 5, 6, 7, 8, 9, 10 feet diameter respectively, at the velocity given above.

"It is immaterial whether the smallest drum is the driving or the driven drum; if the diameter of the smallest drum remains constant, the width of the belt will remain constant; if the diameter of the other drum should be increased indefinitely."

Example.

91. A horizontal "Corliss" engine, having a 23-inch by 48-inch cylinder, making 52 Rpm, has an 18-feet fly-wheel pulley; upon this runs a double leather belt, 28 inches wide and 80 feet long, driving a 6-feet diameter pulley, whose centre is about 18 feet horizontally distant, and whose bottom face is about on a level with the top of the fly-wheel pulley face; both pulleys of iron, smoothly turned.

The lower fold of belt drives; the top fold runs quite freely, with considerable sag.

The maximum load of engine is 217 indicated horse-power, the minimum about 150. These figures give 31.6 and 45.75 square feet of belt per minute per horse-power respectively.

Comparison of Single and Double Belt.

92. A 34-inch pulley on a line shaft running 200 Rpm, drives a 44-inch pulley on a grindstone shaft. The grindstone is 72 inches diameter, its shaft nearly on same level as line shaft, and 7 feet 4 inches away. About midway between these pulleys, a 10-inch diameter tightener, weighing 90 lbs., rests upon the top fold of belt, bearing it down 14 inches from straight line of pulley faces. This tightener is carried by a horizontal swinging frame, having radius arms 4 feet 6 inches long. A 7-inch *single* leather belt, of best make, was completely worn out in four months, another lasted seven months, while a 7-inch *double* oak-tanned leather belt lasts about four years.
— *Samuel Bevan, at H. Disston & Sons' Keystone Saw-Works, Philada.*

RULES FOR BELTING.

From "Treatise on Mill Gearing," by Thomas Box. E. & F. N. Spon, London, 1869.

93. "The laws by which the proportion of the entire circumference embraced by the belt governs the ratio of the weights $T t$ are very complicated. Let

$F =$ the co-efficient of friction.
$L =$ length of circumference embraced, in feet or inches.
$R =$ radius of the pulley, in the same terms as L.
$T =$ the greater weight in Fig. 15.
$t =$ the lesser weight.

Then $T = t \times (2.718)^{\frac{FL}{R}}$

$t = \dfrac{T}{(2.718)^{\frac{FL}{R}}}.$

"These formulæ cannot be worked except by logarithms, and they then take the following forms:

$Log.\ T = Log.\ t + \left(.4343 \times \dfrac{FL}{R}\right)$ | $Log.\ t = Log.\ T - \left(.4343 \times \dfrac{FL}{R}\right)$

"In the table on page 120, the cases of failure are particularly instructive; column 11 shows that in all the cases failure might have been expected. Thus, No. 1 required a 10½-inch double leather belt, where a 6-inch gutta-percha one failed to do the work. In No. 2 a larger pulley was substituted, a 7-inch double leather belt should have been used, and the 6-inch gutta-percha one did the work badly. No. 4 failed with a 9-inch single belt to do the work for which a 14-inch single or a 7-inch double belt was required. No. 6 required a 10-inch single or a 5-inch double belt, and failed to do the work with a 6-inch single belt. No. 8 required a 13-inch single or 6½-inch double belt, and failed with an 8-inch single belt. It will be observed that in cases Nos. 1 and 11 the difficulty was overcome by using larger pulleys; and in cases No. 4, 6, and 8, by converting the single belt into a double one. Circular saws and some other kinds of machinery require extra strength of belt, as shown by No. 18.

"The rules and tables we have given apply strictly to *leather* belts only; leather is in every way the best material, and is not likely to be permanently superseded by the new materials, gutta-percha, India-rubber, etc.

"The table shows that the power of a gutta-percha belt $\frac{5}{16}$ or $\frac{3}{8}$-inch thick is from 25 to 50 per cent. greater than that of a single leather one. We found in practice that leather belts bear about 310 lbs. per square inch of section, and we may allow that gutta-percha will bear about 400 lbs. From direct experiments, the cohesive strength of gutta-percha is 15 cwt. or 1682 lbs. per square inch."

Driving Power of Belts, from Cases in Practice.

No.	Nominal Horse-Power.	Diam. of Pulleys. Driver.	Diam. of Pulleys. Driven.		Distance between Shafts.	Belt Contact on smaller Pulley.	Speed of smaller Pulley.	Particulars of Belt. Folds.	Particulars of Belt. Width.	Particulars of Belt. Material.	Calculated Width of Belt.	Remarks.	Square Feet of Belt per Min. per Horse-Power.
		Ft.	Ft.	Ins.	Ft.								
1	10	5	2	6	8	.446	80	Open.	6	Gutta-percha.	10½ double.	Failed entirely; pulleys made larger, see No. 2.	31.4
2	10	7	3	6	8	.424	80	"	6	"	7 "	Badly; required resin, and gave trouble.........	44.
3	10	10	2	5	30	.472	165	"	6	"	4½ "	Drove well; circular saw, 3′ 6″ diam.; same engine as No. 1..................	62.7
4	12	7	7	0	32	.58	36	Crossed.	9	Single leather.	14 single.	Failed; strap altered to double one, see No. 5..	50.
5	12	7	7	0	32	.58	36	"	9	Double "	7 double.	Drove well.	
6	6	5	5	0	9¼	.5	40	Open.	6	Single "	10 single.	Failed; strap altered to double one, see No. 7..	52.3
7	6	5	5	0	9¼	.5	40	"	6	Double "	5 double.	Drove well.	
8	18	12½	3	6	20	.424	152	"	8	Single "	13 single.	Failed; strap altered to double one, see No. 9..	61.89
9	18	12½	3	6	20	.424	152	"	8	Double "	6¼ double.	Drove well.	
10	18	6¼	1	3	19½	.457	770	"	6½	Gutta-percha, $\frac{5}{16}$ thick.	7¾ single.	Drove, but not well; same engine as No. 8, etc.	91.
11	3	1¼	2	0	12	.49	90	"	6	Single leather.	9½ "	Failed; pulleys made larger, see No. 12.........	58.9
12	3	2	3	2	12	.48	90	"	6	" "	6 "	Drove well....................................	94.2
13	20	9	5	0	16	.457	65	"	9	Double "	11 double.		
14	12	10	3	0	16½	.43	113	"	9	" "	5¼ "	Drove well; pulleys lagged with wood...........	66.5
15	12	3	4	9	11	.473	100	"	8	" "	7½ "	" "	52.5
16	12	10	3	0	14	.42	133	"	6	" "	6 "	" "	52.2
17	10	6	4	5	14½	.6	49	Crossed	7	" "	7¼ "	" "	40.4
18	10	4¼	2	2	11¼	.6	463	"	4	" "	3¼ single.	" " extra strong for 4′ 6″ circ. saw......	104.9
19	10	8	5	0	20	.476	64	Open.	9	Gutta-percha.	5½ double.	" "	75.3
20	6	7	3	6	13	.46	106	"	5¾	Do. ¼″ thick.	6½ single.	" "	93.1
21	6	9	3	6	11	.416	128	"	5½	Single leather.	5½ "	" "	107.5

Driving Power of Belts.

Let A, in Fig. 15, be a pulley fixed so as to be incapable of turning, and T t weights suspended by a belt E, which passes round the pulley, and may be caused to embrace it more or less by a small guide-pulley D. Let now the weight T be increased until the friction of the belt is overcome, and it slips on the pulley, the weight T descending.

Fig. 15.

The ratio between T and t varies —

1st. With the co-efficient of friction of the material of the belt E, sliding on the material of the pulley A. 2d. With the *proportion* which the arc of the pulley embraced, bears to the whole circumference of the pulley.

"It is independent of the breadth of the belt, *so long as* T *and* t *remain the same*, but inasmuch as T and t, or the strain on the belt, may increase with the breadth, this must not be understood to mean that a narrow belt will drive as much as a wide one; for other things remaining the same, the strain, and therefore the driving power, varies directly and simply as the breadth.

"The ratio between T and t is also independent of the *diameter* of the pulley, other things remaining the same; thus, for instance, a strap which slips on a pulley 1 foot in diameter, with a weight of 1 cwt. at one side, and 2 cwt. at the other, would do the same on a pulley 10 feet or any other diameter, the surfaces being similar.

"This appears contrary to our instinctive notions, but is quite correct, as I have proved by experiment. But this must not be understood to mean that a small pulley will carry as much power as a large one, for obviously, if both are set in motion, making the same number of Rpm, the relative speeds of belt would be proportional to the diameters, and the power would vary in the same ratio.

"From Morin's experiments the co-efficients of friction are as follows:

.47 for leather belts in ordinary working order on wooden pulleys
.28 " " " " " cast-iron "
.38 " " soft and moist " " "
.50 for cords or ropes of hemp on wooden pulleys.

"It appears from Morin's experiments that with cast-iron pulleys the driving power is the same whether they are turned or not, the adhesion of the belt to the polished surface generating as much friction as with a rough surface.

"If we take the case of a belt in ordinary working order on a cast-iron pulley, the co-efficient of which is .28, and calculating for four cases in which the circumference is successively ¼, ½, ¾, and wholly embraced, we find that while $t = 1$ in all cases, T becomes successively 1.553 — 2.41 — 3.77 and 5.81.

"The following table is calculated in this way, and gives throughout the value of T when $t = 1$ for different kinds of surface of pulley and states of belt. Decimal parts of circumference of pulley are given instead of fractions named above.

"When a rope is used, and it is wound more than once round the drum, the frictional power is enormous; thus with a rough wooden pulley and a rope 2.5 times round it with $t = 1$, T is 2575.3.

Table showing Ratio of Strains on the Belts of Driving Pulleys,
$$Q = T - t.$$

Ratio of the Arc embraced by the Belt to the entire circumference.	t	New Belts on Wooden Pulleys.		Belts in the Ordinary State on				Soft Belts on Cast-Iron Pulleys.		Ropes on Wooden Drums.			
				Wooden Pulleys.		Cast-Iron Pulleys.				Rough.		Polished.	
		T	Q	T	Q	T	Q	T	Q	T	Q	T	Q
.2	1	1.87	.87	1.80	.80	1.42	.42	1.61	.61	1.87	.87	1.51	.51
.3	1	2.57	1.57	2.43	1.43	1.69	.69	2.05	1.05	2.57	1.57	1.86	.86
.4	1	3.51	2.51	3.26	2.26	2.02	1.02	2.60	1.60	3.51	2.51	2.29	1.29
.5	1	4.81	3.81	4.38	3.38	2.41	1.41	3.30	2.30	4.81	3.81	2.82	1.82
.6	1	6.59	5.59	5.88	4.88	2.87	1.87	4.19	3.19	6.58	5.58	3.47	2.47
.7	1	9.00	8.00	7.90	6.90	3.43	2.43	5.32	4.32	9.01	8.01	4.27	3.27
.8	1	12.34	11.34	10.62	9.62	4.09	3.09	6.75	5.75	12.34	11.34	5.25	4.25
.9	1	16.90	15.90	14.27	13.27	4.87	3.87	8.57	7.57	16.90	15.90	6.46	5.46
1.0	1	23.14	22.14	19.16	18.16	5.81	4.81	10.89	9.89	23.90	22.90	7.95	6.95
1.5	1									111.31	110.31	22.42	21.42
2.0	1									535.47	534.47	63.23	62.23
2.5	1									2575.30	2574.30	178.52	177.52

Pulleys in Motion.

"We have so far considered the pulley as fixed; we will now apply the foregoing facts to the case of pulleys in motion. The mechanical conditions of a driving pulley, with half its circumference embraced by the belt, are shown by Fig. 16, in which we have, as before, the pulley A and the weight T and t as in Fig. 15, where we found them to be respectively 1 and 2.41. But in this case, the pulley A being free to turn, the weights T and t being unequal, there would be no equilibrium without an additional weight at Q, and, supposing the drum J to be the same diameter as the pulley A, it is self-evident that the sum of Q and t must be equal to T; therefore $T - t = Q$; or $2.41 - 1.0 = 1.41 = Q$.

RULES FOR BELTING.

"The mechanical power transmitted by the belt, supposing Q to be raised by a rope coiled around the drum as a hoist or windlass, is the *difference* between T and t, and Q might be increased indefinitely, if we could increase T and t indefinitely in the normal proportion; there is, however, a limit to which this can be done, namely, the cohesive strength of the strap by which the heaviest weight, T, is carried. Where leather is used we can obtain the requisite cohesive strength by increasing the width of the belt, or by making it a double or treble one, and this width must in all cases be proportional to T, and not to t or to Q.

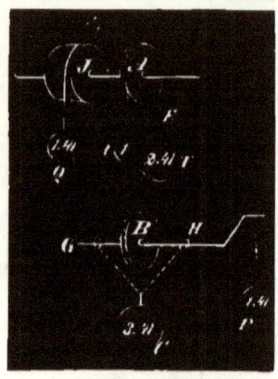

Fig. 16.

"In Fig. 16 G may represent the engine shaft, H its crank, and P the power which is equal to Q. It will be observed that the weight C, or pressure on the bearings due to the tension on the two straps, and also the maximum tension T, is much greater than the power P or the weight Q.

"If the weight Q had been 1.0, the maximum tension T would evidently have been $\frac{2.41}{1.41} = 1.71$, and the minimum tension t have been $\frac{1.0}{1.41} = 71$, and thus we obtain the strain as shown in Fig. 17; this is the most useful form in which the question can be put, as we thus obtain the proportional maximum strain or width of belt for a unit of power at P.

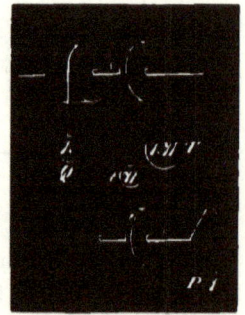

Fig. 17.

"With a wooden pulley the friction of the surfaces is greater, and the strains for the weight Q are different. Here for $t = 1$ we find by the table above that T is 4.38, and hence Q = 4.38 − 1 = 3.38. For Q or P = 1 we should have $T = \frac{4.38}{3.38} = 1.29$, and $t = \frac{1}{3.38} = 29$; so that with the same power, P, a belt 1.29 inch wide, on a wooden pulley, would do as well as one 1.71 inch wide on a cast-iron one.

"In the case of a pulley of cast-iron with $\frac{2}{10}$ of the one embraced, the table shows that $T = 1.42$, and t being 1.0, Q will be $1.42 - 1 = .42$. For $Q = 1$ we have $T = \frac{1.42}{.42} = 3.38$, and $t = \frac{1.}{.42} = 2.38$.

"With a crossed belt on cast-iron pulleys, the arc embraced being $\frac{7}{10}$ of the circumference, we have $T = 3.43$, by table $T = 1$, and $Q = 2.43$; and hence with $Q = 1$, we obtain $T = \frac{3.43}{2.43} = 1.41$, and $t = \frac{1}{2.43} = .41$.

"Comparing all the cases presented it will be seen that, with the same engine power, the breadth of belt would be in the ratio 1.71, 1.29, 3.38, and 1.41."

The following Article from Vol. 3, for 1859, "Publication Industrielle," par Armengaud Ainé, relates to Belts employed for the Transmission of Power:

94. "Several years prior to this date, M. Laborde, M. E., presented to the Industrial Society of Mulhouse a paper on the subject of belts, in which he made the following observations:

"1st. The resistance to be overcome must be less than the power required to slip the belt on its pulley.

"2d. The tension must not permanently elongate the belt.

"3d. The tension must not uselessly increase the friction of the shaft bearings.

"4th. The belt must be flexible, in order to allow of an easy folding in all its parts.

"The first three conditions named are self-evident, while of the fourth it may be said that a belt never requires doubling, but should always be composed of a single thickness of leather.

"The webs of a single leather are extended and compressed in passing over the pulleys without in any way injuring their texture, while the two leathers composing a double belt are subject to such a friction upon each other that their destruction follows rapidly, notwithstanding the numerous points of connection uniting both; it is therefore best to abandon double belts altogether.

"In order to maintain the durability and flexibility of belts it is advised to apply to them, as they need, fine grease, or ordinary grease mixed with tallow, which may be done while they are running. They are apt to slip for a few minutes after greasing, but soon adhere again, and finally drive the better for the application

"Extended experimental observations have proved the superiority of smooth-faced pulleys over such as are rough, or ribbed in the one or the other direction: in that increased area of surface contact with the belt is presented by the former.

"Upon the above considerations as a basis, M. Laborde develops his formulæ.

"1st. The width of belts must be in direct proportion to the power to be transmitted, while the speed remains uniform.

"2d. The width of belts vary inversely to the speed.

"Consequently the products of the widths and speeds of belts are proportional to the power transmitted by them.

"Experience demonstrated to M. Laborde that a belt $3\frac{1}{4}$ inches wide, running 533 Fpm, readily transmitted one horse-power of 33,000 foot-pounds, having the usual tension, and without deforming itself, when the pulleys are smooth-faced and of equal diameter, in order that the belt may embrace their semi-circumference.

"This is equivalent to 144.35 square feet of belt per minute per horse-power, and 19 lbs. strain per inch of width.

"The author has used this rule a number of years, and expresses himself well satisfied with the result.

"M. Carillon, of Paris, a mechanical engineer of no less reputation employs a rule based upon the following statement: A belt can transmit one horse-power, if it have a surface velocity of 96.9 square feet per minute, providing not less than one-third of the circumference of either pulley be embraced.

"Notwithstanding our great confidence in M. Carillon's deductions — believing that in most cases his allowance of driving surface of belts will be sufficient, since in many cases belts are run at a higher velocity—we yet think it preferable to adhere to the base established by M. Laborde.

"The reduction of driving surface may be made with more security by employing well-worked leather, as that of Messrs. Sterlingue & Co., who condense it under the hammer, or that of Mr. Bérendorf, in whose machine it becomes strongly compressed.

"Tables 1 and 2 (not given here) are developed from the following example: If a belt travel 100 metres per minute, it should be 132 millimetres wide in order to transmit one horse-power, which is equivalent to a 5.2 inch belt travelling 328 Fpm; or, in other terms, it is equal to 142.13 square feet of belt per minute per horse-power.

"It is easy to understand why the sizes of belts, as indicated by the preceding figures, must be modified in several particulars. Firstly, when the pulleys are of very different sizes, or, if expressed in more general terms, when the pulleys are embraced by the belt less than the semi-circumference; and, secondly, when the belt is crossed, or when more than half the pulley surface is encircled.

"M. Paul Heilmann presented very judicious observations on this subject to the Society of Mulhouse, which results are reproduced in the Society's Bulletin, No. 40, 1835, and may be expressed thus:

"The friction of a belt upon a pulley depends:

"1st. Upon the pressure or tightening.

"2d. Upon the number of degrees of contact.

"3d. It is independent of the diameter of the pulley.

"4th. It is independent of the width of the belt.

"It is evident that the less the pulley is surrounded by the belt, the tighter must be the belt in order to transmit a given power, because the power which can be transmitted to the pulley is always less than, or at best equal to, the friction produced on its surface; and if the resistance offered by the machine be greater, the belt will slip. Thus the width of the belt has no other purpose than to give it a resistance — a power sufficient to withstand a certain tension without being injured or broken.

"M. Heilmann says that this tension, and with it the width of the belt, must necessarily be an inverse proportion to the numbers as represented in the following table, which table has been calculated after the formulæ and by the aid of the hyperbolic logarithms.

$$\text{Friction} = P \, e \left(\frac{fS-1}{R} \right)$$

"In which, $P =$ resistance to be overcome.

$e =$ base of hyperbolic logarithms, $= 2.718$.

$f =$ proportion between friction and pressure.

$R =$ radius of pulley.

$S =$ lineal contact of belt with pulley.

"This formula is the one taught in the Mechanical Engineering Department of the Polytechnic College.

"In the table, the first column represents the angle of contact of belt, in degrees and minutes.

"The second column represents the fractional part of the circumference corresponding to the angle.

"The third column shows the ratio between friction and pressure following the angle of contact.

"The fourth column contains the result of the division of the ratio 0.4670, which corresponds to the half-circumference, by the successive ratios of the friction and the pressure.

1	2	3	4
° ′			
22.30	$\frac{1}{16} = 0.0625$	0.0491	9.511
30.	$\frac{1}{12} = 0.0833$	0.0660	7.075
45.	$\frac{1}{8} = 0.1250$	0.1005	4.646
60.	$\frac{1}{6} = 0.1667$	0.1363	3.426
67.30	$\frac{3}{16} = 0.1875$	0.1545	3.023
90.	$\frac{1}{4} = 0.2500$	0.2112	2.211
112.30	$\frac{5}{16} = 0.3125$	0.2706	1.725
120.	$\frac{1}{3} = 0.3333$	0.2911	1.604
135.	$\frac{3}{8} = 0.3750$	0.3330	1.402
150.	$\frac{5}{12} = 0.4166$	0.3763	1.241
157.30	$\frac{7}{16} = 0.4375$	0.3983	1.172
180.	$\frac{1}{2} = 0.5000$	0.4670	1.000
202.30	$\frac{9}{16} = 0.5625$	0.5390	0.866
210.	$\frac{7}{12} = 0.5833$	0.5674	0.823
225.	$\frac{5}{8} = 0.6250$	0.6145	0.760
240.	$\frac{2}{3} = 0.6667$	0.6669	0.700
247.30	$\frac{11}{16} = 0.6875$	0.6937	0.673
270.	$\frac{3}{4} = 0.7500$	0.7769	0.601
292.30	$\frac{13}{16} = 0.8125$	0.8642	0.510
300.	$\frac{5}{6} = 0.8333$	0.8941	0.522
315.	$\frac{7}{8} = 0.8750$	0.9551	0.489
330.	$\frac{11}{12} = 0.9163$	1.0190	0.458
337.30	$\frac{15}{16} = 0.9375$	1.0515	0.444
360.	$1 = 1.0000$	1.1522	0.405

"From the preceding observations it will be easy to determine the width of a belt in all cases that may occur in practice whenever the maximal force in horse-power is given, which is to be transmitted and the speed of belt known.

"If the pulleys are of equal diameters, all that is needed is to find the width of the belt, in accordance with the examples from which tables 1 and 2 are constructed, corresponding to speed and power required.

"If the pulleys are of different diameters, then use the following

Rule.

"Determine the number of degrees of contact with the smaller pulley; find, in the third table, the number in fourth column corresponding thereto; multiply the number thus found into the width of belt given in tables 1 or 2.

"In all the preceding we constantly admitted the belt of single thickness, and, consequently, the same power of resistance.

"Although this is generally the case, yet for transmitting small powers at great speed it is better to reduce the thickness and augment the width of the belts, because they will then develop better on their pulleys, which are usually of small diameters.

"In such cases belts of inferior quality may also be employed, by determining their width from a less co-efficient of resistance. On the contrary, for the transmission of great powers at slow speeds, it is advisable to use the thickest possible leathers, in order to avoid great width.

"We have, as yet, taken no account of the belt's own weight, which, in certain cases, is to be added, wholly or in part, to the resistance to be overcome, whilst in others it will have to be deducted from said resistance; but as this has a slight influence on the practical results, it may be left out of consideration.

"Belts should be calculated to meet the maximal resistance, not the average.

Belts of Gut.

"In speaking of belts, it will not be superfluous to announce that an English inventor, Mr. John Edwards, conceived the idea of making belts of gut, prepared in endless flat bands of different lengths and widths, for use on pulleys, and united evenly.

"The filaments of gut are woven into ribbons on looms similar to those used in manufacturing metallic gauze, and the joints made by splicing, care being taken to cut or burn the extremities of the interlaced filaments, in order to obtain perfectly united bands. It is known that experiments were made to manufacture belts from fibrous substances, such as hemp and wool, but it is thought, up to this date, that they will not endure the same wear and tear as leather. The gut was, and is yet, employed with advantage in the shape of cords running in grooved pulleys.

Belts of Wool.

"Another patent has been issued in England to Mr. J. Heywood for a system of belts or bands of wool, which the inventor prepares by soaking in a mixture of linseed oil and rosin. He boils, for instance, $6\frac{2}{3}$ lbs. of oil, adds $4\frac{1}{2}$ lbs. of powdered rosin, and agitates the mixture to a perfect union. After having the bands soaked, he submits them to the action of a pair of rolls, and afterwards exposes them to dry, when they are ready for use.

Belts of Gutta-Percha.

"M. M. Rattier & Co. introduce, in a great measure, gutta-percha belts, which give good satisfaction. They also make these belts with wire-gauze cores, which prevent stretching. Notwithstanding this precaution, we believe it best to employ such belts for light transmission only, with slow speed, and to subject them to slight tensions, in order to avoid injury by heating."

From "Publication Industrielle," par Armengaud Ainé, Vol. 9, 1860.

95. The application of pulleys, cones, and drums for the transmission of power has become so general that, with cog-wheels, they constitute a large part of the stock of patterns carefully kept for use.

There does not exist, in fact, an organic means of transmission more simple and inexpensive than that by the agency of belts.

In most cases the belt and pulley form a mechanical agent at once the most convenient, the most easily erected, and requiring the least combination of parts; it suffices only that they be in exact proportions: 1st, to obtain the necessary speed, and 2d, to communicate the required power.

This mode of transmission has the advantage of smooth and quiet action, of light weight in comparison to the power transmitted, and of less liability to destructive wear and tear, and consequent accident, as with the use of gearing.

In accordance with these facts, gearing has been replaced, of late, by belts and pulleys, even where considerable powers are transmitted.

To gain all the advantages which such a system is expected to furnish, it is absolutely necessary to fulfil several essential conditions, without which the best results cannot be obtained; for instance, if the pulleys be not of proper diameter, the speed would not be in the ratio desired; again, if the pulley faces be too narrow for the power transmitted, the belt will slip; or if, on the contrary, all the dimensions be augmented beyond the requirements of each case, material would be uselessly wasted, and power continually lost, in giving motion to needless weight of parts.

The principal questions concerning the belt and pulley arrangement are the following:

1st. Determine the diameters of the pulleys according to the number of revolutions their respective shafts are to make.

2d. Calculate the dimensions of the belt according to the power it is to transmit, and decide the diameter of one of the pulleys.

3d. Ascertain, also, the proportions of the different principal parts of each pulley in conformity to the width of the belt.

The speeds of pulleys connected by belts is in the inverse ratio of their diameters.

The width of belts is calculated from the tensile resistance of the leather. We merely examine here the leather belts most generally used in machine shops and factories.

In closely observing the action of belts on pulleys, it is found that the power which they transmit depends on the amount of friction developed on the surface of the pulley, and upon a certain degree of tension applied to the belt when put on.

M. Morin has furnished us with the following:

1st. If the belts are sufficiently tightened they do not slip, but transmit the speed in a constant ratio, and inversely to the diameter of the pulleys.

2d. In the transmission of power by endless rope or belt, on pulleys, from one shaft to another, the sum of the tensions in both folds remain constant, in a manner, that if the driving fold is overburdening itself, the driven fold is relieving itself to the same amount, and that the sum of both tensions is the same when the machine is stopped.

3d. The ratio between friction and traction, exercised by the primitive tightening, is very nearly proportional to the degrees of contact with the pulleys when within the ordinary limits, varying, in practice, but little from $\frac{1}{4}$ to $\frac{3}{4}$ the circumference.

Belts should not be subjected to working strains over 284 lbs. per square inch.

Action of Belts.

1st. The friction developed at the circumference of pulleys is proportional to the primitive tightening of the belt, and depends also on the angle in which the belt envelopes the pulley; and, further, on the nature and condition of the surfaces in contact.

2d. The friction is independent of the diameter of the pulley and of the width of the belt.

Most of the belts in actual use for transmitting small powers have larger dimensions than the calculations would give, for the reason that belts are frequently overloaded, and the quality of the leather not always of the best. They sometimes are made to carry 280 lbs. to the square inch, instead of 140 to 210, to which latter strains they are generally admitted in practice.

For the transmission of great power, there is much interest felt in employing the very best leather, in order to reduce the width of belt and pulley as much as possible.

From "Designing Belt Gearing," by E. J. Cowling Welch, we transcribe the following:

96. "The ultimate strength of ordinary leather belting is about 3086 lbs. per square inch; thus with belts $\frac{7}{32}$ thick we have a breaking strain

    Through the solid part..................675 lbs.
       "   "   riveting...........................382 "
       "   "   lacing ....210 "

"Taking a safe working strain of say one-third of each of these, we have

    Through the solid part..................225 lbs.
       "   "   riveting...........................127 "
       "   "   lacing.............................. 70 "

"The working strength of the belt must be taken as that of its weakest part, which is the lacing."

"In order to ascertain the greatest *actual* or *indicated* horse-power ($I.HP$) capable of being transmitted by any particular belt, whose velocity (V) in Fpm and breadth (B) in inches are known, we ascertain the force (R) transmitted to the surface of the pulley; then

$$I.HP = \frac{RVB}{33,000}.$$

"From the foregoing it will be seen that each unit of breadth of the belt carries its own tension, and when this is at its maximum safe amount, and still we are not able to transmit the required power by it, and we cannot increase the angle of contact, or use a belt of sufficient width to effect the same, either from the shaft being unable to withstand the total tension of the broader belt, or from any other cause; we can only overcome the difficulty by increasing the velocity of the belt itself — that is, by increasing the diameters of the pulleys over which it runs; not that we get any greater adhesion by so doing, for increasing the two pulleys in the same proportion, the angle of contact in both cases remains the same, so also does the tension; and as the adhesion is independent of the surfaces of contact, therefore the adhesion remains the same, whether the larger or smaller pulleys are used; but with the larger pulleys we get greater leverage to overcome the resistance, and a correspondingly greater velocity of belt."— *E. & F. N. Spon, London, 1875.*

By Z. Allen, Providence, R. I., from "Proceedings of N. E. Cotton Manufacturers' Association," No. 10, April 19, 1871.

The Relation of Small Shafting to Hollow Shafting.

97. "On small shafting, pulleys are used very much less in diameter, consequently the friction of the shafting is as its velocity. If the circumference of the pulley is used for the bearing, the friction is very much increased. The torsion of the hollow shafting is as the cube of its diameter; if you use a 3-inch shaft, the torsion of that is as its cube. If you take from the centre of the shaft, you have left the outside shell only, so that the amount of power gained theoretically is as the amount of iron taken from the inside of the shaft, running upon the bearing of the outer.

"One difficulty in small shafting is in properly fastening the pulleys. The construction of this shafting is such that it requires a larger diameter to hold the pulleys than to transmit the power, consequently I have taken $1\frac{3}{16}$-inch diameter as the smallest shafting that it is practicable to run in mills. An inch shaft, well sustained, will drive 100 looms; but you have to fasten to it the couplings, the pulleys, and set-screws; and if the holes are not properly drilled, the set-screws will cramp the shaft. After putting up a line 150 feet long, it is necessary to straighten it. You cannot straighten the line of shafting in the shop, because the pulleys are not made in the shop. In order to straighten the shafting we take a lever, put it under the rail, and spring it into place.

"We are using a line of $2\frac{3}{16}$-inch shafting, driving 16,000 ring spindles. The quantity of oil required to lubricate this shafting is so small, that, if I tell you, it will seem almost impossible. I asked the overseer how much oil he used in oiling this shafting, and he told me only two or three drops to a bearing once a week, and said that he would run the whole line a year with a half-pint of oil.

"I found one shaft had been running eighteen months with no dripping pans underneath, the overseer giving as a reason that the quantity of oil consumed was so small that none were required. We all know that it requires oil to run a shaft, and we can form an idea of the amount of friction by the quantity of oil consumed.

"The line of $2\frac{3}{16}$-inch shafting, which drives 16,000 spindles, runs through a mill 350 feet long, and is fitted up with bearings 8 feet apart, and carries a reasonable share of the pulleys which drive the machinery.

"I have adopted this method of having the main driving pulleys about once in 150 feet, and counter lines about 150 feet long, and belted in the middle. The middle shaft is made $2\frac{3}{16}$-inch diameter.

"If the pulleys are very small, it requires, to run at a slow speed, more power than it does through gears, on account of the strain which you are obliged to put upon the belt. In England it is the custom to use gears mostly to transmit the power, which requires a stiff shaft to hold them in place; consequently, a gear never yields; it must go. With a belt it is very different. In making the formula for a shafting, I adopted as a standard one-fifth of the breaking weight. There is no pulley put on strong enough to run more than this; but with gears it must go.

Gearing.

"There are many ways of transmitting power from the motor to the machine, or place where it is to be utilized. I will invite your attention to the three that are commonly in use among our manufacturers, viz., Gearing, Shafting, and Belting. Gearing and shafting transmit a uniform motion, that is, a certain number of revolutions, but not always a uniform revolution, owing to the elasticity of the shaft or imperfect construction of the gearing. Power transmitted through pulleys by belts or straps is variable, and cannot be relied upon when uniform motion is required, owing to the elasticity and thickness of the belts, and their liability to slip. Power transmitted through gears and pulleys may have an increased or diminished velocity by having gears and pulleys of different diameters. But with shafting the velocity is positive, as by construction both ends of the shaft must run with the same number of revolutions. Each of these methods has its advantages, but neither motion in all cases can be made to supply the place of the others. When a positive ratio is required, between the driver and driven, it must be through gears; and as gears are universally used to transmit power from the water-wheel or water-wheel shaft to the second mover, let us for a few minutes consider gears and their formation. Possibly no part of mechanical science in common use is so poorly understood or wretchedly abused as the formation of gearing. Each draftsman or mechanic has his favorite tooth or form of tooth. It is his pet child, and there is no other like it. To ask him to demonstrate or explain why it is better, would be considered almost an insult; but however perfect it may be in theory and construction, if the gears are not properly adjusted to each other, and made to run as designed, the whole theory and mechanism becomes useless, as the teeth are formed for a definite pitch and cannot be used for any other, either theoretically or practically, when a smooth motion is required.

"In forming teeth for gears we first draw what is called the *pitch line*, or circumference of uniform motion, which is the working diameter of the gears. The teeth are formed from this line, and it is indispensable to the smooth running of the gears that these lines should run together, otherwise there would be a grumbling noise or jar, like the rolling of a fluted roll over a plain hard surface. I think I can demonstrate to any geometrician, that a tooth similar to the epicycloidal and hypocycloidal tooth is the only one that can be made to run smoothly.

"This tooth is formed by having two circumferences run together, corresponding to the pitch line or diameter of the required gears. However, as this is not the proper place to discuss theories, I will not occupy your time by doing so. Within the last month I have started a new Turbine water-wheel of about 350 horse-power. The crown gear 7 feet diameter, and jack gear 4 feet diameter. The teeth in these gears are parallel below the pitch line, and when started they did not run smoothly. I had them ground together with tallow and emery, and they at once commenced forming a tooth similar to the epicycloidal tooth.

"In discussing the properties of gears, I have come to the following conclusions: First, That the loss by transmitting power through gears is $1\frac{1}{2}$ per cent. in the driver, $1\frac{1}{2}$ per cent. in the driven, and $1\frac{1}{2}$ per cent. in the teeth, in all $4\frac{1}{2}$ per cent.; *i. e.*, when the diameter at the pitch line is eight (8) times that of the *bearing*. If the diameter is only four to one, then the loss is double, or 9 per cent.; *i. e.*, the friction or loss of power is inversely, as the ratio of the diameter of gears to their bearings. In this statement I have not considered the weight of the gears or shaft. In horizontal shafting the *weight* has no effect, as the weight of the gear seldom is equal to the pressure upon the teeth.

"Secondly. If intermediate gears are used in transmitting power, and the three axes are in the same plane, the friction is double, or 9 per cent. in lieu of $4\frac{1}{2}$ per cent. If the driver and driven have different diameters, the opposite sides of the teeth in the intermediate must be of a different shape, *i. e.*, made to conform to the different diameters of the driver and driven.

"Thirdly, for the same reason, the driver cannot admit of two driven gears of different diameters at the same time and run smoothly.

"As the destroying force or concussion is as the square of the velocity, and velocity of contact is to the pitch of teeth, as verse sine to sine. I have therefore adopted, to transmit the greatest amount of power with regard to durability, the following formula for first drivers, and made tables to correspond. Let $d =$ diameter in feet, $p =$ pitch in feet, and $HP =$ horse-power: then

$$\frac{d}{(6\sqrt{d}+1)} = p,$$

and to find the velocity of the periphery in feet, multiply the square root of the diameter by 750 feet, or $750 \sqrt{d} = v$, and

$$d^2 \sqrt{d} \left(\frac{1}{6\sqrt{d}+1} \right)^2 \times 2200 = HP,$$

i. e., if the pitch and velocity are obtained by the above rule, and the breadth is $2\frac{1}{2}$ times the pitch, which I think will be found correct for spurs, and $2\frac{1}{4}$ for bevel gears.

"Usually, I think the pitch of gears is too large for the diameter to insure good results. An increased pitch on the same diameter will not transmit more power, as the velocity will have to be diminished to make it run smoothly. These formulas are intended for the smaller gear, the larger is not to be considered.

"There are advocates for the rolling of gears together, *i. e.*, the teeth can be so formed that one tooth can be made to roll into the other; but I think this can be shown to be theoretically and practically impossible.

"If spur gears are firmly sustained and well adjusted, and the teeth actually cut in the epicycloidal form, 33 per cent. can be added to the velocity indicated in the following table. This will increase the horse-power in the same ratio.

Table showing the Diameter, No. of Teeth, Pitch, Velocity, Revolutions and Horse-Power of Gears.

Let $D =$ diameter in feet. $T =$ No. teeth. $P =$ pitch in inches. $V =$ velocity of periphery in feet; and $HP =$ horse-power.

Diameter in Feet.	No. of Teeth.	Pitch in Inches.	Velocity of Periphery in feet per minute.	Revolutions per minute.	Horse-power.
1	22	1.72+	750	238.8	44
1½	26	2.15+	915	194.7	86
2	30	2.53+	1057	168.3	137
2½	33	2.86+	1187	151.1	197
3	36	3.16+	1299	137.8	263
3½	38	3.43+	1403	127.5	337
4	41	3.69+	1500	119.3	414
4½	43	3.93+	1591	112.5	499
5	45	4.16+	1677	106.7	590
5½	47	4.38+	1759	101.7	682
6	49	4.59+	1837	97.4	783
6½	51	4.74+	1912	92.6	993
7	53	4.98+	1984	90.2	1004
7½	55	5.16+	2053	87.1	1115
8	56	5.34+	2121	84.4	1233
8½	58	5.52+	2186	81.8	1357
9	59	5.68+	2250	79.5	1483
9½	61	5.85+	2311	77.4	1607
10	62	6.01+	2371	75.4	1738
10½	64	6.16+	2430	73.6	1876
11	66	6.31+	2488	71.9	2020
11½	67	6.46+	2543	70.3	2167
12	68	6.61+	2598	68.9	2311
12½	70	6.75+	2651	67.5	2462
13	71	6.89+	2704	66.2	2610
13½	72	7.02+	2755	64.9	2773
14	74	7.16+	2806	63.8	2932
14½	75	7.30+	2856	62.7	3099
15	76	7.44+	2904	61.6	3262
15½	77	7.55+	2953	60.6	3429
16	79	7.68+	3000	59.6	3604

Shafting.

"We will next consider shafting and the transmission of power through the same, the theory of which, I presume, is well understood by you all; it is, therefore, only in the adaptation that I may differ with some or all of you.

"Wrought-iron shafting of one inch diameter will transmit from

14 to 15 horse-power at 100 Rpm before there is any set twist. You will observe by this that a shaft is seldom twisted off, but is usually broken by jar of gears, or being out of line, or by transverse pressure. A shaft 2 inches diameter, 100 revolutions, will transmit 100 horse-power before there is any set twist. A shaft 4 inches diameter, 100 revolutions, will transmit 800 horse-power before twisting, but will frequently be broken with very much less power if out of line; while 1 inch to 2 inch shafting, being flexible, will hardly be influenced by small variations. You will perceive from this that torsion is hardly to be considered in shafting a mill, as it will require larger shafting to prevent springing by transverse pressure than it does for torsion. With prime movers, or wheel shafts, we can afford to pay an extra insurance in loss of power and weight of iron, as there is usually but one or two in the mill, and should any accident occur to these it would cause the stopping of the mill, and the loss might cost the price of a dozen shafts. I have, therefore, taken one-fifteenth ($\frac{1}{15}$) of the twisting weight, or the cube of the diameter, etc., $\frac{d^3 \times R}{100} = HP.$

For second movers we have the formula $\frac{d^3 \times 2 \times R}{100} = HP.$

For third movers, or mill shafting, $\frac{d^3 \times 3 \times R}{100} = HP.$

"In advocating small shafting I do not pretend that, theoretically, there is any saving of friction in transmitting the same amount of power. It requires the same amount of friction for a 1-inch shaft as it does for a 6-inch shaft, if both are equally strained, as a 6-inch shaft, of course, would run very much slower to transmit the same amount of power, but that in most cases the diameter is larger than is required, as the transverse pressure requires a larger diameter than the torsional, as before stated.

"This led me to consider if there might not be some way devised to meet this difficulty. In most of our mills the bays are about 8 feet, and require shafting of about 2 inches diameter to sustain the lateral pressure of a card or loom belt; yet this same shaft has torsional strength, at 150 revolutions, to run 900 looms before twisting, although it may not be running more than 8 to 10 looms or cards when near the end of the line, while a shaft ¾ inch diameter is all that is required to perform that amount of work, if well sustained. To meet the difficulty I have made a cast-iron rail, so constructed that the hangers slide along the whole length of the line without regard to the beams. By this arrangement there can be as many hangers as are required — one to each pulley, if necessary. The number of bearings do not increase the friction if properly arranged, as it is by this rail. I use this rail for all shafting less than $1\frac{7}{8}$ inches diameter, where the bays are 8 feet. I have running 2 lines, 160 feet each, each line driving 60 breaker cards and lap bead. Its diameter is $1\frac{3}{16}$ inches to $1\frac{5}{16}$ inches, and runs 280 Rpm; driving pulleys on shafting for cards, 7 inches diameter. I have also about 1500 feet more driving cards and looms; about half of it has run 16 months without any repairs. I would here state shafting might be much smaller but for the difficulty of having it made thoroughly in our workshops. The pulleys must be well balanced and nicely bored, or the set-screws or keys will spring the shaft.

"I use this rail in connection with shafting for cards and looms. It is not so necessary for spinning and other machinery, as the machines or pulleys on them are a greater distance from each other. I have, in one of the Lawrence Manufacturing Co.'s mills, a shaft $2\frac{3}{16}$ inches diameter, running 416 Rpm, in common Babbitt boxes, driving 14,000 ring spindles, $1\frac{5}{8}$-inch ring. This shaft has run 18 months without any repairs or extra labor whatever. It has no self-oilers, but is oiled once a week with a common oil-can, using a mixed oil of 2 parts sperm and 1 part Downer's paraffine.

"We have another line of shafting 300 feet long, $2\frac{1}{4}$ inches diameter, running 433 Rpm, driving 15,000 throstle and mule spindles, with full complement of machinery. This shaft has run about 10 months; about $\frac{1}{2}$ of it was not under cover, being exposed to the cold weather of last winter. This line has given no trouble. I have yet to see a shaft less than $2\frac{1}{4}$ inches diameter twisted off, and hope if any one present has they will state the fact and circumstances to the meeting. I have often seen larger ones broken by being out of line, and I think this is one of the strongest arguments in favor of small shafting. I think $\frac{1}{2}$ of the friction and $\frac{3}{4}$ of the weight of the shaft can be saved over the old system of small and quick shafting well arranged. One can hardly afford to waste a large amount of power to drive the heavy shafting of a large mill to prevent an outlay once a year or so of some small accident that *may* possibly occur. And, furthermore, I claim it is better that a small shaft should break than hold so firmly, as in case of a large shaft it would do, as to cause injury either to life or machinery, as the case may be. Self-protection is the first law of nature, we are told, hence civil engineers always construct mills with a view that nothing shall give out in the future — no matter what its present cost in material and power — as they know full well their reputation is at stake; and should any of their work need renewing in a year or so, they would be condemned. Furthermore, they construct with the knowledge that inferior capacities may run the machinery, and we all know by experience what and how great those difficulties are.

Horse-Power of Shafts — Speed 100 Revolutions per Minute.

FIRST MOVERS.		SECOND MOVERS.		THIRD MOVERS.			
Diam.	Horse-Power.	Diam.	Horse-Power.	Diam.	Horse-Power.	Diam.	Horse-Power.
Inches.		Inches.		Inches.		Inches.	
3	27.00	$2\frac{1}{2}$	31.25	1	3.00	$2\frac{13}{16}$	66.35
$3\frac{1}{4}$	34.33	$2\frac{3}{4}$	41.59	$1\frac{1}{8}$	3.59	$2\frac{7}{8}$	71.29
$3\frac{1}{2}$	42.87	3	54.00	$1\frac{1}{8}$	4.27	$2\frac{15}{16}$	76.04
$3\frac{3}{4}$	52.73	$3\frac{1}{4}$	68.66	$1\frac{3}{16}$	5.02	3	81.00
4	64.00	$3\frac{1}{2}$	85.74	$1\frac{1}{4}$	5.85	$3\frac{1}{16}$	86.16
$4\frac{1}{4}$	76.76	$3\frac{3}{4}$	105.46	$1\frac{5}{16}$	6.78	$3\frac{1}{8}$	91.38
$4\frac{1}{2}$	91.12	4	128.00	$1\frac{3}{8}$	7.79	$3\frac{3}{16}$	97.15
$4\frac{3}{4}$	107.17	$4\frac{1}{4}$	153.52	$1\frac{7}{16}$	8.91	$3\frac{1}{4}$	102.98
5	125.00	$4\frac{1}{2}$	182.24	$1\frac{1}{2}$	10.12	$3\frac{5}{16}$	109.04
$5\frac{1}{4}$	144.70	$4\frac{3}{4}$	214.34	$1\frac{9}{16}$	11.19	$3\frac{3}{8}$	115.33
$5\frac{1}{2}$	166.37	5	250.00	$1\frac{5}{8}$	12.87	$3\frac{7}{16}$	121.10
$5\frac{3}{4}$	190.10	$5\frac{1}{4}$	289.40	$1\frac{11}{16}$	14.41	$3\frac{1}{2}$	128.62
6	216.00	$5\frac{1}{2}$	332.74	$1\frac{3}{4}$	16.07	$3\frac{9}{16}$	135.63
$6\frac{1}{4}$	244.14	$5\frac{3}{4}$	380.20	$1\frac{13}{16}$	17.86	$3\frac{5}{8}$	142.90
$6\frac{1}{2}$	274.62	6	432.00	$1\frac{7}{8}$	19.77	$3\frac{11}{16}$	150.42
$6\frac{3}{4}$	307.54	$6\frac{1}{4}$	488.28	$1\frac{15}{16}$	21.81	$3\frac{3}{4}$	158.20
7	343.00	$6\frac{1}{2}$	549.24	2	24.00	$3\frac{13}{16}$	166.24
$7\frac{1}{4}$	381.07	$6\frac{3}{4}$	615.08	$2\frac{1}{16}$	26.32	$3\frac{7}{8}$	174.55
$7\frac{1}{2}$	421.87	7	686.00	$2\frac{1}{8}$	28.78	$3\frac{15}{16}$	183.13
$7\frac{3}{4}$	465.48	$7\frac{1}{4}$	762.14	$2\frac{3}{16}$	31.40	4	192.00
8	512.00	$7\frac{1}{2}$	843.74	$2\frac{1}{4}$	34.17	$4\frac{1}{16}$	201.12
$8\frac{1}{4}$	561.51	$7\frac{3}{4}$	930.96	$2\frac{5}{16}$	37.09	$4\frac{1}{8}$	210.56
$8\frac{1}{2}$	614.12	8	1024.00	$2\frac{3}{8}$	40.18	$4\frac{3}{16}$	220.28
$8\frac{3}{4}$	669.92	$8\frac{1}{4}$	1123.02	$2\frac{7}{16}$	43.44	$4\frac{1}{4}$	230.29
9	729.00	$8\frac{1}{2}$	1228.24	$2\frac{1}{2}$	46.87	$4\frac{5}{16}$	240.60
$9\frac{1}{4}$	791.45	$8\frac{3}{4}$	1339.84	$2\frac{9}{16}$	50.47	$4\frac{3}{8}$	251.22
$9\frac{1}{2}$	857.37	9	1458.00	$2\frac{5}{8}$	54.26	$4\frac{7}{16}$	262.14
$9\frac{3}{4}$	926.86			$2\frac{11}{16}$	58.23	$4\frac{1}{2}$	273.37
10	1000.00			$2\frac{3}{4}$	62.39		

The Transmission of Power from Motors to Machines.

"It will be instructive to you, as superintendents of the present improved machinery of mills, to look back on the imperfect modes of transmitting power early used in the commencement of the cotton manufacture in New England.

"So cheap and abundant was water-power then that steam-power was not at first resorted to.

"It was deemed necessary to locate a mill directly over the water-wheels, so that a main upright shaft might be arranged upward through the several stories, to transmit the power more directly to a main horizontal shaft in each room, to distribute power to each machine.

"The shaftings were all made square, to receive the cast-iron wheels fastened by wedges. The pulleys were made of wood, by clamping together pieces of joists, notched to fit the shafts, by means of screw-bolts. Instead of the numerous light pulleys now used, long wooden drums were built around the shafts, and made of boards nailed upon circular plank heads. With the slow speed of 40 or 50 turns per minute, some of these drums were necessarily made 3 or 4 feet diameter and several feet long, darkening the rooms by their ponderous magnitudes, and requiring very high ceilings to admit them.

"These great drums being frailly nailed together and unbalanced, could not be used with quick-revolving movements without shaking them to pieces, and also shaking the floors intolerably.

"This was the style of mill shafting and pulleys in use when I first commenced building a mill at Allendale, in the year 1822.

"Cast-iron pulleys, clamped upon the shafting, were soon after introduced, with their faces smoothed by grinding on stones. Round iron shafts were deemed a great improvement, with drilled wheels and pulleys fitted to them. With these advantages, the speed of the horizontal shafts was augmented to produce 80 or 90 Rpm. Early experiences of the troublesome difficulties in operating a manufactory, resulting from the imperfect modes of mill gearing then in use, excited an impatience of longer enduring.

The Disadvantages of Transmission of Power from Motors to Machines with Slow Speed.

"In all calculations of the strength of shaftings, wheels, and pulleys there is a certain relationship between the velocity of their movements and transmission of power. A belt, shaft, or a pulley that makes one revolution to do the same work which is done by another making 2 revolutions, has double the stress imposed upon it, and must have double the strength. By doubling the speed there is an opportunity of economizing the weight and costs of the materials employed for mill gearing in a somewhat corresponding ratio. These are strong incentives to incite us to attempt improvements of transmission of power in all manufactories.

"In the functions of the mechanisms of all animals the most admirable scientific skill is displayed in adapting the proportion of bone and muscle to the speed of movement designed to be accomplished. These are models for study presented to the engineer for copying, in artificial mechanisms for transmitting power. The slender limbs of the deer, of the greyhound, and race-horse — all designed for fleet movements — show an impressive contrast with those of the elephant or the heavy draught-horse.

"Instead of increasing strength of wheels and pulleys by additions of more metal, when they are to be used with a higher speed, they become less fit for turning with swift velocities. They may even fly into pieces by the tendency of this excess of matter to move in a straight line, recognized as 'centrifugal force.'

"To duly apportion the formation of the parts of all machines for the transmission of power, is the scientific task assigned to engineers. In addition, there is requisite a knowledge of the various properties of strength, elasticity, and hardness to endure wear, and other qualities. Even changes of temperature, with consequent expansion and contraction of metallic parts, is not to be overlooked. The imperfections of settling foundations, by disarranging the best fitted lines of strong shafting, may cause them to fail of durably transmitting power from motors to machines. The very rigidity of the parts, resulting from the great size of shaftings requisite for transmitting power with slow velocities, is a principal cause of their failure, where light and flexible shafting, with high velocities, might durably perform the service. I will here give you a remarkable illustration of the actual results of transmitting power by mill gearing, with slow speed, by the strongest shafting, made without regard to cost for securing durability of service.

"In persistently carrying out the old system of heavy shafting with slow velocities, in a large cotton-mill built in Connecticut, in the year 1857, the attempt was made to transmit the power from four large water-wheels through a line of cast-iron shafting of the great diameter of 12 inches, made in sections of 10 feet in length, each piece weighing 3500 lbs., with couplings of the weight of 3800 lbs. each. So great was their weight and rigidity, that the settlings of the foundations, changes of temperature, and continual jar caused them to break so frequently as to render necessary the replacement of them by new wrought-iron shafts. Quite recently these, in turn, have been discarded as unsatisfactory, and are superseded by light, quickly-revolving shaftings, driven by belts from pulleys 20 feet diameter and 24-inch face, with a surface velocity of over 5000 Fpm. All that was practicable to perfect the old system of slow speed of mill gearing was here done to maintain a dying struggle for its prolonged existence.

Experiments for Testing the Advantages of Transmitting Power with High Velocities.

"Having realized, from practical experiences, the disadvantages of the slow speed system, more than twenty years previously to this persistent attempt to perpetuate it, an entirely opposite system was commenced by me in the construction of a second mill at Allendale, in the year 1839. The idea was there carried out of more than doubling the then existing speed of mill shafting, from 90 revolutions to over 200 per minute, for the special purpose of reducing the size and weight of all the shafting and pulleys in nearly a corresponding ratio, with the economy of costs and motive power.

"To accomplish this object, several important innovations were necessary upon the old modes of transmitting power. A wheel-pit was requisite outside of the mill building, in a separate wheel-house, for the double purpose of obtaining more space for larger cog-wheels, to get up the requisite speed, and of excluding the noxious steamy dampness arising from all water-wheels shut up within the mill walls.

"The pulleys in previous use, with ground or turned surfaces, would not operate quietly, without being turned inside as well as outside, to balance the rims, and prevent the tremor consequent on the use of all unbalanced pulleys revolving rapidly. This improvement also reduced the weight of the pulleys to correspond with the reduced weight of the shafting and wheels, made of half the previous diameters, excepting those used in the wheel-pits of larger size to get up speed.

"Before this systematic balancing of pulleys was commenced, no inconsiderable portion of the power was transmitted to shake the floors, and even to cause some of the old wooden mills to rock to and fro. This experiment, deemed somewhat wild at the time, proved successful. It has gradually been adopted as an economical system of transmitting power from motors to machines by high velocities of shafts and belts.

"In constructing a third mill in Georgiaville, in the year 1853, a further attempt was made to more than double the speed again, and to still further economize the weights and cost of materials and power used.

"The details of the experiments there made and practically applied, you have requested me to give an account of, as facts that may serve usefully for guidance hereafter to others, in further perfecting the transmission of power for operating the machinery of mills.

System Pursued in Transmitting Power from Motors to the Manufactory.

"In the location of mills on water-courses, it has commonly been deemed necessary to place the main building directly over the spots where the wheel-pits are unavoidably located, on some steep hill-side, or rocky precipice, however unfavorable the site may be for grading and for costly foundations, with dark and damp basement-rooms, unsuitable for occupancy by workmen.

"One of the most important advantages derivable from the new system of transmitting power economically to a distance from water-wheels as motors, is practically available in selecting a good level site for the location of a manufactory.

"In carrying out this system at Georgiaville, the power has been transmitted several hundred feet from a bluff, where a fall of water of 36 feet descent was available by two successive falls of 20 and 16 feet each. To accomplish this task, with the massive shafts and couplings then in common use (1852), appeared to be too costly and difficult of execution with satisfactory results.

"Encouraged by previous experiments for practically transmitting power by swiftly-revolving shafts and belts, the attempt was boldly made to carry the power to the manufactory, instead of carrying the manufactory to the power, which was necessarily located on a hill-side, where the wheel-pits were to be excavated.

"The motors were a pair of water-wheels, 24 feet diameter and 18 feet long, with a fall of water 20 feet, and a second pair of water-wheels, 50 yards above them, 18 feet diameter and 19 feet long, under a fall of 16 feet.

"A small shaft only 3 inches diameter, if revolving with 200 Rpm, was deemed sufficient to transmit all the power of the upper pair of wheels; and by transmitting this power to another lower line of shafting of the same size, but with the velocity doubled to 400 Rpm, it was also deemed sufficient to receive the additional power of one of the lower pair of 24-feet wheels. A driving pulley of 10 feet diameter on the upper line of shafting transmitted the power, by a belt 12 inches wide to a 5-feet pulley on the lower shaft, to its double speed.

"This idea was more readily conceived than executed. The movement of a pulley of the dimensions of 10 feet diameter, with a surface velocity of over 6000 Fpm, had never before been attempted practically. Doubts were suggested of the safety of using belts with this velocity in mills. But after having trusted my own body to travel with the speed of a mile a minute, over English railways, with numerous other passengers, drawn by a ponderous locomotive engine of 35 tons' weight, whirled around curves, over precipitous embankments, and uncertainly fastened rails, it seemed very rational to trust a leather belt to travel with the same speed. Thus reassured, the doubter might smile at the suggestion of danger of risking a light belt to journey at the same rate. But there had been no light pulleys made suitable for this use. Those previously in use, made of two iron rims, covered with wooden lags bolted thereto, were rejected as unfit.

"Although the superior convenience of belts over wheel-work and shafting for transmitting power had induced many attempts to use them 30 years ago, yet the experimenters had commonly failed of successfully operating them with the low rate of speed then used. Pulleys had not been made sufficiently light and well balanced for any one to adventure to use them with the high speed required for leather belts to operate advantageously. With the slow speed it was necessary to strain the belts so tightly on the pulleys, to produce sufficient adhesion, without slipping around on the smooth surfaces, that the lacings and texture of the leather yielded; and so frequent repairs were required, that the superintendents of mills nearly all abandoned the use of them for transmitting the power from the motors to the mill shafting. They fell back on the old system of slowly revolving heavy shafting and wheels.

"To carry out the proposed system new patterns of pulleys were therefore made. The first pulley, 10 feet diameter, proved to be imperfect, and, when tested with a velocity of about 8500 Fpm, the rim soon made its exit through the roof of the wheel-house, and continued its course in a parabolic curve through the air several hundred yards, until it finally transmitted its motive power to plough a furrow in a meadow. A remodelled pulley, made to take the place of the wandering one, stood the test and has continued faithful, without deserting its post, to perform the duty assigned to it ever since, during a period of 16 years. The same belt has also remained in use, in good order, after travelling about a quarter of a million of miles every year in its daily circuits, with a velocity of 6000 Fpm.

"As a test of the efficacy of this small 3-inch shaft to transmit the power from 3 water-wheels, it may be stated that not a single shaft or coupling has required renewal or repairs, and they appear still capable of a much longer service. This same 3-inch shaft has also served to transmit all the power of the steam-engine used in times of drought.

"The contrast between the two systems of high and low rates of speed of shafts and belts, for transmitting power from motors to manufactories, is instructively exhibited in these two narrated instances of the practical application of each of them, with conclusive results of the failure of the latter.

"To avoid the use of the brittle teeth of wheels, quite recently the adhesion by friction of the surfaces of wheels has been employed for the transmission of power. To intensify the friction, the adhesion has been increased by turning grooves and ridges on the faces of the wheels brought together by pressure.

"This arrangement has been found advantageous for engaging and disengaging heavy washing machines and other apparatus in dye-houses and bleacheries, where leather and India-rubber belts fail.

"It may be questioned whether there is not a great loss of power by transmission through a length of 300 feet of a 3-inch shafting.

"Undoubtedly some power is lost by friction in this case, but not so much as by the massy shafts described, or by a similar line of main shaftings in mills, where the friction is much augmented by loading the shafts with numerous pulleys, and by the tension of numerous belts.

"If kept in line and properly attended, the friction of a naked shaft of the length of 300 feet is so small as to be easily turned by the hand of a man. The friction being caused by the weight of the shaft only, is not affected by any extent of power transmitted by it while revolving, whether it vary from one to an hundred horse-power.

"Where high velocities of mill gearing are used, it is desirable to limit any accidental extreme acceleration that might prove injurious.

"This was readily accomplished by the simple arrangement of a latch, to be lifted by a touch of the whirling arm of the ball regulator, whenever the accelerated speed causes it to rise to a certain prescribed limit. The lifting of this latch disengages the connection of the regulator with the gate of the water-wheel, and simultaneously engages another adjacent revolving wheel, that instantaneously shuts the gate more quickly than can be done by hand. By this automatic action the water-wheel itself is made a self-regulating machine.

"A wire extended from the distant mill, like a bell-wire, serves to communicate with the same latch by a slight pull of the hand, and to shut the gate of the water-wheel by the same automatic arrangement. This was devised for use only in case of accident, requiring the immediate stoppage of the machinery.

"This same system should be applied to automatically shutting off steam from an engine, whenever the velocity may accidentally become accelerated beyond a prescribed limit, to endanger the machinery.

"The very important advantage of combining together the action of the several motors of a manufactory to co-operate in concert for equalizing the regulation of the speed of the looms, self-actors and other machines, requiring great uniformity of movements, is really available, where the velocity of a mile per minute of a connecting belt is adopted. While the strongest shafting and cog-wheels fail to accomplish this work, even with the most massive materials, as has been described, all 4 of the water-wheels of the Georgia mill have been very satisfactorily and successfully made to act in unison by a single belt of only 8 inches in width. This belt serves to transmit back and forth between the motors any excess of power that either may receive, and to return any surplus, to an extent of 60 horse-power. A range of variation of 120 horse-power is thereby available for maintaining an equable movement of all the machines of a large manufactory with admirable regularity. The elasticity and slight slipping of the belts relieves the shocks of more than 100 tons of water-wheels, which break the teeth and shafting made of rigid, unyielding iron.

"This system was necessarily introduced to prevent the waste of water that ensues when two mill regulators are used to control the flow of water successively from one pair of wheels above another lower pair. The two regulators cannot be made to act harmoniously, for each one is governed by the varying load of machinery imposed on each motor. The annoying waste of the surplus water in times of drought, shut off by the regulators, and flowing past without useful effect, can be prevented entirely by using only one regulator on the upper wheels for controlling the whole of the machinery.

"Thus a small leather belt, of only 8 inches width, has been successfully employed for many years, and is still employed, with the velocity of a mile a minute, to control the speed of 4 water-wheels, like leather reins to bridle 4 steeds.

"In a manufactory operated by two independent motors, with distinct lines of shafting, the two systems may be connected even by an inch belt moving with high velocity, to modify, wonderfully, the sudden extremes of speed, so disadvantageous where machines are operated, requiring nice adjustments of power."

"In the present experimental state of the introduction of pulleys and belts, moving with high velocities for the transmission of power to a distance from motors, a few facts may be briefly stated to inspire confidence in the operators of mills to adopt new arrangements by learning what has been found practically successful.

"A good leather belt, one inch wide, has sufficient strength to lift 1000 lbs.

"The speed of a mile per minute for main driving leather belts has been found both safe and advantageous for practical use.

"The capability of belts to transmit power is determined by the extent of its adhesion to the surface of pulleys.

"The extent of adhesion of belts varies greatly under varying circumstances of the use of them, and is very limited in comparison with the absolute strength of the leather.

"The adhesion and friction, causing the belt to cling to the surface of a pulley without slipping, is mainly governed by the weight of the leather, if used horizontally.

"If belts are strained tightly on the pulleys, then the adhesion is increased in proportion to the increased tension produced.

"The weight of leather in vertical belts tends to produce a sag beneath the under side of the under pulley; and, if loosely put on, might not touch it at all, to transmit power by adhesion. For this reason it is necessary to strain on more tightly all vertical belts, with a dependence on the elastic stretch of the leather for producing adhesion.

"A vertical belt of single leather of the width of 6 inches, and with a velocity of 5200 Fpm, has practically been used very satisfactorily at the Georgia mill, during several years, to operate 10,400 self-acting mule-spindles, and the spoolers and warpers for the same; and another belt of similar width and velocity, 110 feet in length, has served to transmit the power from a 24-feet water-wheel, 18 feet long, under a fall of 20 feet, with the same velocity of 5200 feet.

"A 24-inch belt of single leather, with the velocity of 4850 Fpm, has transmitted all the power of a steam-engine of 6-feet stroke, 30-inch cylinder, making 40 Rpm, and with so slack a tension on the returning side as to flap and wave with an undulating movement.

"These statements are specified simply to show what has been done by belts running with certain velocities, not for the purpose of holding them up as models for imitation.

"No fixed rule can be given for calculating the actual adhesion of belts; for this adhesion depends upon so many contingent facts of their relative positions and weights, as affected by greater or less lengths and breadths, and lightness. As the result of experimental observations, it may safely be calculated that, with a properly slack belt, the effective adhesion of a horizontal belt may be taken at 30 lbs. to each inch of width of short belts, and double of this on long belts, with threefold or more if tightly strained on the pulleys, which never should be done, for this increases the friction of the bearings and waste of power, in addition to injuring the durability of the leather for service.

"Clamps with powerful screws are often used to put on belts with extreme tightness upon the pulleys, and with most injurious strain upon the leather. They should be very judiciously used for horizontal belts, which should be allowed sufficient slackness to move with a loose undulating vibration on the returning side, as a test that they have no more strain imposed than what is necessary simply to transmit the power.

"Rather than to continue to use horizontal belts with overstrained tightness to obtain the necessary adhesion, it is often better to use larger pulleys, which require less adhesion to transmit an equal extent of power.

"On the scientific principle that the adhesion, and consequently the capability, of leather belts to transmit power from motors to machines, is in proportion to the pressure of the actual weight of the leather on the surface of the pulley, it is manifest that, as longer belts have more weight than shorter ones, and that broader belts of the same length have more weight than narrower ones, it may be adopted as a rule that the adhesion and capability of belts to transmit power is in the ratio of their relative lengths and breadths. A belt of double the length or breadth of another, under the same circumstances, will be found capable of transmitting double the power. For this reason it is desirable to use long belts. By doubling the velocity of the same belt, its effectual capability for transmitting power is also doubled."

From Daniel Hussey, Lowell, Mass., in "Proceedings of N. E. Cotton Manufacturers' Association," No. 10, April 19, 1871.

98. "A leather strap or belt an inch wide will sustain 1000 lbs. before breaking. I have, therefore, taken 8 per cent. of the breaking weight, or 80 lbs. to the inch, or about 400 feet to the horse-power, as a tension that will not materially injure the leather for a long period by overstraining or stretching. This is used for single belts — main drivers only. A double belt will give $\frac{1}{3}$ more equally well.

"As regards the velocity of belts, this subject admits of a wide margin. Ordinarily, counter-belts, where the centres are not more than 12 feet apart, will require 1000 feet to horse-power per minute, and card and loom belts from 2000 to 3000 horse-power per minute. When at the Nashua Co.'s mills, I ran a 20-inch single belt 7200 Fpm from a 14-feet diameter to a 4-feet diameter pulley, which ran successfully on the 14-feet diameter, but the centrifugal force on the 4-feet diameter pulley caused it to jump or fly from the surface and run a little uneven, owing to the uneven weight and thickness of the leather.

"I think it would have run well on a 6-feet diameter pulley. When it was running 6000 Fpm it ran very satisfactorily indeed, and I could not have asked to run it better. From this experiment I have come to the conclusion that 6000 feet is as fast as a belt should run when the small pulley is not over 4 feet diameter. Taking this as a basis of calculation, a 10-feet pulley may run a belt 10,000 Fpm with safety. It is, however, seldom in practice that we should need such quick speed. Some three weeks since I commenced running a single belt 5400 Fpm — the smaller pulley being about 4 feet diameter — which gives excellent satisfaction. I know of no definite rule for running belting; everything depends upon surrounding circumstances.

"A horizontal belt, running on not less than a 7-feet diameter pulley, 50 feet from centre to centre, and working side at bottom, will run well with 400 feet to a horse-power, the slack being taken up by its own weight. The same belt, at an angle of 45°, will require 500 feet to the horse-power, and with a vertical belt it will be almost impossible to run it any length of time without a binder (which, of all things, we most dread in a mill). I will now mention one law of belting that may not be known to you all, *i. e.*, the hug or adhesion is as the square of the number of degrees which it covers on the pulley, or, in other words, a belt that covers $\frac{2}{3}$ of the circumference of a pulley, requires 4 times the power to make it slip as it does when it covers ⅓ of the same pulley.

"Belts, like gears, have a pitch-line, or a circumference of uniform motion. This circumference is within the thickness of the belt, and must be considered if pulleys differ much in diameter and you must get a required speed.

"Owing to the slip, elasticity, and thickness of the belt, the circumference of the driven seldom runs as fast as the driver. With two pulleys of equal diameters, one may be made to run twice as fast as the other without slipping, if you use an elastic belt of India-rubber.

I simply mention this to show the effect of elasticity in belts. As the power of a belt is as its velocity, it is well to run it as fast as possible to avoid lateral pressure, and, consequently, friction of the shaft."

Pulleys.

"One of the greatest objections to the fast running of shafting and belts is the want of pulleys properly constructed. My experience leads me to the conclusion that it is not safe to run a cast-iron pulley 4 feet diameter 400 Rpm, owing to the unequal shrinkage of castings in cooling and other imperfections. Running slow, the centrifugal force has but little effect; but as the centrifugal force is as the square of the velocity, it is not so easily overcome in rapid motions.

"If you make the rim of the pulley thicker, the centrifugal force increases with the thickness, and, consequently, nothing is gained by the extra iron. I have, therefore, substituted white pine felloes made of one-inch boards, breaking joints for the rim, built on cast-iron hubs and arms. The centrifugal force of material is as the specific gravity, and the specific gravity of cast-iron is 13 times that of pine, hence the centrifugal force must be 13 times greater; but the tensile strength of cast-iron is only two to one of that of pine, therefore the rim of a pulley made of white pine felloes will sustain from 4 to 6 times the centrifugal force of a rim made of cast-iron; that is, the same diameter with white pine felloes will run more than double the velocity without being torn asunder. It is less likely to be broken by jar or blow, and is less than half the weight, and, of course, takes less power to run it. I have run a pulley made in this way 16 feet diameter, 4 feet wide, 90 Rpm for 18 months. I have just started another, 17 feet diameter, 62 inches wide, 100 Rpm, driving on to one made the same way 4 feet diameter, and running 425 Rpm. Both of these are working well. I am fully convinced that, with quick shafting, *wood* must take the place of *cast-iron* for the rims of pulleys 3 feet diameter and above.

"No. 2 section of Lawrence Manufacturing Co. has been running with gears, shafting, pulleys, and belts, conforming as nearly as possible to the above rules, and is driving the shafting for 38,000 spindles (throstle, ring, and mule) with the same amount of power as it formerly required for 19,000 spindles."

From Leffel's "Mechanical News."

99. "But not only have band-saws been used with success for heavy work, but another application of a metallic band of essentially the same nature has been made, and, it is said, with satisfactory results. We refer to the employment of a belt of sheet-iron instead of leather or rubber for transmitting power from one pulley to another. This has been done in at least one instance on record (see Art. 146), and the operation of the belt is reported to have met all the requirements of the work. The pulleys in this case were of cast-iron; and it is suggested that if pulleys with an elastic surface were employed, still better results would be obtained. The substitution of steel belts in place of cast-iron is also recommended as, at least, an experiment worth trying; and the fact that in the case of a band-saw the saw itself transmits the power by its friction on the lower pulley. As high as 15 horse-power being in some cases effectively transmitted by such a saw, is pointed out as a proof of the entire adaptability of steel to belting purposes.

"The tensile strength of low steel is such that it is calculated that a belt of this material one foot wide and $\frac{1}{16}$ inch thick could, with a safe working strain — the same in proportion to actual strength which is allowed in ordinary belts — transmit 900 horse-power. So far as ability to bear tension is concerned, this is certainly enough and to spare. (See Art. 79.)

"The various mechanical difficulties which may suggest themselves to the reader as liable to occur have, in a great degree, been overcome in the manufacture of band-saws, for which almost precisely the same conditions — facility of joining and general management — are required as would be called for in belting for any class of work."

The following, from "Overman's Mechanics," 1851, we reproduce here.

100. "For the transmission of rotary motion belts are generally used; iron chains have also been employed, but they are now almost universally abandoned for wire ropes. If an India-rubber, leather, or any other description of belt passes around a pulley it adheres to it with a certain force, which may be called adhesion. A certain tension of belts is always required to prevent slippage; besides which the angle of contact is an element of adhesion. The formula for the force F, which is to be transmitted by a belt of the tension, t, is:

$$log.\ F = log.\ t + .434 \times C \times \frac{S}{R},$$

in which C is the co-efficient of friction, $log.$ the common logarithm, S is the arc of the pulley covered by the belt, and R the radius. The common co-efficient of friction cannot be applied in this case; it is .47 for greased leather upon wood, .50 for dry leather upon wood, .28 for dry leather upon cast-iron, .38 for oiled leather upon cast-iron, and .50 for new hempen ropes upon wood. India-rubber belts may be classed with oiled leather. To increase the arc on the driving pulley, that which is driven may be made smaller, and to increase the arc on both the belt may be crossed. In many instances the arc as well as the tension is increased by a tension pulley.

In cases where all these are insufficient to produce the adhesion required, the rope may be put around the pulley more than once, to afford it a longer time of contact. This is particularly resorted to where ropes are to pull heavy loads, as up inclined planes. This arrangement is here represented:

If the pulley A is grooved, of which at least two are fastened to the same shaft, the rope is directed on one of these pulleys, and, passing around it, goes to B, which revolves on an inclined axis, such that the rope will be received from A' and delivered to A in the plane of the grooves. The number of pulleys may be multiplied to gain adhesion. This method of augmenting friction is preferable to the tension roller, as no increase of tension is required; and it has the additional advantage of bending the rope in the same direction, which makes it more durable.

To determine the strength and size of a belt, find first the amount of labor to be performed by it. This labor is its tension with velocity.

If a belt passes over a 3-feet pulley which makes 100 Rpm, its velocity will be:

$$100 \times 3 \times 3.1416 = 942.48\ Fpm.$$

If this belt is to transmit 2 horse-power, its tension on the pulling side is:

$$\frac{2 \times 33000}{942.48} = 70\ \text{lbs}.$$

In this case it is assumed that one side of the belt is slack; if this is not the case, which in the average of practical instances may be depended upon, the tension on the following side of the belt is subtracted from the above. We here see of how much more service the horizontal belt is than the vertical, for it increases the tension by its own weight, and also the arc of contact.

Fig. 18.

In most of these cases, we may neglect the width of the pulley in the calculation of friction; for the strength of the belt, if sufficient to resist the tension, makes the belt wide enough for adhesion. In all cases it is advisable to make the belt sufficiently wide: no other loss arises from too wide a belt than that of first cost and the loss in rigidity. If a belt is too narrow, or the arc of contact too short, the tension must be increased, in order to afford sufficient adhesion to the pulleys.

Short belts are very disadvantageous, and so are vertical ones: they always require more tension than either long or horizontal belts. Those which are too narrow will stretch, in consequence of which tension and adhesion are diminished. The adhesion of leather upon iron and smooth surfaces is greater than upon wooden and rough surfaces; for these reasons pulleys ought to be made of iron, and perfectly round and smooth. Frequently we see the surface of the pulleys convex, in order to prevent the running off of the belt; this convexity must be very small, or it will diminish adhesion. The most perfect is the cylindrical form of pulleys for flat belts.

Round ropes, or strings, are conducted by grooved pulleys, in which the adhesion of the rope is increased by the wedge-form of the groove into which it is squeezed; the adhesion of these ropes to the pulleys increases, therefore, as the angle of the groove diminishes.

Round grooves are disadvantageous, because they are destructive to the rope, caused by its sliding on the sides of the groove. The best form for the groove is angular, so that the rope touches but in two places tangential to its circumference.

CHAPTER II.

METHODS OF TRANSMISSION BY BELTS AND PULLEYS.

Main Driving Belts.

101. Mr. S. S. Spencer, Superintendent of Conestoga Mills Nos. 2 and 3, at Lancaster, Pa., has kindly furnished me with the following particulars of the main driving belts now in use in these two mills:

Mill No. 2 is driven by a horizontal condensing Corliss engine, having a 30-inch cylinder, 6-feet stroke, and running 52½ Rpm. The fly-wheel is 22 feet in diameter, and weighs 25 tons. It has teeth on its periphery of 5.183 inches pitch, 18-inch face, and drives a "jack" 9 feet 7½ inches diameter on a counter-shaft 9 feet 10 inches below the engine shaft. On this counter-shaft are three 9-feet 6-inch pulleys, each driving a 5-feet pulley on the main lines as here shown.

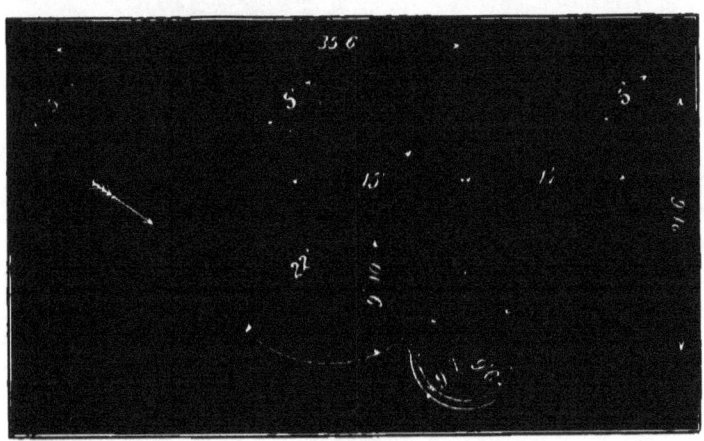

Fig. 19.

The belts running to right and left are each 23½ inches wide, and each transmits 125 horse-power. The middle belt is 29 inches wide, and transmits 175 horse-power. All are of double leather.

Mill No. 3 is driven by a horizontal non-condensing Corliss engine of 28-inch cylinder and 5-feet stroke, running 50½ Rpm, having a 22-feet diameter fly-wheel pulley of 17 tons' weight.

Two 14½-inch double leather belts on the fly-wheel run to right and left, driving 7-feet pulleys on the line shafts, thus: (Fig. 20.)

Fig. 20.

The belts transmit 250 horse-power, have been in use since 1852, and are doing well. If we suppose each belt does half the work, we will have 33.64 square feet of belt in motion per minute per horse-power.

Another case worthy of note is that of a 16-inch cylinder, 48-inch horizontal Corliss engine, with 14-feet diameter fly-wheel pulley, making 65 Rpm, carrying a 12-inch double leather belt over a 5-feet pulley on the line shaft, which is located under the second floor of the factory in the usual way, and 17 feet horizontally distant from the engine shaft. This engine is doing 90 horse-power, and has been running 8 years. The velocity of belt surface here is 2860 Fpm, and gives 31.77 square feet of belt per minute per horse-power.

Mr. Spencer further says: "Whether a belt is more or less liable to slip on the larger pulley, I say less in all cases. There is great advantage in covering the smaller pulley with leather where much work is required. It is much better to use narrow belts and pulleys of large diameter, than wide belts and pulleys of small diameter. This fact is probably less understood and appreciated than any other in connection with belts and pulleys. Belts require care in their application and management. I have belts that have been in use 22 years under rather unfavorable circumstances, and look to-day (1872) as if they would last 10 years longer."

Main Driving Belts.

102. Mr. W. B. Le Van presents this and the next case of driving belts:—At A. Campbell & Co.'s factory, Manayunk, Philada., the motive power is transmitted by 3 belts to 3 separate line shafts, as shown in Fig. 21.

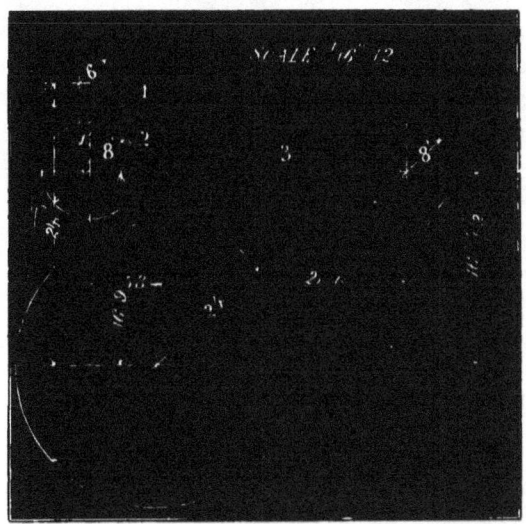

Fig. 21.

Belt No. 1 to the upper 6-feet pulley is 17 inches wide. Belt No. 2 to the 8-feet pulley beneath is $21\frac{1}{2}$ inches wide. Belt No. 3 to the 8-feet pulley, to the right in the cut, is $26\frac{1}{2}$ inches wide. The belts are all double leather, and run from the pulley fly-wheel on the engine shaft, which is 24 feet diameter, 6 feet width of face, and weighs, with its shaft, 43 tons. The engine is a horizontal Corliss, with 30-inch cylinder, 5-feet stroke, making 52 Rpm, driving about 20,000 spindles and the usually connected machinery.

In Fig. 22 we give an indicator card taken from this engine on March 2, 1874, under a boiler pressure of 85 lbs. Scale of card 30 lbs. to the inch; average pressure 45.8 lbs.; horse-power exerted by 1 lb. of pressure 11.14; estimated power to run engine only 16.71: hence we have $(11.14 \times 45.8) - 16.71 = 493.5$ horse-power. Some of the cards ran as high as 47 lbs. average pressure. The average of 8 cards was 456.74 horse-power.

METHODS OF TRANSMISSION. 157

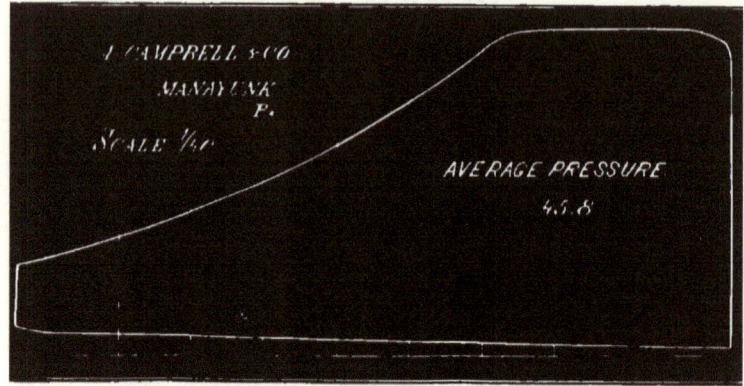

Fig. 22.

Example.

103. At Great Bend, Indiana, an 18-inch cylinder, 48-inch stroke, Corliss engine, under 90 pounds of steam, at 65 Rpm, transmits 190 horse-power usually, and at times 222 horse-power, through the medium of a 22-inch single leather belt over a 12-feet fly-wheel pulley, to a 42-inch pulley on the line shaft.

This belt was originally 24 inches wide, but using the figures above we get 23.64 and 20.23 square feet of belt in motion per minute per horse-power respectively. This is the hardest worked belt of any yet noted.

Belt at J. & J. Hunter's Print-Works, Hestonville, Philada.

104. A 5-ply gum belt, 24 inches wide and 79 feet long, runs on the two pulleys 15 feet 2 inches and 6 feet 6 inches in diameter, situated as shown in Fig. 23, the lower fold drawing.

The engine has a 20-inch cylinder, 36-inch stroke, and makes 56 Rpm, works under 80 lbs. boiler pressure of steam, and gives off 60 to 70 horse-power.

The centre of the 6-feet 6-inch driven pulley is 5 feet 6 inches above, and 17 feet 9 inches horizontally distant from the centre of the engine shaft.

The data given above show that when 60 horse-power are transmitted by this belt, there are 88.97 square feet of belt travelling per minute per horse-power, and that when 70 horse-power are transmitted, 76.26 square feet are travelling per minute per horse-power.

This case is interesting as showing the amount of power a belt is capable of transmitting when adhesion is produced on the pulleys by its own weight. At the time these dimensions were taken the belt folds were within 4 feet of each other.

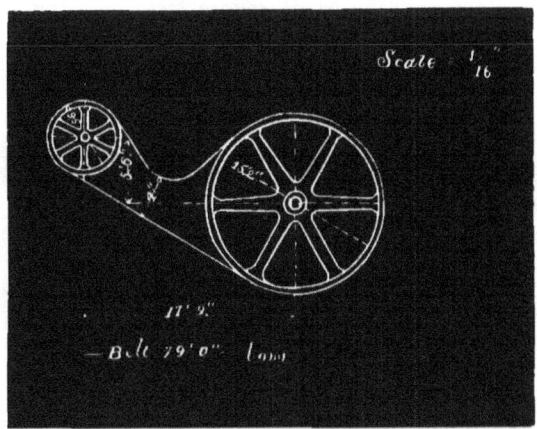

Fig. 23.

Mr. H. W. Curtis, of this city, furnishes the following on Main Driving Belts.

105. "We have at our works, now Hale, Kilburn & Co., a 20-inch single leather belt 167 feet long, having a 3¼ inch single leather belt cemented and riveted on its outside face at either edge, and transmitting the power of a 20-inch diameter, 48-inch stroke, horizontal Corliss engine. It is arranged to give power directly to 3 shafts, each in a separate room, from which the power is further conveyed by means of vertical belts to the other parts of the factory. The lower fold of this belt extends from the fly-wheel over a pulley 4 feet in diameter (see cut No. 24) situated 28 feet from and 23 feet above the horizontal line of the engine.

"The upper fold is carried 12 feet 3 inches higher, and over a pulley 3 feet in diameter, situated directly above the 4-feet pulley.

"The main receiving pulley is 6 feet diameter, situated in an adjacent building, 48 feet horizontally distant, having its centre on a level with that of the 4-feet pulley.

"The fly-wheel is 18 feet diameter, and runs 60 Rpm, giving a velocity to this belt of 3392 Fpm, and was calculated to give 125 horse-power on the 3 pulleys collectively, in the proportion of 80 on the 6-feet pulley, 24 on the 4-feet pulley, and 21 on the 3-feet pulley.

"This would make 70.6 square feet of belt per minute per horse-power on the 6-feet pulley, 235.55 square feet on the 4-feet pulley, and 292.2 square feet on the 3-feet pulley. If considered as a 27-inch belt, it would be working at 61 square feet per minute per horse-power.

"This belt has worked up to 125 horse-power, as proven by indicator cards taken from the engine, which subjects the lower or draw-

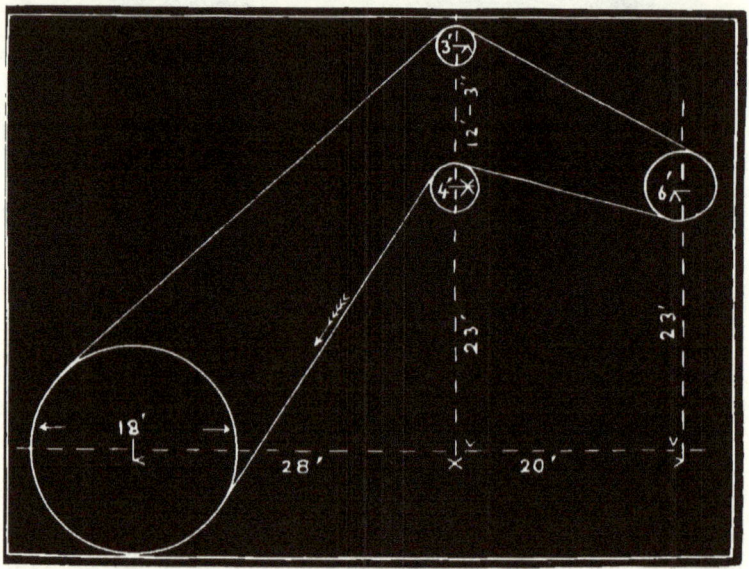

Fig. 24.

ing fold of the belt to a tensile strain of 1216 lbs., or 45 lbs. per inch of width, allowing 7 inches for 3½-inch strips, making the belt equal to one of a single thickness 27 inches wide.

"This belt has been running 9 months; its upper fold is very slack, the longest span is 50 feet at an angle of 45°, having a sag of 20 to 24 inches, and it has given entire satisfaction during that time. We have also 3 other belts, with similar strips on their outer faces. These belts were all tried single at first, but would not do the work required of them. The first is 20 inches wide, taking power from a pulley 4 feet in diameter to one 3 feet in diameter, situated 12 feet

10 inches directly above. The next is 16 inches wide, taking the greater part of the power from the 20-inch belt by means of a 40-inch pulley to one 24 inches, situated 12 feet 6 inches directly above. These belts were put on at the same time as the main belt, and after trying them 5 days, running only part of the machinery, and that with insufficient power, we thought it best to try the strips. The result was that with the strips they have driven all the machinery connected with them, giving no trouble whatever, and have not been tightened more than once in the time named above.

"The next is an 8-inch belt driving a pump, the piston of which is 4 inches diameter, 12-inch stroke, is double-acting, and makes 24 strokes per minute. The pulley on the crank shaft of pump is 22 inches diameter, and this is driven by a 7-inch pulley, situated 9 feet below, and 5 feet from the perpendicular line of the pump; both pulleys are covered with leather. The pump lifts water 12 feet, through a $2\frac{1}{2}$-inch pipe, and forces it 80 feet more of vertical height through 123 feet of 2-inch pipe. A single leather belt 8 inches wide was first applied, but it would not drive the pump at all. It was thoroughly tried by being drawn so tightly that it parted at one of the splices in a few minutes. It was then provided with strips, one on each edge $1\frac{1}{2}$-inch wide, and put on again, driving the pump successfully. It runs about 3 hours each day, and has not been tightened in 5 months.

"From the above results it is plain to see that our experience with belts of this character has been very satisfactory thus far, and we do think that belts made heavier and stronger on their edges conform to the convexity of pulleys better, and that the same weight of leather will drive more and keep straighter than when put in any other form. We do not, however, recommend belts strengthened by narrow strips on their outer faces for running at a high speed over very small pulleys; in such places, only light belts, of an even thickness, should be used."

Main Driving Belts.

106. Referring to Fig. 25, which represents the driving arrangement of the "Patent Metal Co.," Philada.,

A is a 12-feet pulley, 18-inch face, running 55 Rpm.
B " 4 " " 18 " " " 150 "
C " 6 " " 32 " " " 50 "
D " 6 ft. 6 in. " 32 " " " 46 "
E " 15-inch single leather belt.
F " 23 " double " "

The pulley D is 28 feet from A, measuring from centre to centre of shafts, which lie in the same level plane, and B is 24 feet horizontally distant, and 12 feet above A, which latter is a heavy main driving pulley on the engine shaft. The cylinder of this engine is 20 inches diameter and 36 inches stroke, the fly-wheel A making 55 Rpm. Indications taken during recent trials showed that 105 horse-power were transmitted by the belt F to the pulley D on the roll shaft, and that 20 horse-power were transmitted by the belt E to the pulley B on the line shaft. During one trial the belt F was found to have sufficient adhesion to hold the engine still with steam on, the amount of steam admitted to the cylinder being unequal to the resistance at the rolls, when the indicator recorded 125 horse-power. Afterwards, with altered valve giving later point of cut-off, this same belt drove the rolls through their work without stoppage, and with an average indication of the same horse-power.

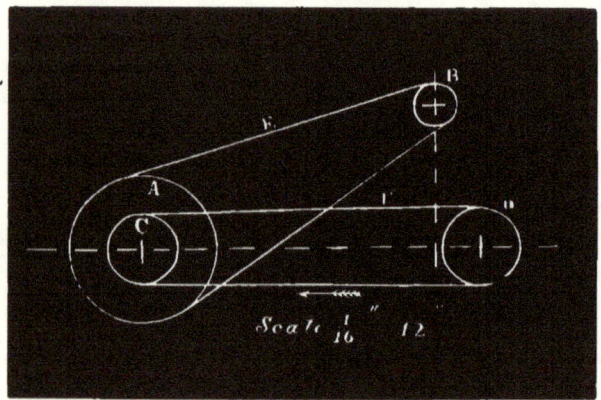

Fig. 25.

Means of Increasing the Adhesion of Belts.

107. Various means have been devised for augmenting the adhesion of belts to the pulleys which they drive, to some of which we now refer. With the ordinary single and double belts, where the usual tightening of the same as well as where the application of the several oily and adhesive matters in common use have proven insufficient, a very simple remedy has been applied with success, in the shape of a narrower belt drawn tightly, with ends laced in the usual way, and allowed to run with, and on the outside, of the same.

This is a remedy almost too simple to refer to, yet it has, in many instances, proven a valuable addition to the driving power of certain belts which fail to transmit the power desired from lack of adhesion. In many cases belts do not possess sufficient tensional strength to impart the power due to their surface velocity; with such the auxiliary belt above referred to is the cheapest and readiest cure.

Double leather belts are frequently employed in place of single ones, to increase adhesion on the pulley surface; but while we are assured of greater tensional strength in the double belt, we have no data to prove its superior adhesive power.

We have, however, the valuable testimony of Mr. F. W. Bacon, of New York, whose large experience in this line of practical engineering gives weight to his conclusions. He says: "I never use double belts under any circumstances. I am satisfied that, other things being equal, they will not do as much work as a single belt, because they do not come in contact with as much surface as the more pliable single belts do; hence there is no advantage in doubling the thickness. If greater tensional strength is required use wider belts."

In Fig. 26 we present the "traction gearing" of Mr. Alonzo Hitchcock, of New York, which consists of 3 pulleys, each on an axis of its own, and a driving belt so placed that the driving pulley, A, touches the driven pulley, C. These two are forced into close contact by the auxiliary pulley B, over which the belt is tightly drawn from the pulley A. The larger pulleys may be leather covered, and all should be straight on the face and have their centres in the same straight line, and, of course, their axes in the same plane.

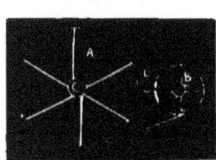

Fig. 26.

It is evident, on inspection, that the belt and pulley surfaces all favor rotary motion in the pulley C, and while the pulleys A and C may vary greatly in diameter, producing rapid increase of speed, the belt passes freely and with full driving effect over comparatively large pulleys. The pulley B being made of any diameter desired.

This combination was patented January 30, 1867. The inventor says: "I claim distributing the power around the shaft to be driven, so that the tendency to displace the shaft on one side is counteracted by that on the other, by the means and in the manner shown."

A more complicated arrangement of pulleys for gaining a high speed of rotation at once, without intermediate pulleys, is shown by Parker's patent belting in Fig. 27.

In this the auxiliary pulley, B, is connected to the driven pulley, C, by an endless belt, F F, but, unlike the usual method, the driving pulley A is set against the outside of the belt to contact with B and C, the belt passing between.

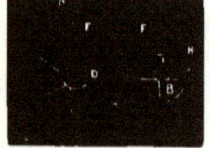

Fig. 27.

The axis of the pulley B turns in the ends of the arms, E, one being on each side, which are jointed with the arms D at H, outside the pulley; the other ends of D turn upon the shaft bearings of pulley A.

These levers form a toggle by which the auxiliary pulley is forced against the driving pulley. The effect of this combination is to draw belt and pulleys into close driving contact, and at the same time to avoid severe lateral strains on the shaft bearings.

The driving and driven pulleys may have diameters of 30 to 40, and even as high as 50 to 1 respectively; the driven pulley may, indeed, be nothing more than an enlargement of its shaft, and the auxiliary pulley must be of such diameter as will prevent contact of belt at F.

The pulley surfaces must all be straight; that of C may be covered with leather, the belt must be made of well stretched leather of uniform texture, must have a permanent joint, and be of equal thickness throughout its entire length, and all the parts must be fitted with great exactness to insure perfect working of the combination.

Many of these have been made and used for driving small circular saws, where hand-power is employed, and for such and like purposes they answer very well.

One of the simplest methods of increasing the efficiency of any belt or cord is with pulleys of given dimensions, to cause it to embrace a greater portion of their circumference, which increases proportionally its adhesion as well as its driving power. The following shows one method of doing this by the cord and pulley arrangement.

In Fig. 28 one fold of the cord is drawn out of its direct course between the peripheries of the driving sheave A and driven sheave C, across the other fold and around a tightening sheave, B, which is so arranged as to be moved to and from the other sheaves, and set at such an angle as to prevent contact of cords at points of cross-

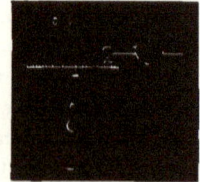

Fig 28.

ing. The movement of B is such as to permit the cord to run in grooves of different diameters on A and C, by which different speeds in C can be obtained when that of A is uniform.

This is an old device, and is frequently applied to the driving gear of foot-lathes.

In this the driving power is limited to the adhesion of the single cord in the sheave grooves, but its circumferential contact is such as to give the greatest effect possible.

In order to greatly increase the adhesion of the cord the plan shown in Figs. 29 and 30 has been devised, and consists of two multigrooved wheels, A and C, the driving and driven sheaves of the system. Into the grooves of these sheaves is wound continuously a single endless cord, in parallel lines from one sheave to the other, such that the cord in leaving the last groove of A is deflected across and above the other cords, and delivered in line of the groove in C by the adjustable single grooved sheaves B and B.

Fig. 29.

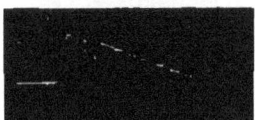
Fig. 30.

The bearings of these sheaves are secured to rods, D D, fixed parallel with the cords, and upon which the sheaves can be slipped and fastened to take up slack of cord.

It is evident that with any cord its adhesion, and consequently its driving power, is increased in the direct proportion of the number of grooves in each driving sheave.

Referring generally to pulleys with angular grooves for round cords of any material, we quote from *Publication Industrielle, par Armengaud ainé*, Vol. IX., p. 428:

"This system is much in use for transmitting power by means of hemp cords or round gut belts of small dimensions. The angular shape of the groove is preferred, in order to increase the adhesion of the cord, which is thus pinched between the surfaces of the groove by the primitive tension it receives similar to ordinary belts. The angle which the two surfaces form with each other is made ordinarily 60°."

A very simple method of driving two pulleys on separate shafts from one pulley is shown in Fig. 31. The driving pulley A has a belt running to B, which it turns in the usual way.

The pulley C may be driven by A also, by simply running its belt over the belt which turns B.

The belts atop of each other on A increase the adhesion sufficiently to drive both B and C.

We have found it convenient, as well as economical, to resort to this expedient for driving lines of shafting in different stories from the prime mover below.

"Where pulleys of very unequal diameter are coupled by a belt, the surface of contact with the smaller pulley is so little that the belt must be very tightly stretched, in order to transmit its full duty. This is especially the case where the two pulleys are very near each other. In the case of the small centrifugal pumps, made by Gwynne & Co. for plantation and farm use, the horse gear is connected with the pump by belting from a large pulley to the riggers of the pump, the relative diameters of the two being about as 6 : 1, while also they are placed but four or five inches apart. The short belt, thus acting upon a very small portion of the surface of the rigger, requires to be very tightly stretched, and in this way a heavy strain is brought upon the bearings. To relieve this strain, Messrs. Gwynne & Co. place a friction wheel between the pulley and rigger, so as to touch both in a line connecting their centres. This wheel, revolving freely upon a fixed centre, receives the strain exerted by the belt. No means of adjustment are provided to compensate for the wear of the journals, nor is such provision believed to be necessary. The arrangement, so simple in itself, appears to possess a considerable advantage over the ordinary practice of throwing the whole strain of the belt upon the bearings."

Fig. 31.

Fig. 32.

The intermediate pulley is recessed in the centre of its face, so as to bear on the others at the edges only. The arrangement is said to work well, and is here shown (Fig. 32).

Double Separate Belts.

108. Mr. J. S. Lever, of Philadelphia, presents this curious performance of a pair of belts, shown in Fig. 33.

"The driving pulley, *a*, on the engine-shaft is 5 feet diameter, and runs 75 Rpm; the driven pulley, *b*, on the line-shaft is 3 feet diameter, and about 20 feet distant in a line near 45° with the horizontal. Both pulleys are wooden drums covered with leather, and have sufficient breadth of face for two 12-inch belts. When this arrangement was started two 12-inch single leather belts were put on side by side, and for some time were run in that way, but never satisfactorily. After many fruitless efforts to obtain a uniform action of the two belts, one accidentally mounted the other, the two running thenceforth as one double belt. In this way they drove the line-shaft better than ever before. Many experiments were tried in the relative tightness of the two belts, which invariably proved that the best driving was always secured when the inside belt, *d*, was very slack, sagging, say 12 to 18 inches, and the outside belt, *c*, quite tight, the working sides, of course, running close to each other, as shown in the cut. This was in use some thirty years ago in a factory near this city."

Fig. 33.

Imparting and Arresting Motion.

109. Mr. Thomas Shaw's dead-stroke power-hammer illustrates the application of the belt for giving to and taking motion from a shaft at the pleasure of the operator (see Fig. 34). The same devices can, however, by an easy transition, be applied to other machines.

In this the driving pulley, carrying a loose belt, is on a line-shaft over the driven flanged pulley, which latter is on a shaft at the top of the hammer-frame. This shaft carries a crank-wheel actuating the hammer, as shown, and is partly invested by a leather band for arresting its motion. One end of this band is secured to a pin in the hammer-frame under the crank-wheel; the other end is fastened to the swinging lever, to which also the tightener-pulley of the driving belt is applied.

The action of these belts is produced by opposite motions of the lever; thus, when the operator pushes it, the arresting-band releases the crank-wheel, and the tightener-pulley presses upon the driving belt, which, being constantly in motion, applies its adhesion to the pulley on the crank-shaft and propels the hammer; and it does this with a varying velocity, according to the pressure upon the tightener. Withdrawing the lever relaxes the driving belt and tightens the

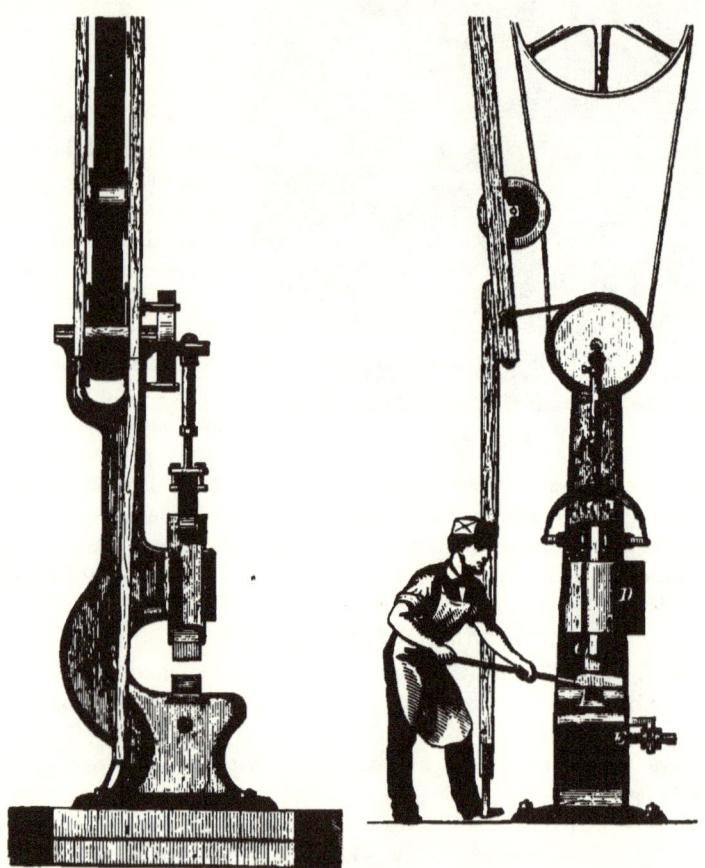

Fig. 34.

arresting-band. These motions are under the easy control of the operator, and such is the nature and action of the belt in this application, that these motions can be repeated rapidly and effectively without destructive wear to any part of the machine.

Combined Fast and Loose Pulley for Round Belts, by John Shinn, of Philadelphia.

110. The round belt, f, fits in a groove formed between two half-pulleys, of which A' is fixed and A slides upon a fixed key on the shaft, B; between A' and A, and running loosely on the shaft, is a flat-faced pulley, C. When A is separated from A' a short distance, the belt, f, will cease to turn them, and will run on and turn C instead. The belt drives the shaft, B, only when pinched between the half-grooves

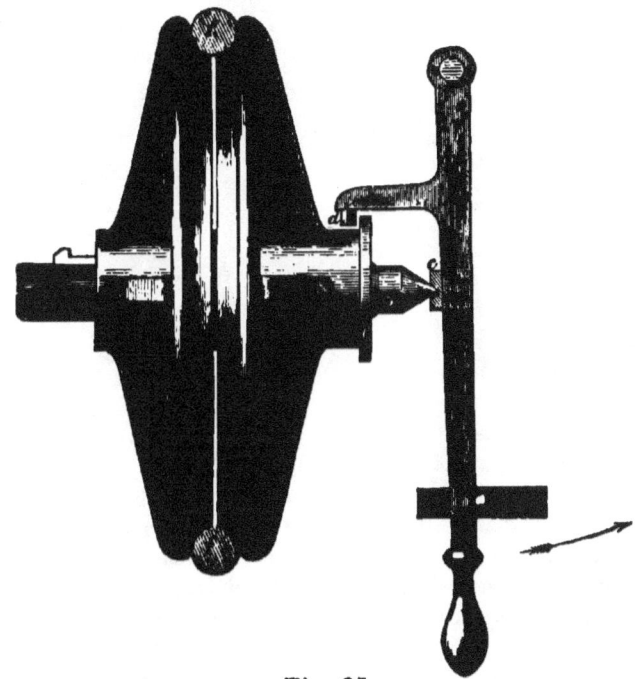

Fig. 35.

of A' and A. The lever, D, when moved in the direction indicated by the arrow, withdraws the half-sheave, A, and permits the belt to run on the loose pulley. Simple and efficient means for holding the parts together and drawing one half from the other are shown in the cuts.

It is not proposed, of course, to drive very large and heavy machinery with round belts, such as are required for this description of shifting pulleys; but, as far as a round belt will go with advantage, these pulleys will be found of the greatest service. Thus, the round belt cuts off less light, occupies less room, makes smaller holes in the floors, needs lighter driving pulleys to carry it, and thus saves power; while, as regards the driving power of round belts, we have seen one of an inch diameter doing for years work which proved too much for a 7-inch flat belt.

Band Links.

111. "Where tension alone, and not thrust, is to act along a link, it may be flexible, and may consist either of a single band or of an endless band

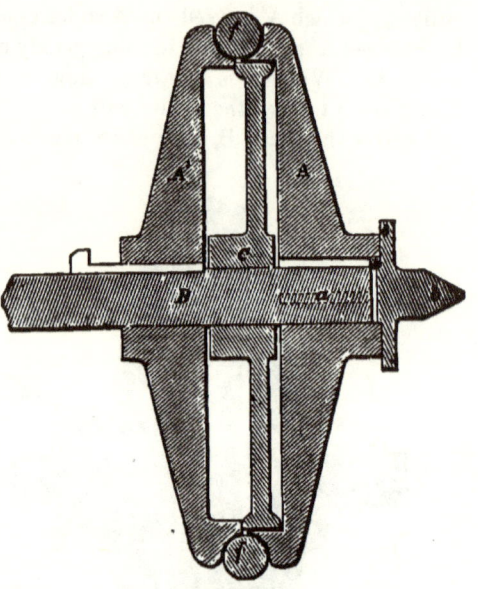

Fig. 36.

passing round a pair of pulleys which turn round axes traversing and moving with the connected points. For example, in Fig. 37, A is the axis of a rotating shaft. B that of a crank-pin. C the other connected point, and B C the line of connection; and the connection is effected by means of an endless band passing around a pulley which is centred upon C, and round the crank-pin itself, which acts as another pulley. The pulleys are of course secondary pieces, and the motion of each of them belongs to the subject of aggregate combinations, being compounded of the motion which they have along with the line of connection, B C, and of their respective rotations relatively to that line as their line of centres; but the motion of the points B and C is the same as if B C were a rigid link, provided that forces act which keep the band always in a state of tension.

"This combination is used in order to lessen the friction, as compared with that which takes place between a rigid link and a pair of pins; and the band employed is often of leather, because of its flexibility." — *Rankine's Mill Work, p. 213.*

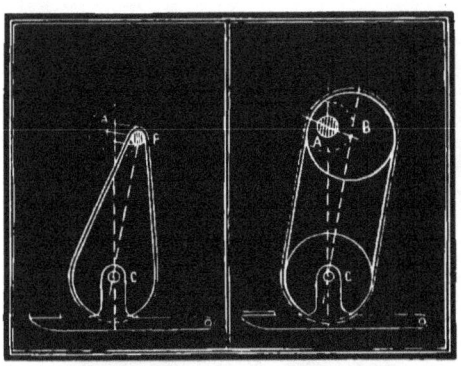

Fig. 37.   Fig. 38.

In Fig. 38 we give a substitute for Fig. 37, in which an eccentric, B, takes the place of the crank, allowing a straight shaft to be used. When the eccentricity of B equals the radius of the crank, the result is the same, but experiment has proven, in the case of the eccentric used in the treadle arrangement of the latter, that the motion lacks freedom, the treadle moving heavily.

Weaver's Belting.

112. The object of this arrangement is to obtain high speed in a shaft directly from a driving pulley without the aid of intermediate counter pulleys or gears, and with reduced lateral stress on the bearings of the driven shaft.

A, B, and C, Fig. 39, show 3 shafts parallel to one another. A and C carry straight-faced pulleys, upon which run 2 belts of equal length and width, separated to prevent contact with each other while running. The lower fold of belt, D, is carried *over* the shaft, B, and the upper fold of belt E is carried *under* B, and each, in running, imparts motion to the driven shaft in the same direction, and at the same time balancing the lateral pressure on its journals.

A is the driving shaft with large pulley; B the driven shaft of comparatively small diameter, and C, a counter shaft, with its pulley of any convenient diameter, is placed in position to carry and return the belts, and may be moved and secured to and from B by screw adjustment or otherwise, to secure proper tension of belts.

Belt as a Friction Clutch.

113. The belt offers a simple and efficient means of producing intermittent effects, as in that of operating stamps, when it is desirable to control by the hand the number of blows in a given time, as

METHODS OF TRANSMISSION. 171

well as to vary the intensity of the same at will. Arranged in this manner it acts as a friction clutch.

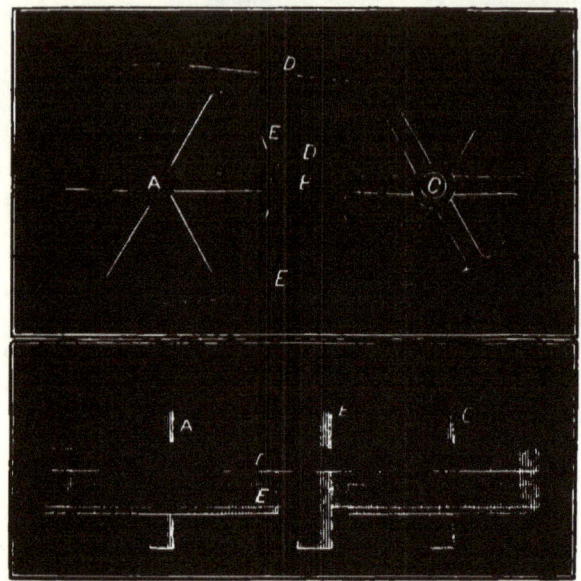

Fig. 39.

In Fig. 40, the shaft J, which may be the "main line" of the shop, or a counter shaft, carries a flanged pulley, H, continuously revolving in the direction of the arrow. Over this pulley is thrown loosely a belt, D, one end of which is fastened to the stamping weight, A, at E, and the other end is secured to the floor at F.

The weight, A, is guided in the parallel uprights, B B, and rests upon the base, C. As the apparatus stands, the weight A is not lifted, owing to slackness of the belt, but by taking hold of the belt at G by the hand, and drawing it forcibly in a horizontal direction and at right angles to the shaft, a severe tension is created in the belt on the pulley, which latter lifts the weight at a velocity nearly equal to that of its rim. Relaxing the pull on G lessens the tension, when the weight falls freely back to the base.

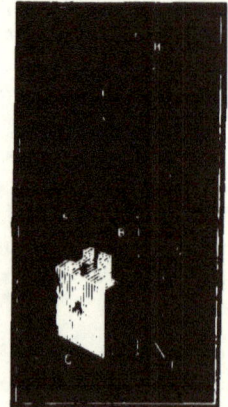

Fig. 40.

It is evident that the number of blows struck may be repeated at will, and that the force of the blows, in so far as they may be due to the height of the fall, may also be regulated by the duration of the lateral pull on the belt.

Since the effect of the lateral pull on G is as G F + G H, and becomes less as it leaves the straight line joining H and F, it is important to make the distance H F as great as may be convenient, and to keep the belt as taut as can be without destructive friction on it at the pulley surface when not in use.

For the lighter hand-stamping operations, this is a simple, cheap, and ready means of obtaining gravity effects on dies and moulds.

Nine Dispositions of the Quarter-Twist Belt.

The Shafts at Right Angles but not in the same Plane, the Belt Running on Two Pulleys.

114. When two shafts are at or nearly at right angles with each other, and not in the same plane (Fig. 41), and it is desired to drive one from the other by two pulleys only and a connecting belt, experience has proved that certain conditions are necessary. In the first place, the distance between the near faces of the pulleys must not be less than four times the width of the belt. The pulleys A and B should be so placed that the belt will lead from the face of one to the centre of the face of the other, that is, so that *a plane passing through the centre of the face of one pulley will be tangent to that part of the face of the other from which the belt is running.*

The following diagram gives the position and proper proportions referred to:

The pulley A, from which the belt deflects, should have a wider face than B, in the proportion of 10 to 6, and should be more rounding on the face than is usual, and the pulleys should be as small as may be to do the work, and should be of nearly equal size.

About 25 per cent. of belt contact is lost when the belt makes a quarter-turn, even when the pulleys are of the same size. We have noticed in the performance of a *leather* belt that the first 90° of lap on the pulley fit closely as in the ordinary straight belt arrangement; but in the second 90°, about half the width of the belt is forced from contact with the pulley by the strain in the substance of the belt, due chiefly to its imperfect elasticity, and primarily to the oblique deflection of the fold which is leaving the pulley.

With a belt perfectly elastic the same amount of contact, if not more, can be obtained, as with the open belt; since the belt would adhere to the face of the pulley up to the line of departure the same in one case as in the other.

A Quarter-Twist Belt.

115. Mr. L. H. Berry, of the Atlantic Works, Philadelphia, gives the particulars of a quarter-twist belt, arranged by him, and shown in Fig. 42, for driving a 54-inch circular saw, the periphery of which travels at the rate of 8400 Fpm, and the mandrel lying at right angles to the driving-shaft.

"On the mandrel is a 12-inch pulley, and on the driving-shaft, which runs horizontally 8 feet above, is a wooden drum 24 inches diameter, 8½ feet long, upon which the belt—a 10-inch heavy single leather, travelling 1800 Fpm—traverses back and forth, following the reciprocating movement of the saw mandrel. The forward movement of the saw when cutting is at the rate of 60 Fpm, and the return movement 120 Fpm.

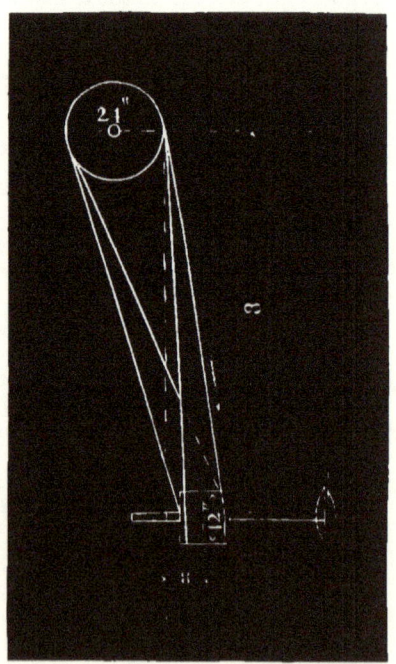

Fig. 42.

"From some cause (centrifugal force, perhaps, or because the belt was new, and, therefore, not as pliable as it would otherwise have been) the centre of the pulley had to be set 8 inches out of the path of the vertical line from the periphery of the drum. (See cut.)"

The Shafts at Right Angles but not in the same Plane, the Belt Running on three Pulleys. Figs. 43 and 44.

116. A is the driving pulley on a horizontal shaft; B the driven pulley on a mill-spindle or upright shaft; C the tightener or guide pulley, which is placed at the proper angle for receiving the belt from B and delivering it to A. It has a short shaft running in bearings secured to a frame which slides vertically in fixed grooves, and may

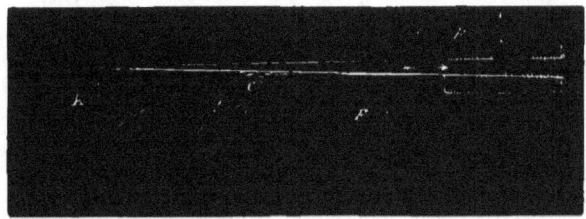

Fig. 43.

be raised to tighten the belt for driving, or lowered to slacken the belt for stopping, B, at pleasure. B is made wide and straight on the face to admit of motion in raising and lowering the stones, as well as to allow of lead of belt by the different positions of C, which are due to length and tightness of belt.

A and C should be rounding on their faces. The cut shows the proper positions of the pulleys and shafts, and also gives good work-

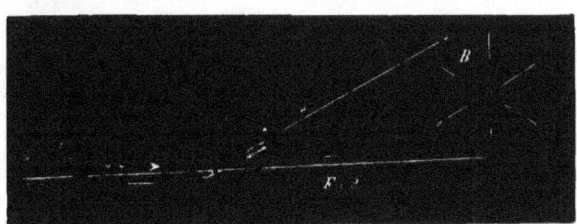

Fig. 44.

ing proportions, the particulars having been obtained from machinery in use; but the motion of the belt, as shown, should be reversed.

The quarter-twist belt, with intermediate guide pulley, like Figs.

43 and 44, will permit of very short distance between the driving and driven shafts. A case in practice may be cited, in which the driving pulley is 40 inches, the driven pulley 18 inches, and the guide pulley 16 inches in diameter; all of them are 8 inches face, and the shafts are 4 feet 7 inches from centre to centre, vertically.

This distance might be even less without injury to the belt. In the erection of this arrangement it was found necessary to set the face of the driven pulley one inch back of the centre of the face of the driving pulley, and to give the axis of the guide-pulley an inclination of 30° to the horizontal line.

The Shafts at Right Angles but not in the same Plane, the Belt Running on four Pulleys. Fig. 45.

117. Let E be the *driving* shaft, with tight pulley, A, and loose pulley, B, and F the *driven* shaft, with tight pulley, D, and loose pulley, C; all the pulleys of same size and with rounded faces, in the usual way.

Let the pulleys be arranged in a square on the plan, whose side is the diameter of pulleys at centre of face, and let an endless belt be put on, as shown, and run in the direction of the arrow. It will be noticed the loose pulleys, C and B, run in opposite directions from that of the shafts on which they turn; but since they carry the slack fold of the belt, they are relieved of heavy strain on the shafts. This is a good plan for wide belts when the shafts are a proper distance apart — say 10 times the breadth of the belt — and solves the sometimes difficult problem of carrying considerable power around a corner by a belt. There is no loss of contact of the belt on any of the pulleys of this system, and no lateral straining and tearing of the fibres of the belt, as in the usual quarter-

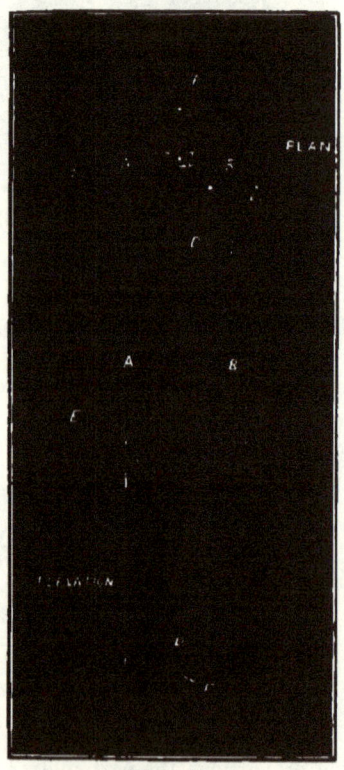

Fig. 45.

METHODS OF TRANSMISSION.

twist arrangement, in which only two pulleys are used. The lower shaft may drive the upper one, as well, by changing the direction of motion, or changing the relative positions of the tight and loose pulleys.

The Shafts at Right Angles but not in the same Plane, the Belt Running on four Pulleys. Fig. 46.

118. A is the driving pulley on a horizontal main line-shaft; B the driven pulley on a mill-spindle or upright shaft; C a tightener on a shaft parallel to the main shaft, with bearings, in a frame, which, with the pulley, can be raised or lowered when required to start or

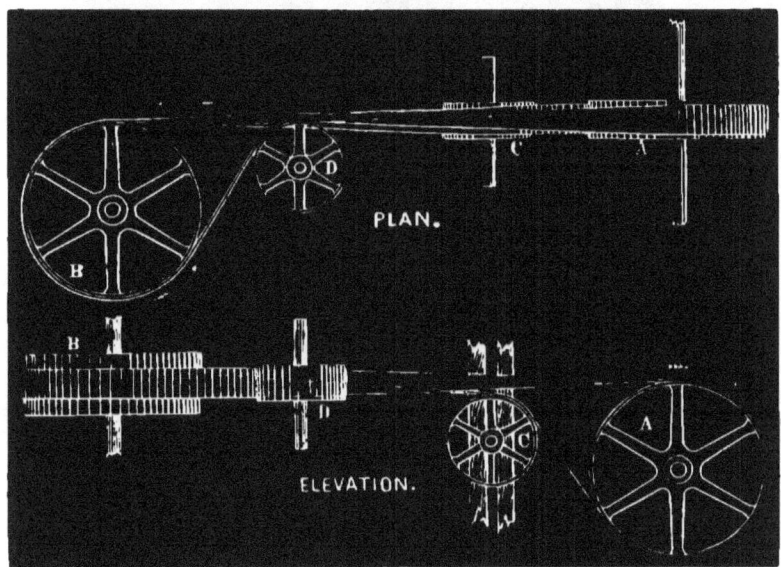

Fig. 46.

stop the pulley B; D a guide pulley on a vertical shaft running in fixed bearings. The course of the belt is indicated by the arrows. This plan may be resorted to when the pulley A cannot be placed on the main shaft in a position to receive the belt directly from B, as in the case shown in Figs. 43 and 44.

The Shafts may or may not be at Right Angles, must be in or near the same Plane, the Belt running on four or five Pulleys. Figs. 47, 48, and 49.

119. Fig. 47 shows the usual method of transmitting power to shafts which are at or near right angles with the driver, and Figs.

48 and 49 show an extension of this method to driving two such shafts from one.

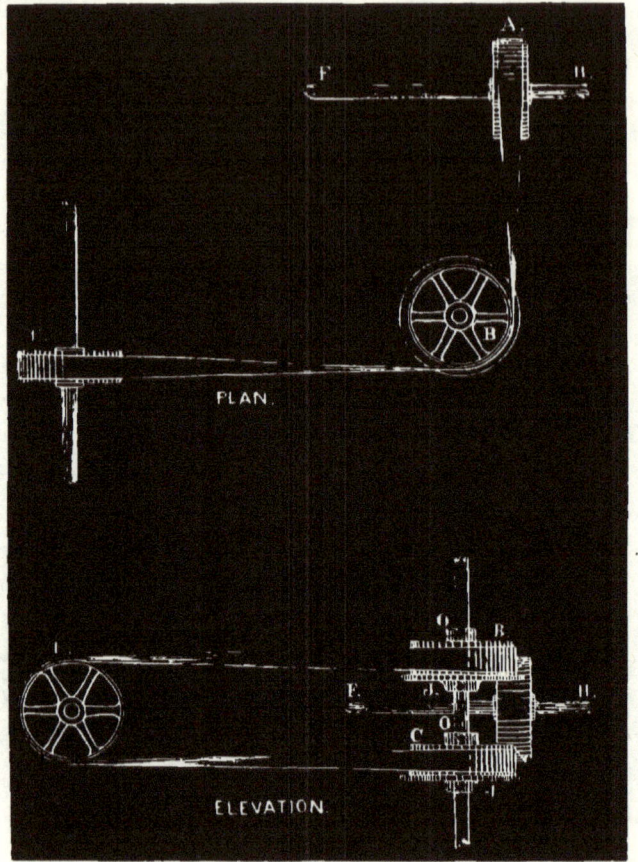

Fig. 47.

Let A be the driving pulley on the main shaft, F H; D and E driven pulleys on the counters, at right angles to the main. Place two upright shafts, each with a loose pulley, so that its face will be opposite the middle of the face of A, one to the right and one to the left, over these pass a belt as shown in the cuts. The belt will run either way in both.

In Figs. 48 and 49 it will be observed that the driving face of the belt is changed between the two pulleys, D and E, which may be avoided by giving the belt a half twist in this part, which we think, however, would injure the belt more than by using both sides of the same.

Collars O and O are placed over the pulleys B and C, and we have added the stationary flanges J and J to the uprights under the pulleys, introduced by Messrs. Wm. Sellers & Co., Philadelphia. This device, whether applied to vertical or horizontal pulleys, is in every

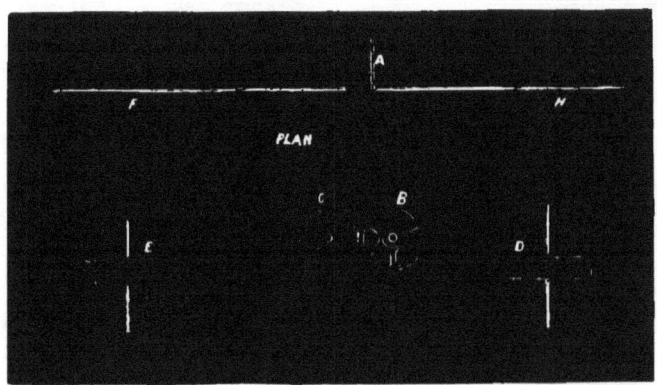

Fig. 48.

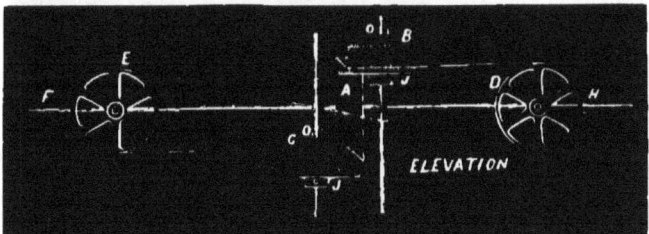

Fig. 49.

way superior to flanges fast to pulleys which tend to lift the edges of the belts, and turn them over. On the other hand, when the belts strike stationary flanges, they are thrown back on the pulley faces again, except, perhaps, in the case of soft, flabby belts, which are liable to curl at the edges and roll up.

The Shafts may be at any Angle with each other, may not be in the same Plane, the Pulleys may differ much in Diameter, and the Belt may be crossed.

120. What cannot be done with the preceding methods of arranging the quarter-twist belt, may be done by the guide pulley devices shown in Fig. 50, in which the vertical cylindrical staff, A, is secured to a flange, J, and held by a brace, G, to an over-head timber, H, or other fixture.

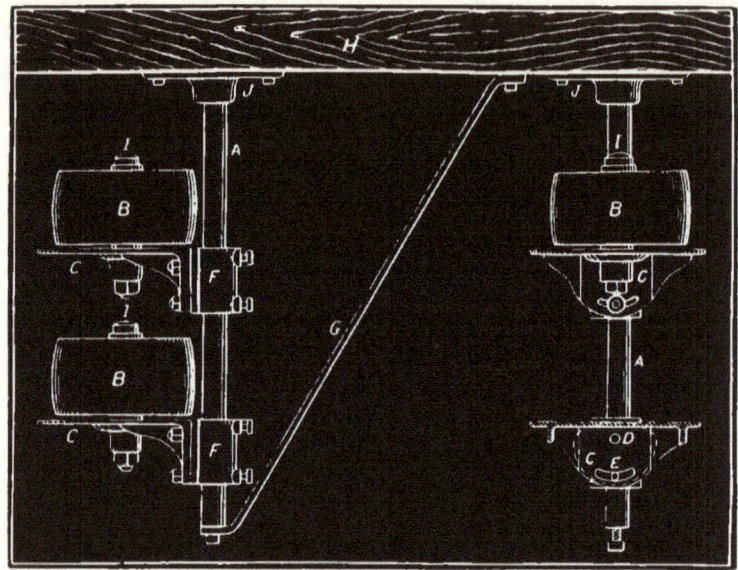

Fig. 50.

Upon this staff are placed two hubs, F, held by set screws in any position, and each formed with a flat face, to which a flanged bracket, C, is bolted.

The upper bolt, D, is utilized as an axis about which the bracket can turn, and the lower bolt, E, in a slot permits the turning and holds the bracket at the inclination required by the belt.

In the centre of each flange, C, is secured a pin, I, upon which the pulleys, B, turn.

The facility with which these pulleys may have their axes inclined to accommodate the angle of a belt passing from a smaller to a larger pulley, from a higher to a lower shaft, and crossing to the opposite faces of pulleys, favors the employment of this combination, which has an all but universal adjustment, in many places where the previously described arrangements will not serve at all.

The writer, early in February, 1871, was required to manage a case of belting two shafts, set at right angles, one above the other, having pulleys of different diameters which were to run in opposite directions, and all of them lying close to one another. This was disposed of successfully by the use of mechanism, exactly like that shown in Fig. 50, and this, with that shown in Fig. 47, are existing methods employed at the People's Works, Philadelphia, for belt driving around corners and in confined places.*

Holes for Quarter-turn Belts.

121. Draw on a level floor, with chalk line and tram, two full-size views of the pulleys and position of the floor through which belts are to pass, or lay them down on paper to a convenient scale: observing that, *that fold of the belt which leaves the face of one pulley must approach the centre of the face of the other in a line at right angles to the axis of the latter.* Completing the figures as shown in Fig. 51, the points of intersection $a\ b\ c$ and d will indicate the places in the floor, E F, where the centres of both folds of the belt will pass when drawn tightly and at rest. The obliquity of the opening can best be obtained by trial of tape line or narrow belt applied to the pulley faces in position, passing through small trial holes in the floor. Allowance in the hole should be made for the sag of slack fold of belt. This is the usual safe and practical shop method.

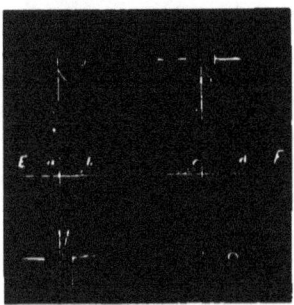

Fig. 51.

Belt for Cooling Shaft Journals.

122. A very ingenious as well as simple method of cooling a journal, consists in placing an endless belt of loose water-absorbing texture on the shaft, as near the heated part as may be, and allowing the lower bight to run in cold water, which may be held in a vessel at a convenient distance below the shaft.

Continuous contact of the liquid band carries away the heat of friction as it is produced, without spilling or splattering of water on and about the machinery, and without contact of the lubricant in the journal boxes.

We have seen this method successfully applied to the shafts of the rolls of calico printing-presses.

---

* Two shafts at any angle with each other may be effectively driven by two belts, each having less than an $\frac{1}{8}$ twist and each running on two pulleys, by placing a counter-shaft above or below and across the main lines at or near equal angles to the main line shafts.

CHAPTER III.

CEMENTS, ADHESIVES, AND FASTENINGS FOR BELTS.

Cement for Cloth or Leather. Molesworth.

123. 16 gutta-percha, cut small.
      4 India-rubber, "
      2 pitch, "
      1 shellac, "
      2 linseed oil, "
} melted together and well mixed.

Water-proof Cement for Cloth or Belting. Chase.

124. "Take ale, 1 pt.; best Russia isinglass, 2 oz.; put them into a common glue kettle, and boil until the isinglass is dissolved; then add 4 oz. of the best common glue, and dissolve it with the other; then slowly add $1\frac{1}{2}$ oz. of boiled linseed oil, stirring all the time while adding and until well mixed. When cold it will resemble India-rubber. When you wish to use this, dissolve what you need in a suitable quantity of ale to have the consistence of thick glue. If for leather, shave off as if for sewing, apply the cement with a brush while *hot*, laying a weight on to keep each joint firmly for 6 to 10 hours, or over night."

Elastic Varnish. Smithsonian Report.

125. "2 pts. rosin, or dammar-resin, and 1 pt. caoutchouc are fused together, and stirred until cold. To add to the elasticity, linseed oil is added. Another varnish for leather is made by putting pieces of caoutchouc in naphtha until softened into a jelly, adding it to an equal weight of heated linseed oil, and stirred for some time together while over the fire."

To Render Leather Water-proof. MacKenzie.

126. "This is done by rubbing or brushing into the leather a mixture of drying oils, and any of the oxides of lead, copper, or iron, or by substituting any of the gummy resins in the room of the metallic oxides."

Water-proof Glue. MacKenzie.

127. "Fine shreds of India-rubber dissolved in warm copal varnish make a water-proof cement for wood and leather."

Another.

Glue, 12 oz.; water sufficient to dissolve it; add 3 oz. of rosin, melt them together, and add 4 parts of turpentine or benzine. Mix in a carpenter's glue-pot to prevent burning.

To Preserve Leather from Mould. MacKenzie.

128. "Pyroligneous acid may be used with success in preserving leather from the attacks of mould, and is serviceable in recovering it, after it has received that species of damage, by passing it over the surface of the hide or skin, first taking due care to expunge the mouldy spots by the application of a dry cloth."

Castor-Oil as a Dressing for Leather. MacKenzie.

129. "Castor-oil, besides being an excellent dressing for leather, renders it vermin-proof. It should be mixed, say half and half, with tallow or other oil. Neither rats, roaches, nor other vermin will attack leather so prepared."

Adhesive.

130. A good adhesive for leather belts is printer's ink. I have the case of a 6-inch belt running dry and smooth and slipping, which latter was entirely prevented for a year by one application of the above.

To Fasten Leather to Metal.

131. "A. M. Fuchs, of Bairere, says that in order to make leather adhere closely to metal, he uses the following method: The leather is steeped in an infusion of gall nuts; a layer of hot glue is spread upon the metal, and the leather forcibly applied to it on the fleshy side. It must be suffered to dry under the same pressure. By these means the adhesion of the leather will resist moisture, and may be torn sooner than be separated from the metal."—*Athenæum.*

Dressing for Leather Belts.

132. "One part of beef kidney tallow and two parts of castor-oil, well mixed and applied warm.

"It will be well to moisten the belt before applying it. No rats or other vermin will touch a belt after one application of the oil. It makes the belt soft, and has sufficient gum in it to give a good adhesive surface to hold well without being sticky.

"A belt with a given tension will drive 34 per cent. more with the hair side to the pulley than the flesh side."— *F. W. Bacon, N. Y.*

Cement.

133. "A cement for joining pieces of leather, one which repeated tests have shown to be very efficient, may be made by dissolving in a mixture of ten parts of bisulphide of carbon and one part of oil of turpentine, enough gutta-percha to thicken the composition. The leather must be freed from grease by placing on it a cloth, and pressing the latter with a hot iron. It is important that the pieces cemented be pressed together until the cement is dry."

Water-proof Cement for Belting. Moore.

134. "Dissolve gutta-percha in bisulphide of carbon to the consistence of molasses; warm the prepared parts and unite by pressure."

"To increase the power of rubber belting, use red lead, French yellow, and litharge, equal parts; mix with boiled linseed oil and japan sufficient to make it dry quick. This will produce a highly polished surface."

Fastenings, Etc.

135. From various authorities we have many ways of securing and treating belts. "Down East" we find shoe-pegs used for joining the ends, which, being scarfed, glued, and pressed together forcibly, have the pegs dipped in glue and driven in, and when dry pared off smoothly. Such joints are warranted to stand as long as the belt, but will not resist water unless joined by water-proof glue.

Dried eel-skin lacings, slit lengthwise of the animal, are recommended for their ever enduring qualities, but these were known long ago on the other side of the Atlantic. (See Arts. 39 and 67.)

A simple adhesive for rubber belts is made by sticking powdered chalk, which has been evenly sprinkled over, to the surface of the belt by cold tallow or boiled linseed oil.

Oil can be taken out of belts without injury to the leather by several applications of 4 F. *aqua ammonia.*

Wilson's Belt-Hooks.

136. We are indebted to Crane Bros. for description and illustration of the above, which we present in Fig. 52.

No. 1. — The teeth go through the belt and clinch. To be used only on belts running at very high speed over small pulleys, and in making up belts, instead of lapping, glueing, and pegging.

No. 2. — Is the only belt-hook in the market that can be used repeatedly and successfully more than once. A hammer is all that is required to secure them to belts. Lay the hook down on something solid, with the teeth up, and *drive the belt down tight* to the plate. None of the belt is cut away by punching holes. The strain comes in *fifteen* places instead of *three,* as with lacings. Belts can be mended in $\frac{1}{4}$ the time required for any other hook or lacing. They will last longer than a dozen lacings. They are the *only good* fastening for rubber or paper belts.

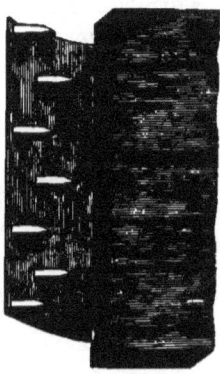 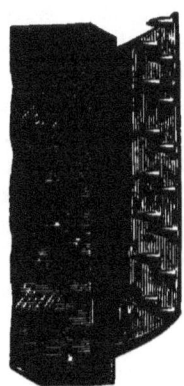

Fig. 52.

These hooks have been thoroughly tried for three years in all places — in machine-shops, cotton, woollen and paper mills — and all who use them admit that they are the *best* and *cheapest* fastening in use, taking into account the *durability* of the hooks, *wear* and *tear*, and *time* in *mending* belts.

Blake's Patent Belt-Studs.

137. This approved device for fastening rubber and leather belting was patented April, 1860, and March, 1861, and re-issued August, 1868, shown in Figs. 53 and 54. Messrs. Greene, Tweed & Co., 10 Park Place, New York, have the sole right to make and sell this invention, and during the last four years have supplied the studs to over 10,000 manufacturing concerns in the United States.

CEMENTS, ADHESIVES, AND FASTENINGS.   185

They are recommended to be better and cheaper than either lacing or hooks. When the above-mentioned studs for fastening the ends of belts were tested on a 6-inch 4-ply rubber belt, the following result was obtained:

Belt-hooks tore out under a strain of	. .	800 lbs.
Lacing " "	. .	1690 "
Blake's studs held up to "	. .	4600 "

Where lacing or other methods of fastening are used, requiring punched holes, which weaken the material of the belt, the whole strain transmitted by the belt must be carried by a width section of the belt less by the sum of the diameters of the holes, which, in usual practice, is about $\frac{1}{3}$ the width of the belt; whereas, by the use of patent studs, no punched holes are made, the material of the belt is not reduced, and nearly the whole section is available for strain, the stud being of such a form and inserted in such a manner as to grip nearly its entire width.

Belts fastened by studs are running now nearly four years without repair. The ends of the belts are kept close together, and the studs do not touch the pulleys.

When bolts have become too rotten to hold lacings, they can be fastened with these studs, which will hold until the belts are completely worn out. For damp places these studs make a secure fastening.

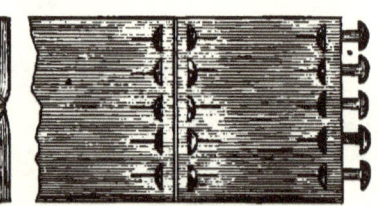

Fig. 53.

In fastening belts by these studs, the slit in a leather belt should be a $\frac{1}{4}$ inch, and in a rubber belt $\frac{3}{8}$ inch from the ends, and the studs inserted $\frac{1}{2}$ inch apart.

The cuts above give the form of the studs, the sizes made, and their position in the joint.

Fig. 54.

The smallest size is used for sewing-machine belts and the like, the next size, say, for 2-inch belts, and the largest for 4 and 5-ply rubber and double leather belts.

Champion Belt-Hook.

138. Fig. 55 will convey a correct idea of the manner of adjusting this hook. It will be observed that the substantial double bearing of each hook precludes the possibility of its "tearing out." Shortening or taking up slack in belts is only the work of a moment when this hook is used. It is conceded that no belt fastening is equal to this for strength. It is less expensive than the *Blake stud* or

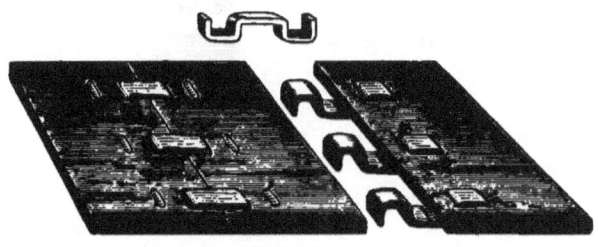

Fig. 55.

than *lace leather;* and, although it costs more than the "C" hook, it is in the end cheaper, because it retains its original shape in the belt, and the same hook can be used over and over again. It can be adjusted with greater ease and in much less time than any other belt fastening.

Machine-riveted Strapping.

139. "Peter McIntosh & Sons, Glasgow, call attention to their machine-riveted strapping, riveted with copper or iron wire. This new joining is highly recommended, being much stronger than hand-sewn, and specially adapted for double strapping."

Fig. 56.

The Lincolne Belt-Fastener.

140. "The disadvantages, such as insufficient strength and loss of time, attending the usual modes of fastening belts by means of sewing or of leather laces, are well known. What worker in a mechanical workshop has not had to do with the comparatively tedious operations of lacing a belt? Several mechanical belt-fasteners are now in use. One of the newest and most ingenious forms, affording a very strong and yet light joint by very simple means, is that which we now illustrate. It is a Canadian invention, and we understand that, though of very recent introduction, it is making rapid headway both in England and abroad. The fastener consists of two pieces of tough curved plate (Figs. 57 and 58), tinned, to preserve them from oxidation. The buckle proper is curved as shown, and formed with a series of teeth at each side, its width transversely to the length of the belt being rather less than the width of the belt itself. The ends of the belt are pierced with an awl, or a special tool for the purpose, from the inside, in a somewhat slanting direction, and the points of the teeth are inserted in these holes through the whole of the belt, so as to project at the opposite side to that at which they are inserted. The plate-cover or clasp proper is then slipped over the projecting teeth, of course tying them securely fast, and making the complete buckle. With very wide belts several such buckles are applied. Such a joint is clearly as applicable to India-rubber, gutta-percha, or woollen belts as to those of leather."—*The Engineer, Jan. 27, 1870.*

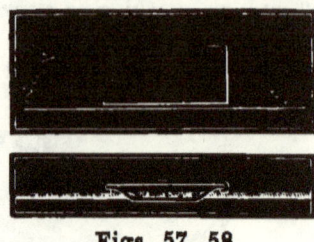

Figs. 57, 58.

Connecting the Ends of Belts.

141. "Two or more oval slots, A A, Fig. 59, are made near one end of the belt to be joined, and in the other end of the same belt D-shaped slots, B B, are made, the material being cut through from the middle of the straight side of the D by an incision parallel to the length of the belt, thus dividing the end into T-shaped parts.

"The ends of the belt are scarfed, so that when engaged they will lie closely to the body of the belt.

"In connecting the ends of the belt, the T-shaped parts are twisted quarter-way round and passed through the oval slots in the other end, and then straightened up again, thus locking the ends without the aid of laces, metal clips, or buckles."—*Howarth, Mech. Mag., London, XCIV., p. 289.*

Fig. 59.

Directions for Lacing Paper Belting.

142. "For narrow belts butt the two ends together, make two rows of holes in each end (thus obtaining a double hold), and lace with lacing leather, as shown in Fig. 60.

"For wide belts, where extra strength is required, rivet pieces equal in length to width of belt on back of each end, and make the connection with lacing, as before. This belting should, in all cases, be put on by the use of clamps, secured firmly to each end of the belt, and drawn together by bolts running parallel with and outside the edge of the belt, making no allowance for stretch.

Fig. 60.

"Wide belts, in dry places, can best be connected with Wilson's belt hooks, riveting down the teeth, thus making a connection that will not wear out."

Patent Belt Fastenings.

143. The Messrs. J. B. Hoyt & Co., present a patent belt fastening in the form of a double-shanked rivet with a link or elongated washer, having a perforation at each end for receiving the end of each rivet.

These rivets are used for making butt as well as lap joints in belts; in either case the belt is punched as for lacing, with holes equidistant from the joint. The perforation, of course, being of such size and distance asunder as to just allow the double rivet to pass tightly in when the whole is secured by laying on the link washer and hammering the ends of the rivets into the countersunk holes of the latter, care being taken to imbed both head and washer in the substance of the belt, flush with the surfaces, in order that they may not wear by contact with pulleys.

They also have the exclusive right to use a patent "rivet and burr." This improvement consists of a conically formed head to the rivet, which gives greater strength of attachment thereto. The burr has a form similar to that of the head, and has in addition a central hole countersunk on top for receiving the spread of metal of rivet end when hammered into it. This is the best form of rivet that can be made, as it gives the greatest strength and durability to all the parts, and is easily sunk into the substance of the belt to a level with its surface.

They recommend for round belts a grooved pulley in which the section of groove shall be neither triangular nor semicircular, but rather that of a spherical triangle or gothic arch.

CEMENTS, ADHESIVES, AND FASTENINGS. 189

This form undoubtedly gives greater belt contact than any other with a given amount of tension on belt.

They also urge the importance of covering pulleys with leather, " 50 per cent. more work can be done by machines without belts slipping if so covered." "The covering of pulleys with leather, in many establishments where there is a deficiency of power, would produce such an improvement as to astonish those not acquainted with its value."

"Large pulleys and drums may be covered by narrow strips of leather, wound around spirally, but narrow pulleys should be covered by leather of same width as pulley face."

"Our pulley covering is cut into strips of required width cemented and made of even thickness, by a machine, then wound in coils like belting."

Good Methods of Lacing.

144. The usual way of joining the ends of a belt — that is, by means of the leather thong — is the best after all, because it is the most convenient; the thong being an article more readily obtained and applied than any other of the numerous and ingenious means devised for securing the ends of a belt.

In the use of thongs, it is the practice of some engineers to cross them in lacing on both sides of a belt; with others, to cross them on the outside only, laying the double strands evenly on each other in the line of motion and on the pulley side of the belt, which experience proves to be the better way.

We present, in Fig. 61, a plan, altered from the original, in which the two ends, *a a*, of the thong are tied in the middle of the belt, as we do not consider any laced joint safe with the tie at the edge. It will be observed that there is no crossing of the lace at all, and that the holes are in two rows fore and aft, or "staggered," as we say, which favors the strength of the belt. This fastening may be a little more tedious to make, and may require rather more contrivance on the part of the party lacing, yet it is one of the best in existence.

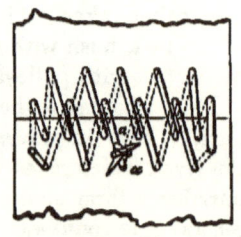

Fig. 61.

The credit of producing this excellent joining is due to Mr. William Annan, of Morrison, Ill., and of proposing to put the tie in the middle to the Eds. of *Scientific Amer.*, in Jan., '66.

CHAPTER IV.

VARIETIES OF BELTING.

Raw-Hide Belt.

145. Mr. Jno. Mason, of Bulkley, Barbadoes, presents a novelty in the way of belting that he has been using for driving centrifugal machines for the past 2 years. He says: "I found leather belting gave so much trouble and was so expensive that I was induced to try a belt made of raw cow's hide, simply dried in the sun, cut perfectly straight, and the joints carefully stitched (square and even) with common saddler's hemp. I find in practice that a belt of this description will last longer than if made of leather, besides being only ¼ the cost. I am driving a line of 3-inch shafting with an 8-inch belt of this description with every satisfaction. They are now used here for driving Weston's centrifugals, as well as those of Manlove and Elliott. I make them $2\frac{1}{2}$ inches broad for driving the latter."—*Engineering, June 19, 1874.*

Sheet-Iron Belt.

146. Mr. John Spiers, of Worcester, Massachusetts, gives us an account of a sheet-iron belt:

"A lathe used for turning rolling-mill rolls, compound geared, has a 48-inch pulley on it; this is driven by an 18-inch pulley on the counter-shaft, which makes 120 Rpm, and is 8 feet from the 48-inch pulley, measured from centre to centre. Both pulleys of iron, smoothly turned on faces.

"A 7-inch double leather belt was used on these pulleys, but would slip when the turning-tool became dull. This belt was replaced by one made of Russia sheet-iron, same as used for stove-pipes and parlor-stoves, and was riveted together in the ordinary way; it was 7 inches wide, and was 2 inches longer than the leather belt. This extra length made up for a want of elasticity in the iron.

"During one year's steady run this iron belt could not be slipped, even when a heavy 'cut' on a 25-inch roll was taken, which broke a

'Sanderson' steel tool having a section of 2 x 2½ inches, a cutting surface of 2½ inches, a feed of ⅛ inch per revolution, and an overhang of 4 inches."

Alexander Brothers' Improvement in Wide Leather Belting. Patented June 15, 1875.

147. We respectfully ask attention to the following description of a patented improvement in the manufacture of wide leather belting shown fully in Fig. 62:

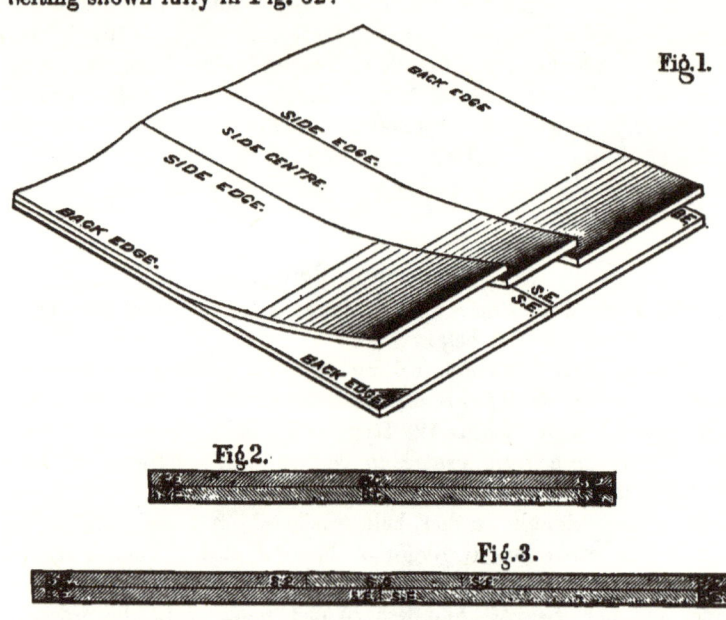

Fig. 62.

The term *hide*, used in this description, means — the skin of the animal tanned whole, the back being in the centre, lengthwise.

The term *side*, means — one of the halves of the hide, made by cutting down the middle of the back.

The ordinary method of making wide two or more ply belts, as shown in cross-section by Fig. 2, is stripping pieces of the width required from the centres of hides, splicing the ends, and on this one ply building up layer after layer, as many as required, breaking joints with each layer, lengthwise of the belt; the width of each layer being, of course, one piece. This brings the back centre (marked B. C. in the cuts), or firmest part of the leather, immediately in the centre of the belt, the edges being composed of leather from the side portions of the hide (S. E.), which is yielding in comparison with the middle. The above construction has three disadvantages:

1st. All pulleys being more or less convex on their faces, the middle of the belt being firm and not conforming to this convexity, and the edges of comparatively loose fibre, the consequence is that the edges of the belt will not bind down to the edges of the pulley, and, after running a short time, will stretch more, owing to their loose fibre and the absence of that lateral support which the central portions have. Thus only a part of the width of the belt is effective, thereby transmitting much less power than if all the surface contact was fully available.

2d. The centre or tight portion of the belt bearing the greater part of the strain, and the other parts not relieving it, will consequently give out proportionately quicker than if the strain was equalized.

3d. Along the portion of the hide over the back-bone full or humpy places are often found, caused by the shape of the animal, and, this part of the leather being more or less hard and stubborn, it is difficult and often impossible, *in the whole hide*, to work them perfectly flat; and, after being made into belting, they present to view an uneven surface all along the centre of the belt, which will never lay down flat to the pulley, thus preventing other parts from touching, and a corresponding decrease in surface contact, even though the edges were supposed to bear.

We overcome the above disadvantages by the following construction:

In making wide double belts we cut the hides along the middle, turn the back edges (B. E.) outward, and the side edges (S. E.) inward, inserting a side centre (S. C.) piece, so as to break joints widthwise, as shown in Fig. 3, Fig. 1 being a perspective view of the same.

In three-ply belts the same method is carried out as shown in Fig. 4.

There are various other arrangements of the pieces which can be used advantageously in certain cases.

The disadvantages of the ordinary method heretofore enumerated are overcome as follows:

1st. The edge portions of the belt being of firm, solid, and unyielding leather, and the middle portions of leather of looser fibre and more yielding texture, it is evident that, after running a short time, the middle will give to the higher part of the pulley, and the edges will not only bind down, but will also afford that lateral support which will prevent the middle stretching as much as it otherwise would, and thus giving an even bearing the whole breadth of the belt, and consequently the greatest amount of pulley contact.

2d. When the middle of the belt becomes stretched, and allows the edge portions to bed themselves down to the pulley, the working strain will be distributed over the entire width, thus preventing wear on any one part alone.

3d. Cutting down the middle of the hide enables the currier to work out any uneven or full places, the surplus being cut away in straightening.

For belts of 16 to 48 inches, or wider, no other plan of making, or material other than solid oak leather, can approach, for effectiveness and durability, the arrangement described and illustrated above.

Vulcanized Rubber Belts.

148. We are indebted to Mr. D. P. Dieterich, Esq., of 308 Chestnut street, Philadelphia, agent for the New York Belting and Packing Company, the oldest and largest manufacturing firm in the United States of vulcanized rubber fabrics adapted to mechanical purposes, for the following valuable collection of facts and statements concerning rubber belts:

"This belting is made of heavy cotton duck, weighing 2 lbs. per yard, woven expressly for the purpose, and is vulcanized between layers of a patent metallic alloy, by which process the stretch is entirely taken out, the surface made perfectly smooth, and the substance thoroughly and evenly vulcanized.

"The superiority of this belting over the best leather belts has been proved by a trial of more than 10 years. It is manufactured by a process peculiar to this company, by which unusual firmness and solidity are obtained, thereby obviating some objections heretofore urged against India-rubber belting made in the old way.

"This, together with the fact that other great improvements have been made in its quality, warrants us in asserting that it is superior to leather, or anything else, for all open belts, particularly heavy or main belts, for the following reasons: It has a perfectly smooth and even surface. It seldom, and scarcely ever, requires tightening more than once. It will always run straight and with perfect bearing on the pulleys, by which we believe that a power of 20 per cent. is gained over the best leather belts. It will stand heat of 300° Fahr. without being affected, and the severest cold will not stiffen it or diminish its pliability. It is much stronger than leather and far more durable. It can constantly be run in wet places or exposed to the weather without injury.

"From information which we have personally collected on the subject of vulcanized India-rubber belting, it appears to us that this material is yet designed to effect an economic revolution in driving machinery.

"In the extensive establishment of Burr & Co., Cliff street, New York, where the manufacture of hat bodies is carried on, and where an immense amount of belting is used, it has taken the place of leather in nearly all work. We instance this case because the machinery in this manufactory is such as to afford a signal test of the quality of belting. One long India-rubber belt, 8 ply and 36 inches wide, is employed to transmit the power from a fly-wheel of 2 horizontal steam-engines of 100 horse-power each.

"The fan-blowers of the 'forming machines,' and those for teasing and cleansing the fur, are driven at the high velocities of from 3000 to 3500 Rpm. This speed wore out the best leather belts faster than those of India-rubber, which have supplanted them.

"India-rubber belting has been for some time used for driving the presses on which the *Scientific American* is printed, and has proved superior in every respect to the leather belts previously employed. It also possesses the qualities of running unaffected under exposure to water, to the open air, and even to a temperature above the boiling point.

"A 5-ply India-rubber belt, 12 inches wide, as now manufactured, is considered equal to a double leather belt of the same width, and can be furnished at $\frac{194}{250}$ of the price of the latter.

"The new variety of India-rubber belting to which we have referred is manufactured by the New York Belting and Packing Company at Newtown, Conn.

"The cotton duck which gives the peculiar uniform and non-elastic character to such material is woven especially for this purpose, with the warp much stronger than the filling, and cut by machinery into strips of a perfectly regular width. Single strips of this duck will bear a tensile strain of 200 lbs. to the inch of width."

Experiments with Belting.

"The comparative adhesion of vulcanized gum and leather belts to the surfaces of pulleys is a question of great interest to manufacturers, and in order to satisfactorily decide it, Mr. J. H. Cheever, of the New York Belting and Packing Co., made a series of experiments, the results of which are here given:

"The apparatus consisted of 3 equal size iron pulleys, with faces turned in the usual way and secured to a horizontal shaft, also fixed. One of these pulleys was used without covering, one was covered with leather, and one with vulcanized gum.

"In the first set of experiments a leather belt was used of good quality, 3 inches wide and 7 feet long, with 32 lbs. weight attached to each end, and the belt thus prepared was laid on the iron face pulley. Additional weights were then attached to one end of the belt until it began to slip, which was in this case found to be 48 lbs.

"When this weighted belt was placed on the leather-covered pulley it required a weight of 64 lbs. to slip it, and when on the vulcanized gum-covered pulley it required 128 lbs. to slip it.

"In the second set of experiments, a 3-ply vulcanized gum belt of the same width, length, and thickness was used, and to each end was attached the same weight as in the other case.

"To cause this belt to slip on the iron face pulley required 90 lbs. additional weight, on the leather-covered pulley 128 lbs., and on the vulcanized gum-covered pulley 183 lbs.

"In the third set of experiments, the shaft, with all the pulleys secured thereto, was permitted to turn freely in its bearings. One end of the belt was fastened to the frame work of the apparatus, and to the other end was attached a weight of 32 lbs. as before.

"A rope was wound several times around one of the pulleys, with one end made fast to the rim, and the other allowed to hang freely downward; to this end weights were attached sufficient to produce rotation of the shaft.

"The results were the same, requiring in effect the same amount of weight on the end of the rope to rotate the pulleys under the belt as it did on one end of the belt to slip the belt over the pulleys."

How to Use Vulcanized Rubber Machine Belting.

"Belts should be cut $\frac{3}{16}$-inch shorter for every foot of length required. After running, say, for 3 weeks, take up the slack and they will never again require shortening.

"To fasten the ends of narrow belts, make 2 rows of holes in each, butt the ends together and unite by strips of lacing leather in the usual way with leather belts.

"To secure the ends of wide belts, lap the joint evenly on the outside with a piece of square gum or leather, equal in width to the belt, and rivet, sew or lace the same firmly to each end of the belt.

"If belts should slip from dust or other causes, they should be slightly moistened on the pulley side with boiled linseed-oil, making several applications if necessary. *Animal oils must never be used*, and belts should be protected, while running, from contact with such oils.

"Should the rubber, from long use, or other cause, be worn from the surface of the belt, give it a coat or two of lead paint, containing sufficient *Japan* to dry it quickly.

"For belts which are shifted, put rolls on the shifter bars with axes inclined towards each other at top and bottom, according to circumstances, which has the effect to press the faces of the belts and relieve the edges from wear. By this plan belts are more easily shifted than by the usual method, and the liability to injure the edges entirely prevented. Use large headed bolts or rivets for securing elevator buckets."

General Statements.

"The more nearly *horizontal* a belt can be applied, the better will the weight of the belt produce a sufficient and uniform friction; and a long belt is better than a short one, inasmuch as the *weight* and '*sag*,' and consequently the friction, are greater.

"To avoid kinks and crooks in the belts, the ends where joined should be cut *exactly* square across the centre line of the belt.

"The pulleys should be perfectly smooth, and the shafts carrying the pulleys perfectly '*in line*,' parallel to each other, and in the same plane.

"Leonard, in his '*Mechanical Principia*,' assumes the ordinary velocity of belts to be 25 to 30 feet per second, and gives a table which may be fully represented by the following:

$$HP = \frac{w\,d}{3.6},$$

$$w = \frac{3.6\,HP}{d}$$

In which $HP =$ horse-power transmitted.
    " $w =$ width of belt in inches.
    " $d =$ diameter of smaller pulley in feet.

"Where belts are not to be exposed to saturation of animal oil, or to frequent abrasion, a combination of rubber and canvas has proved to be fully equal if not superior to leather, and is much cheaper. For large belts rubber is preferable, because the belt, whatever its length or width, is one — not pieces joined by mechanical means or connected temporarily — but solid, and, to all intents and purposes, one continuous fabric.

"For purposes where unusual strength is required — equivalent to double leather — 5 and '6-ply' belts are made. Endless belts of any width or length are made to order, costing, in addition to cost of belt, the price of 3 feet of the belt joined, for joining. For belts less than 6 inches wide, the 3-ply is sufficiently strong, unless the work is unusually heavy. Belts wider than 6 inches should not be less than 4-ply, unless the work is light.

"Wherever double leather belts of 14 inches wide and upward have been used, we recommend our 5 and 6-ply belts, and will warrant them to do more work at about *one-half the cost*.

"As an evidence of the capacity of this establishment, we refer to the 'Champion' belt, the largest belt ever made of either leather or rubber. It is 4 feet wide, 320 feet long, and weighs 3,600 lbs.

"This company is prepared to make any width of belt not exceeding 50 inches.

"The 3-ply vulcanized gum belt weighs $1\frac{2}{3}$ lbs. to the square foot, the 4-ply weighs 2 lbs. Thicknesses as follows: 2-ply, $\frac{3}{16}$-inch; 3-ply, $\frac{5}{24}$-inch; 4-ply, $\frac{5}{18}$-inch; 5-ply, $\frac{5}{12}$-inch; 6-ply, $\frac{7}{16}$-inch."

Edge-laid Belt.

149. "A better plan of making a broad belt, than the usual American *double* leather belting sewn together, is made with the greatest ease, of any thickness or width, perfectly equal in texture throughout, and alike on both sides. It is made by cutting up the hides into strips of the width of the intended thickness of the belt, and setting them on edge. These strips have holes punched through them about $\frac{1}{8}$ of an inch diameter, and one inch apart. Nails made of round wire, clinched up at one end for a head, and flattened at the other, are used for fastening the leather strips together.

"Each nail is half the width of the intended belt, and after the strips are all built upon the nails, the ends of the latter are turned down and driven into the leather, thus making a firm strap without any kind of cement or splicings. When the strap is required to be tightened, it is only necessary to take it asunder at the step lines of splice, cut off from one end of the strap at each step what is required, and piece up again with wire nails or laces, going entirely through the strap.

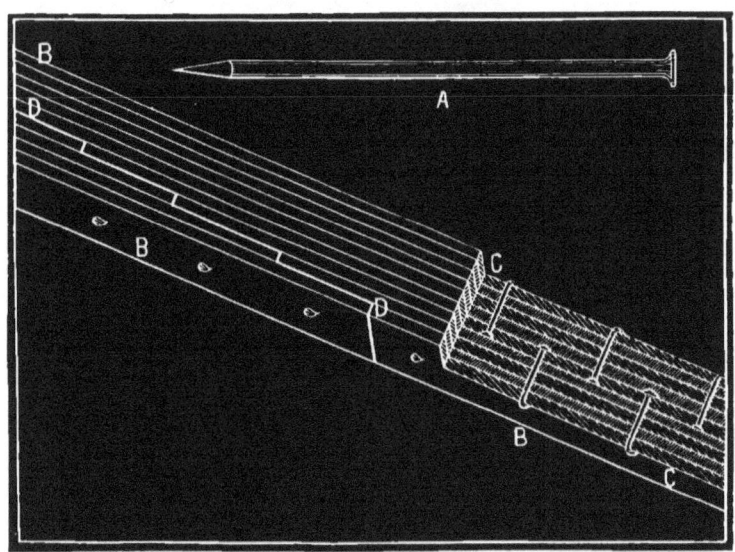

Fig. 63.

"In Fig. 63, A represents one of the nails used in this form of belting, B B the belt in perspective, with the part C C cut half way to show the disposition of the nails, and D D the step lines of the splice."—*E. Leigh.*

Paper Belting.

150. The Messrs. Crane Bros., of Westfield, Mass., who manufacture Crane's Patent Paper Belting, have kindly communicated the following facts concerning their new fabrics for driving belts: — "Our belts are manufactured from pure linen stock, and can be made of any desired thickness, width, and length. We guarantee equal driving power from equal surface, as from leather or gum, and we recommend them only for straight and unshifted belts, making none less than 5 inches wide. They will not stretch nor change shape, and being made all in one piece, of even thickness, will run smoothly and straight. We have proved them equal in durability with leather, and equal also in strength, where they have been arranged for one to pull against the other. They adhere to the pulleys very closely, and generate no electricity while running. They are quite flexible, and do not crack in passing over pulleys even as small as 6 inches in diameter. They are not affected by heat at ordinary temperatures, nor by dust or oil, but will not run in water. Being very tough, they would answer a good purpose for elevator belts, holding the bolts well and running in a direct line without swinging from side to side.

"Compounds similar to those used for stuffing leather belts, or black lead, mixed with sperm oil, are very good to apply to our belts, when dry and slipping.

"For *endless* and for *heavy belts*, our belting material is not to be surpassed, and when fitted to the pulleys at the proper length, it must remain so until worn out. It is 40 per cent. cheaper than leather, and for all heavy and expensive belts must come into general use."

We present some facts derived from testimonials furnished to the makers of the paper belting.

One party says the 15-inch paper belt gives us entire satisfaction. It drives all our machinery with perfect ease; runs very straight, without any swaying or sagging; is not affected by temperature; and there are no indications of electricity, which we consider a great gain.

Another party has had a 15-inch belt, 70 feet long, in use during 15 months; the surface next to pulley is hardly marked, and shows no indications of wear; belt not touched since first put on, for repairs of any kind.

Another party says: "We have been using one of your 14-inch paper belts for the last 4 months to drive a paper engine, which requires about 13 horse-power. It works to our entire satisfaction, and gives us less trouble than any other belt in the mill."

Another party has in use a paper belt 12 inches wide, 50 feet long, and runs from the fly-wheel of a 25 horse-power engine to a pulley on line-shaft, driving machinery which nearly exhausts the power of the engine. It gives equal satisfaction with a leather belt formerly used in same place. A narrower and shorter belt, crossed also, has done quite as well as a leather belt under like circumstances.

Another party has thoroughly tested the paper belt, and is entirely satisfied therewith. The belts are 12 inches wide; one 120 feet long, the other 92 feet; they do not stretch or slip.

Another party says: "The paper belt continues to give good satisfaction. It has already been at work on our engine as long as the majority of leather belts last in the same place, and as yet we can see no signs of wear. It does not stretch nor slip, and running in the open air, as it does, it saves us a great deal of trouble which we have had heretofore with a leather belt. Having had a great deal of experience, we do not hesitate in saying that the paper is a superior belt."

Water-proofed Leather Belting.

151. A company in Philadelphia is at present very successfully engaged in water-proofing various materials by patent process. They operate upon lighter fabrics of all kinds, and also upon leather, and claim to add considerably to the durability and efficiency of belting so prepared. The belting is rendered perfectly water-repellant, and can therefore be employed under circumstances where the ordinary leather would rapidly stretch and become worthless, namely, whenever it is exposed continuously to the wet.

A line of belting thus prepared has been for some months employed in transporting from the bed to the shop damp clay, for subsequent working, the clay being simply heaped upon the belt, and thus traversed. The water-proofed belt has thus far, we are told, not appreciably stretched or deteriorated, which would indicate that the process is a successful one, and of much practical value.

The process, from the account of one who is conversant with the operation, is stated to be as follows:

"Leather bands, having the joints cemented and riveted in the usual manner, are steeped in an alkaline solution, which permeates them and forms a coating in the surfaces of the cells and pores.

"By a subsequent treatment in a solution of metallic salts (sometimes accelerated by pressure), the coating is rendered insoluble and repellant to water.

"The water-proofing effect appears to be thorough, newly cut surfaces being equally repellant to water with the original. Belting so treated possesses greater flexibility and improved adhesion to the pulleys; about 5 per cent. more force is requisite to slip a band after water-proofing it than before. In good leather, the tenacity of fibre is not impaired; poor leather, or leather of unequal texture, cannot be water-proofed without so distorting it as to render it unsalable; the purchasers of water-proofed belts are reasonably sure of good material.

"More force is required to stretch the belting after water-proofing than before, and a considerable degree of elasticity is imparted by the process.

"The belting shortens or shrinks in length and increases in width, thus showing a tendency during the process to resume the original form of the leather before the stretching operation of the belt manufacturer."

Underwood's Patent Angular Belting.

152. This driving belt consists of a number of narrow leather bands, laid a-top of one another, lapping and breaking the joints, in

Fig. 64.

order to secure the greatest combined strength. Shown in Figs. 64 and 65.

To the under side of this compound band are fastened short piles of leather, of equal length, forming blocks, each secured to the band by two iron rivets, as shown above.

The 4 bands, J, are continuous; the 5 pieces, K, and shorter pieces, I, interposing, are held together by the rivets, F; all are shaped to the angle, C D E, which is the correct angle of grooves for the wheels.

The length of the blocks is made as short as construction will permit, in order to in-

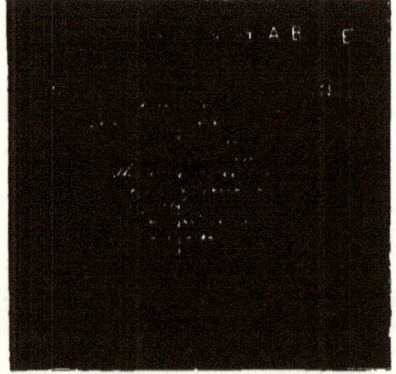

Fig. 65.

crease the surface of contact while bending in the groove of the wheels; the pieces, I, are made shorter to give more flexibility to the band, for which also the blocks are separated by a narrow space.

The elasticity of the leather is sufficient to allow of the necessary bending of the 4 united bands without injury to the fibre, as it is not intended to use this belt over pulleys of small diameter.

The 4 bands, J, thus constituting the "wrapping connector" and tensional strength of this system, while the blocks, K, form the frictional wedges, so to speak, and the sloping edges of all in the angle of the groove contribute to the great adhesive power possessed by this driving belt.

The ends of this belt are joined by bevelling the opposite faces of the part J, for 18 or 20 inches of its length, and then uniting the whole by bolts having washers and nuts at F, and with heads inside, similar in size and position to the rivets; or the separated strands may be joined, the top one say at O, the next one at P, and in no case having more than one joint between any pair of rivets.

We have a practical illustration of the driving capacity of this kind of belt in the N. H. and N. Y. Railroad shops at New Haven, Conn., where a 22-inch double leather belt of Hoyt's make, weighing $2\frac{1}{2}$ lbs. per square foot, running on pulleys of 6 feet and 4 feet diameter, and $16\frac{1}{2}$ feet distant between centres of pulleys, with a quarter-twist, and slipping under a load of 75 to 80 horse-power, with noise which could be heard $1\frac{1}{4}$ miles away, was replaced by $2\frac{1}{2}$-inch angular belts, on V-grooved wheels of same diameter. After 16 months of running, one of the angle belts was removed. This, of course, put all the work on the remaining one, and this one carried the whole load with apparent ease. Afterwards one-third more work was done by this belt without slipping, visible straining, or injury.

20 inches length of the $2\frac{1}{2}$-inch belt weighs 2 lbs. 21 inches of the 3-inch belt weighs 3 lbs. M N is the breadth.

European Compound Leather Belts.

153. An examination of the different leather departments, and the varieties of belting in actual use, reveals a tendency on the part of manufacturers to improve the quality of wide belts by securing 2 inch strips along their edges. Specimens of this character are exhibited by Messrs. Webb & Son, Stowmarket, England; Mr. William Ruland, of Bonn, Prussia; H. Lemaistre & Co., Brussels, Belgium; Placide Peltereau, 32 Rue d'Hauteville, Paris; Poullain Brothers, 99 Rue de Flandre, Paris; and others of less note.

The material forming these strips is (with a single exception) leather of the same quality as the belt. The methods of attachment are variable, as laces, threads, rivets, eyelets, and brass screws.

The English use the threads, Prussians the laces, and the French all the varieties enumerated. Mr. P. Pelterau, proprietor of one of the largest houses in France, makes a remarkable display, not only of belts and their mountings, but of different kinds of leather, such as tanned elephant hide, varying in thickness from $\frac{1}{4}$ to $\frac{1}{2}$ an inch, and hippopotamus hide from 1 inch to $1\frac{1}{2}$ inches in thickness.

His 8-inch and 10-inch belts have leather facings 2 inches wide on their edges. Each of these facings is attached by two leather laces, whose stitches have $\frac{3}{4}$ of an inch span, and run in parallel lines separated by $1\frac{1}{4}$ inches.

The "inextensible belt," for which, at a previous exposition, he received a gold medal, has steel instead of leather edging strips. These strips, for a 10-inch belt, are 2 inches wide by $\frac{1}{84}$ of an inch in thickness, and are attached by 2 riveted rows of copper tacks. These tacks are $\frac{1}{8}$ inch diameter and placed $3\frac{1}{2}$ inches between centres.

Messrs. Poullain Brothers join their single and compound their double belts with headless $\frac{1}{8}$-inch brass screws. This is accomplished by a very ingenious machine, of which there are several types in the French department. It carries a coil of plain brass wire, which, while being fed to the work, passes through a die of 28 threads to the inch. The screw thus formed enters the belt at a point closely clamped by a foot-lever, and, having passed through, is cut off. Finally, the belt being placed on a surface-plate, the points of all the screws are slightly riveted. The most compact and expeditious of these machines is the invention of Mr. Cabourg, 74 Rue St. Honoré, Paris.

Mr. E. Scellos, of 74 Boulevard du Prince Eugéne, exhibits what he terms a "homogeneous belt," for 150 horse-power. This belt is $19\frac{1}{2}$ inches wide by $\frac{3}{4}$ inch in thickness.

It is composed of 104 leather strips $\frac{3}{4}$ inch wide, laid horizontally with reference to the belt, and laced transversely; the distance between laces is $1\frac{1}{4}$ inches, and diameter of lace $\frac{3}{16}$ of an inch. The advantage of edge-bound wide belts, where frequent slipping is an essential, we think will be readily conceded; and to what extent they can supplant double belts is a subject worthy of experimental inquiry. The use of very wide belts is seldom resorted to in the machinery department. One of the stationaries has two central-ribbed pulley-rims bolted to the arms of its fly-wheel, on these run 4 belts 6 inches in width; another has two 12-inch edged belts, and so on. The inclination was always to increase the number rather than the width of the belts.— *Paris Exposition, 1867.— W. S. Auchincloss.*

English Belts of Leather and Iron Mixed.

154. "A contemporary says that the improved steel wire has a strength of 160 to 175 tons per square inch of actual section; that 176 No. 14 wires have a total section of one square inch, and that each wire will bear from 2000 lbs. to one ton breaking strain; and that the ropes made from these wires run readily around 4-feet or 5-feet drums, coil perfectly, and last for a long time. Such being the case, it becomes important to us to look to steel, in a measure, to substitute leather driving-belts." . . .

"While the substitution of leather by other substances, such as the vegetable gums, gutta-percha, and India-rubber, impregnated into strips of coarse woven fabrics, has often been tried, and used, too, with a certain measure of success. Speaking, as we now do, from an experiment as to the value of such a combination for driving-belts, we can certainly assert that we never found them one-quarter as durable as leather; their use was more costly than the older used substance."

. . . "In some, though few, cases iron and steel wire belts have been used, the pulleys on which they run being covered with buckskin or some other leather, to increase the adhesion."

"We have cited these few remarks to throw out the hint that, now, when cheap and strong steel wire can be purchased, there promises to be a fruitful field for inventive talent to devise some means of so weaving steel wire and gutta-percha into flat belting — producing a stronger and better adhering driving band than leather ever can be."
—*Prac. Mech. Jour., Nov., 1867, p. 237.*

Chas. Sanderson, of Sheffield, England, has taken out a patent — dated Dec. 8, 1862 — "For making driving bands of thin sheet metal, coated with rubber, to prevent oxidation." "The bands are first well cleaned with acids, then coated, by electro process, with brass, after which they are coated all over with gum vulcanized thereon, and which adheres tenaciously to the metal coating. Bands of great strength may be made by cementing together several made as above, with a layer of gum between each, the gum imparting flexibility and adhesion to the compound band in passing over the pulleys."

George and Daniel Spill, of Middlesex, England — under date of Nov. 9, 1859 — have taken out a patent for "the manufacture of bands by weaving together covered strips of metal with ends of hemp or other fibrous material."

A strip, or band, or wire of steel is covered with one or more strands of hemp cord, previously passed through a solution of caoutchouc, gutta-percha, glue, drying oils, gums, resins, tar, pitch, or other glutinous, gelatinous, or siccative materials. After the strands have been applied, the strip or wire is passed between rollers, in order to solidify the covering.

Any required number of metal strips or wires thus covered are used as warps in a loom, and hemp cord or other fibrous material — previously covered with a solution of caoutchouc or any of the other before-mentioned materials, or not — is employed to weave the whole together.

The fabric thus produced is passed between rollers, to render it flat and smooth, and before or after so doing a solution of caoutchouc, gutta-percha, or a coat of paint, or any other desired material, is applied thereto.

M. J. Haines, of Stroud, England, has taken out a patent — bearing date Feb. 14, 1860 — for making driving-belts.

"This invention consists in cutting leather or hides into narrow strips of equal width, each strip width representing the thickness of the intended driving-belt, and placing the same side by side, breaking joints with the lengths, to make the whole of uniform strength, and with the cut edges of the leather coming to the upper and under surfaces of the intended belt, until the desired width is obtained. The whole are fastened together by wire, rivets, or screws passing transversely through the strips, and secured on the opposite sides."

"An interesting description of American belting is made chiefly of wool, and the surface of the belt covered with a resinous cement. We saw a small piece that had been in use for $2\frac{1}{2}$ years on a heavy cloth loom in the States."—*Lond. Mech. Mag., Mar., 1863.*

For description of a peculiar form of driving-belt, the invention of W. Clissold, see *Frank. Inst. Jour.*, Aug., 1863, p. 121. It consists of double links of leather, or other similar material, connected by intermediate links of metal, the whole series running in grooved pulleys, the leather only touching the sides of the grooves, and driving by adhesion, in the same manner as ordinary round belts.

J. B. Hoyt & Co.'s Patent Angular Belting.

155. "This invention consists of a novel belt of a trapezoidal form, to be used, in connection with a V-shaped or angular grooved pulley, for driving all kinds of machinery where belting is required to transmit power.

"The angular belt is a great improvement on the round, square, or flat belt; a much greater surface of the belt is brought in contact with the pulley than with any other kind. It will wedge itself into the groove and resist any slipping action during the rotation of the pulley, so that the greater the strain put on one side of the belt the tighter will it be held in the groove; not being liable to slip on the pulley, it may be used very loose, causing less friction, consequently requiring less power to drive machinery, and giving it greater certainty and regularity of motion. As the angular band is not liable to slip, it can be used with greater economy and certainty than either the round or flat belting; its power is limited only by its strength. Made of the same materials and the same width of belt, this form has more strength than any other.

"This belt has been used with perfect success when other bands have failed entirely to impart motion to machinery. They are made without a joint in their length; and when the width requires more than one thickness of leather, the belt is connected, then riveted or screwed, so that the fastening will not come in contact with the pulley."

Friction Wheels.

156. "Wheels acting upon each other are the instruments by which the transmission of force from one part of a system of machinery to another is commonly and conveniently effected. The due connection of the moving parts is accomplished either by the mutual action of properly formed teeth, by straps or endless bands, or by the friction of one face of a wheel against another. The latter method has, when adopted, been generally in small, light works, where the pressure upon the different parts of the machinery is never considerable. Mr. Nicholson saw a drawing of a spinning-wheel for children, at a charity school, in which a large horizontal wheel, with a slip of buff leather glued on its upper surface near the outer edge, drove 12 spindles, at which the same number of children sat. The spindles had each a small roller, likewise faced with leather, and were capable, by an easy and instantaneous motion, of being thrown into contact with the large wheel at pleasure. The winding bobbins for yarns at the cotton-mills operate on the same simple and elegant principle, which possesses the advantage of drawing the thread with an equal velocity, whatever may be the quantity on the bobbin, and cannot break it.

"We are not aware that the same mode of communication has been adopted in large works, except in a saw-mill, by Mr. Taylor, of Southampton. In this the wheels act upon each other by the contact of the end grain of wood instead of cogs. The whole makes very little noise and wears very well; it has now been in use nearly 20 years. There is, of consequence, a contrivance to make the wheels bear firmly against each other, by wedges at the sockets or by levers. This principle and method of transmitting power certainly deserves every attention, particularly as the customary mode, by means of teeth, requires much skill and care in the execution, and, after all, wants frequent repairs."—*Treatise on Mechanics.* *Olinthus Gregory, London, 1806.*

English Leather Belts.

157. Mr. W. T. Edwards, of No. 20 Market Place, Manchester, England, presents, October 24, 1875, the following interesting particulars of English leather and belts: "All our belts are made from the best English leather — that is, *English hides, oak-bark, tanned;* and to make them any length and width without cross joints, we select all the large hides (which are curried on the premises); we then cut the hide *round* (taking off the *bellies* and *shoulders*); they are then cut spirally in narrow strips, 1½ inches wide or more, sewn together longitudinally, in the same manner as the stitching of a cricket-ball cover. Some of the strips are 160 feet long. By this patent (Sampson's) we build the belt any width; the double belts are made by putting two singles back to back, and then pegging them together; by these means we get the greatest amount of strength with the least possible weight, an even thickness throughout, and warranted to run perfectly straight.

"I have just supplied a 38-inch wide double, 90 feet long, for driving direct from the fly-wheel, 14 feet diameter, to a 6-feet pulley. Speed of belt, 2800 Fpm, transmitting 350 horse-power indicated, the belt running horizontally, and driving the whole of the machinery in a cotton-mill. At another cotton-mill I have two 29-inch wide double (121 feet and 145 feet long), driving pulley 28 feet diameter, 5 feet face turned up for the two belts; driven pulleys, 7 feet 6 inches diameter; speed of belts, 4500 Fpm, running at an angle of 75°, and transmitting 600 indicated horse-power.

"I have in another similar cotton-mill double belts 28 inches wide. I have a 36-inch wide double turning about 30,000 throstle spindles in one room, transmitting 350 horse-power indicated; pulleys 16 and 8 feet diameter; speed of belt, 4300 Fpm, running at an angle of 45°. At another spinning-mill I have 8 double and treble belts transmitting 2000 horse-power indicated; pulleys 10 and 7 feet diameter; speed of belts, 4200 Fpm on second motion shaft. I am now making two 26-inch treble belts, which are intended to transmit 800 indicated horse-power. I have a 24-inch double at Sir Joseph Whitworth & Sons' works here transmitting 290 horse-power indicated; pulleys each 10 feet diameter; speed of belt, 4200 Fpm, running horizontally. I have another 24-inch double transmitting 230 horse-power indicated; pulleys 19 feet 6 inches and 6 feet, running horizontally at a speed of 2700 Fpm. I have them up to 26-inch wide trebles working in rolling-mills here in England. I had a 30-inch double turning 350 horse-power indicated, but they have been induced to put in a wider belt. I have one on hand now 34 inches wide. I have plenty of large belts working in this country, and am only giving you a few of the principal ones. To do the same amount of work, I generally allow a *third* more width for single than for double belts."

Various Driving Belts.

158. "The North British Rubber Company, Edinburgh, exhibited *India-rubber belting* in various widths. This belting consists of a number of plies of cotton fabric cemented together by India-rubber, and is said to possess the advantages of durability and superior adhesion as compared with leather.

"The machinery in motion in the Western Annex was driven by India-rubber belts supplied by the exhibitors, which worked in a satisfactory manner.

"Messrs. G. Spill & Co., London, exhibited *machinery belting* manufactured from flax-yarn, saturated with a compound substance, said to be incapable of decomposition.

"The following results of the tests to prove the tensile strength of this belting in the chain-cable testing machine at Rotherhithe appear to show that it is much stronger than leather belting:

Belting.	Width in Inches.	Tensile Strength in Lbs.	Tensile Strength per Inch Wide in Lbs.
1	5	6272	1254
2	5	7448	1489
3	10	16632	1663
Stout leather band of good quality....	4	2100	525

"Messrs. Nobes & Hunter, London, exhibited *compound leather belts*, manufactured of a strong hempen web, sewed between 2 plies of leather.

"Messrs. Bryant & Cogan, Bristol, exhibited *edge-laid leather bands*, consisting of thin strips of leather laid side by side, breaking joint, and united into one band of the requisite width. It is considered that by the mode of working the strap on the edge of the leather instead of on the face, the risk of cracking and breaking the grain of the material is avoided; whilst edge-laid bands may be used more slackly than ordinary bands, as the edge surface 'hugs' the drum or pulley more than the face. (See Arts. 149 and 153.)

"Messrs. C. J. Edwards & Son, London, exhibited *untanned leather belts*. The fibres of tanned leather, he contends, are weakened by the ordinary artificial means used in swelling the hides, which produce heavier and thicker, but spongy leather. Mr. Preller's leather is twice stretched, and is of greater density, as it contracts, in drying, to the original thickness of the hide, and from the results of experiments made at Woolwich Dockyard, to which he refers, it is said to have been ascertained that his leather is at least 50 per cent. stronger than tanned leather.

"Mr. W. Potier, London, exhibited *gut wheel bands*, also bands of twisted leather."—"*Exhibited Machinery of 1862.*" D. K. Clark, C. E., London.

Tool for Putting on Belts.

159. We copy from an advertisement in "Engineering" a very ingenious device for putting on belts while the pulleys are in motion, shown in Fig. 66.

The conical pin and flange, A, revolve on the end of the staff, B, which, being provided with a socket, may have a rod, C, of any suitable length fitted to it.

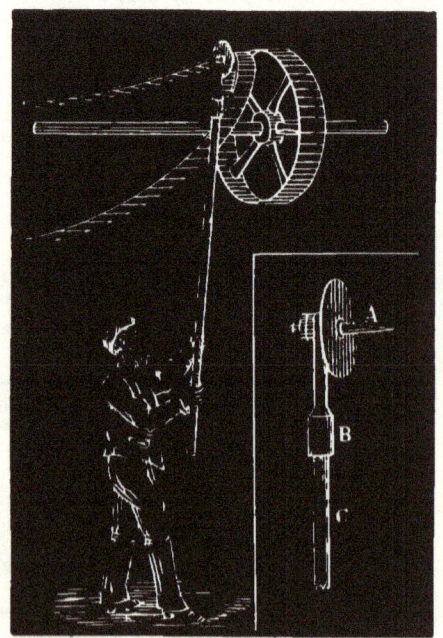

Fig. 66.

The illustration shows how it is to be used.

CHAPTER V.

ON THE STRENGTH OF BELTING LEATHER.

160. "We herewith call the attention of those who use belting to the results of some tests made for us by Riehle's testing machine, as to the relative strength of different portions of a side of leather when submitted to a breaking strain. The experiment was made with half of one of our regular butts, from which 48 pieces were cut, each $11\frac{3}{4}$ inches long by 2 inches wide, on which they are respectively marked, with the breaking strain (in lbs.), stretch (in fractions of an inch), and weight of each sample (in ounces and drams)."

"It will be seen in table (Fig. 67) that the textile strength of leather varies widely, according to the portion of the hide from which the leather is taken, and the test shows that the most valuable leather, or that which insures an enduring, straight, and well running belt is not found in that part of the side which has the greatest textile strength. The long fibre found near the lower edge of a side of leather or in a hide that is not well filled (and still stronger in a raw hide), will stand a much greater breaking strain, but it does not possess those qualifications required for belting and sole leather purposes, which are found in the firmest part of well-filled oak-tanned leather." — *J. B. Hoyt & Co.*

"This test is of importance principally to makers of belt or band leather, but when we compare the results, and notice how great is the variation of tensile strength — those parts which are concededly the best for sole leather, giving such decidedly inferior qualities for belts or for harness — it is quite possible that the experiment may have as much of value to the makers of boots and shoes, as it has to the manufacturers of belting; for, generally speaking, those portions of the hide which give the greatest tensile strength will most readily soak water, and present the least resistance to wear by attrition, when used in the soles of boots and shoes. The leather which is most carefully made from the best selected hides, and in which the tanning is most thorough, and nothing neglected in the finish and trim, will undoubtedly continue, as heretofore, at all times to command the highest prices and the readiest sale."—*Shoe and Leather Reporter, July 6, 1876.*

I have endeavored to present in this chapter as complete an exhibit of the tensile resistance of the beltings in general use for machine driving as could be obtained from standard books: these were found having little to offer in variety, and less to give in the way of details of tests, but the isolated figures derived from these sources, when taken in connection with the interesting and valuable experiments of the Messrs. Hoyt, together with those made at the Centennial, make up a tabular statement which will enable the reader to know *about* what strength there is in leather and in some other beltings.

A glance at either table will show the user of belts that, if he employs such as are well made from good stock, he has little to fear from the breaking of them in the solid part, even when severely used. To prove this, refer to the articles in which the driving tension for continuous service is stated. Take Morin's figures, for example, in Art. 1, p. 17, indicating 355 lbs. to the square inch of leather, being equal to from $\frac{1}{8}$ to $\frac{1}{12}$ the breaking strength of the same, which is ample factor of safety even when the belt has lost part of its original tenacity by wear and tear, and by the weakening effect of oils and adhesives applied to its substance.

Again, take the strain of 100 lbs. to the inch of width, as proposed by Thurston, in Art. 40, which is beyond the limits of usual practice; but single leather belting has been submitted to such a working load without appreciable injury, as represented in Art. 15. Even this is only $\frac{1}{5}$ the rupturing strain of the poorest leathers in the table.

Account should be taken, in these estimates, of the loss at the holes for rivets and laces, for the tightening to gain adhesion, and for the weakening of the belt at the splices and fastenings, which are favored, of course, by the superior tenacity of the fibre of the belt.

Hoyt's tests of a side of leather have awakened an increasing interest in the need of reliable experiments for determining the breaking strain of the various beltings in use. This table tells us plainly that *strength, stretch,* and *show* of leather are not equal and convertible terms, and shakes our faith in the ancient saying that there is "nothing like leather," by proving it to be not *all strong*, but, like everything else, having its weak points. It shows, also, why belts become crooked in use.

The table on page 213 shows the strength of various beltings — the first group presenting the tests made on July 3, 1876, at the Centennial Exhibition, in Messrs. Riehlé Bros.' testing machine; the others are derived from various sources. Most of them will be found in the articles throughout this treatise.

STRENGTH OF BELTING LEATHER.

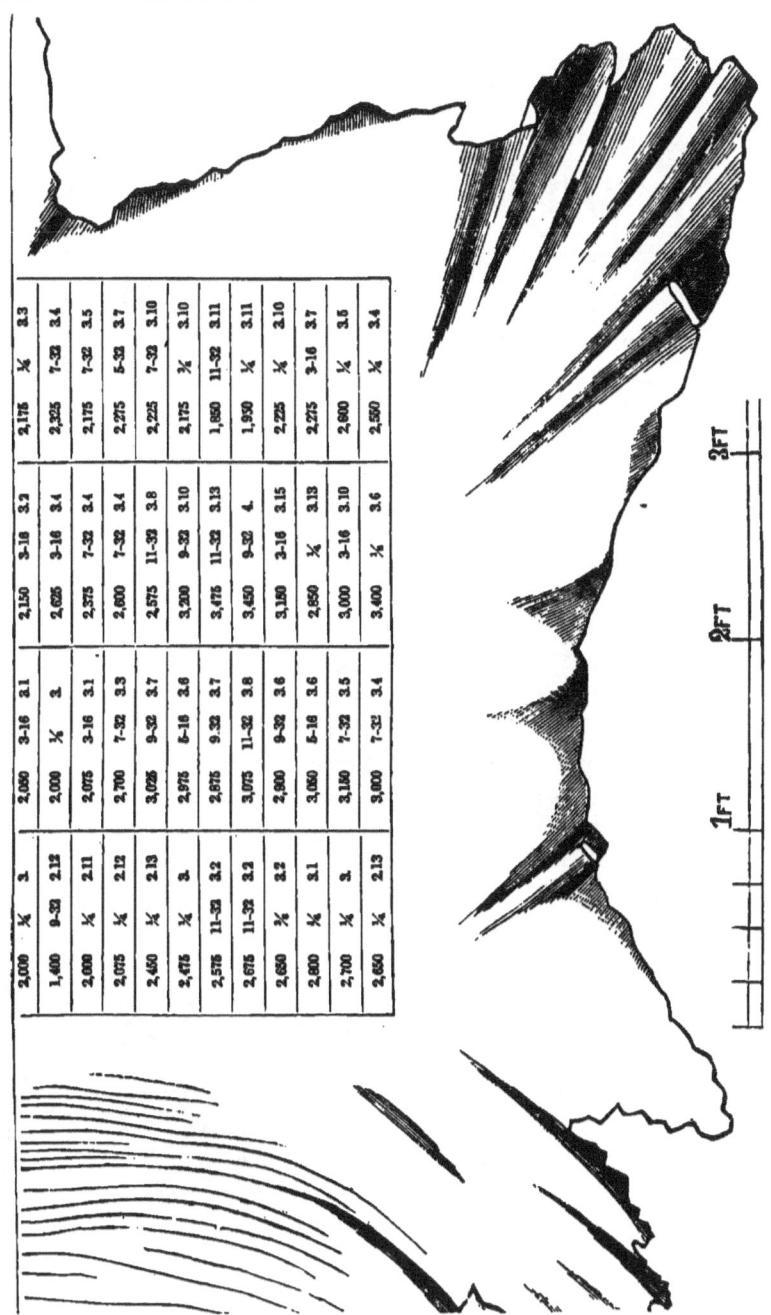

Fig. 67.

STRENGTH OF BELTING LEATHER.

Material in Belt.	Size of Belt Tested.		Force Breaking the Belt.	Force Required to Break One Inch Width.	Force Required to Break One Square Inch.	Remarks.
	Width in Inches.	Thickness in Inches.				
			Lbs.	Lbs.	Lbs.	
Leather	3	¼	3750	1250	5000	Centennial Tests. Oak Tanned.
"	3		3625	1208	4833	" " " "
"	3		3500	1166	4666	" " " "
"	3		3375	1125	4500	" " " "
"	3		3250	1083	4333	" " " "
"	3		3000	1000	4000	" " " "
"	3		2250	750	3000	" " " "
Raw Hide	3	5/32	2875	958	6131	" "
Sugar Tanned	2¾	¼	2000	727	2909	" "
Rubber	3	...	3500	1833	......	" "
" 3-ply	3	7/32	3000	1000	4571	" "
Leather	1	3/16	552	552	2944	Mean of 5 experiments ord'y leather.
"	2	3/16	1077	588	2872	" 5 " " "
"	3	3/16	1522	507	2705	" 3 " " "
Rubber 3-ply	2	...	1211	605	......	" 5 " " "
"	3	...	1763	587	......	" 5 " " "
Leather	1	3/16	530	530	2836	
Rubber 3-ply	1	...	600	600	......	
Leather	1½	...	1050	840	......	Oak Tanned. 86*
"	1¼	...	1850	1480	......	Page Tannage. 86
"	....	...	......	......	3200	Good quality. Many tests. 86
"	....	...	......	......	4000	Good quality. 88
"	....	...	......	......	4278	English. Rankine. 82
Raw Hide	....	...	......	......	6417	" " 82
Leather	1	1/16	930	......	5000	Good new English. 54
"	1		1000	1000	......	
"	3	7/32	2025	675	3086	Towne. American. 163
"	....	...	......	......	4200	Ox. English. Rankine.
Flax	5	...	6272	1254	......	
"	5	...	7448	1489	......	G. Spill & Co., London, flax yarn cemented. 158
"	10	...	16632	1663	......	
Leather	4	...	2100	525	......	
Calves' skin	....	...	......	......	1890	London Mech. Mag., 1863.
Sheep "	....	...	......	......	1610	Brazil " " "
Horse "	....	...	......	......	4000	White " " "
" "	....	...	......	......	3200	Russian " " "
" "	....	...	......	......	1680	Cord. " " "
Cow "	....	...	......	......	3981	Mech. Mag., 1863.
Cotton Duck.	1	...	200	200	......	CENTENNIAL TESTS.
Leather	3	...	3000	1000	......	"Union." Two shaved leathers with cloth cemented between. No rivets in stronger piece.
"	3	...	5625	1875	......	
"	1	...	......	1000	......	Good quality. 97

* These figures refer to the Articles from which the tabular numbers are taken.

CHAPTER VI.

TRANSMISSION OF FORCE BY BELTS AND PULLEYS.

By Robert Briggs, accompanied by Experiments of Henry R. Towne, from the "Journal of the Franklin Institute," January, 1868.

161. There are few mechanical engineers who have not been frequently in want of tabular information or readily applicable formulæ, upon which they could place reliance, giving the power which, under given conditions and velocity, is transmitted by belts without unusual strain or wear. The formula of the belt or brake is well known and simple, and it is only necessary to acknowledge and adopt a value for the co-efficient of friction (or of adhesion, which is perhaps the better term), to allow this formula to be applied in daily use. And this co-efficient of friction has been carefully established by the experiments of General Morin and M. Prony, and has been made available to English and American engineers, by the translation of Bennett.*

With every point needed, therefore, at the command of the engineer, it is somewhat surprising that a more extensive publication and general use of the data has not followed.

But, notwithstanding the existence of this correct mathematical and experimental information, the numerous tables which have been given by mechanical engineers appear to have had only that kind of practical basis which has come from guessing that an engine or a machine, either the driving or the driven, with a belt of given width, was producing or requiring some quantity of power which might be expressed in terms (foot-pounds) generally without any stated arc of contact. Three rules, given by practical mechanics, vary so much as to give as bases for estimate (without regard to arc of contact) 0.76 horse-power, 0.93 horse-power, and 1.75 horse-

---

* Bennett's *Morin's Mechanics*, New York, 1860.
   It must be remarked that there are some mistakes in the text of Bennett's translation, which will lead to serious errors, unless read by a careful investigator.

power, respectively, for the power of a belt one inch wide running 1000 Fpm.*

It was the requirement to know the exact useful effect of a novel disposition involving an unusual small arc of contact of the belt upon the pulley, where much embarrassment would result if the application proved itself unsatisfactory, that led to the present inquiry. As the writer was not able to give the time demanded for making such experiments as would establish the practical co-efficient of adhesion, he suggested what was desired to Mr. H. R. Towne, and the numerous experiments, of which he gives the accompanying report, are the result of his labor and care.

It was not until after the experiments were completed, that either Mr. Towne or the writer knew of the publication of M. Prony or General Morin, although Bennett's translation rested upon the shelves of the writer's library; but, aside from the gratification which we feel at the corroboration, we think the reader of this article will be pleased to know that our data is founded upon the ordinary pulleys and belts of the workshop, and our experiments were not impaired by any niceties which common workmen would not apply.

Even the crudeness of our experimental apparatus, and the general, not over-exact, method adopted, will serve to demonstrate to the minds of practical men the possibility of relying upon figures which have been established so nearly in accordance with the customs of the workshop.

We take from Rankine's *Applied Mechanics* the following formula of a belt, only changing the words, in the hope to make it comprehensible to the general reader in an elementary way:

Let A be a pulley upon which the belt passes from T_2 to T_1. Let $r =$ the length of radius of the pulley A. Let $T_1 =$ the tension of the belt (or the strain) on the tight side.

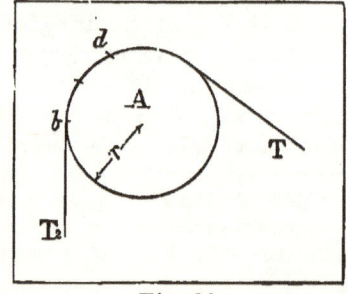

Fig. 69.

Let $T_2 =$ the tension of the belt (or the resistance on the loose side). Then the pull on the belt by which it transmits power $= P = T_1 - T_2$.

---

* It will be shown, by the deductions from experiments, that 1.33 horsepower, nearly, is easily given in continuous work by a belt one inch wide running 1000 Fpm.

and this difference represents the adhesion or friction, resulting from the contact between the belt and the pulley.

If we suppose the tension on a unit of width of belt at $b = T_2$, then it follows that the normal pressure per unit of surface at the point $b = \dfrac{T_2}{r}$. The units of dimension, either linear or superficial, may be inches and square inches, feet and square feet, metres and metres square, or any other units of relative value. Thus we may say the normal pressure upon a square inch of surface at b equals the tension on an inch-wide belt, T_2, divided by the length in inches of the radius, r.

We will endeavor to make this understood by comparing the case with the well-known instance of the relation of pressure to tension in the shell of a cylindrical boiler. If we take an example of a boiler having 10 inches radius (or 20 inches diameter), and with an internal uniform pressure of 100 lbs. per square inch, it will be recognized that the tangential strain per inch of length of shell will be equal to the pressure multiplied by the radius, or 1000 lbs.; and this tangential strain is uniform at all points of the circumference.

The tension, T_2, in like manner, corresponds to the tangential strain just stated, and the resulting normal pressure corresponds to the internal uniform pressure. And as, in the instance of the boiler, the tangential strain is exerted at all points of the circumference, so the normal pressure proceeding from the tension in the case of a belt, is independent of the length of arc of contact on the belt, and refers to the point of contact, b, wherever that point may be taken on the pulley.

Admitting, therefore, that the pressure at the point $b = \dfrac{T_2}{r}$, we have the friction resulting from the contact of the belt on a unit of surface $= f\dfrac{T_2}{r}$ (when f is the co-efficient of friction of the leather of the belt on the pulley). This gives a new value for the tension of the belt at some point, c (very near b), (or tangential strain at that point), $= T_2 + f\dfrac{T_2}{r} = T_2\left(1 + f\dfrac{1}{r}\right)$. And it follows that the pressure at $c = \dfrac{T_2}{r}\left(1 + f\dfrac{1}{r}\right)$, and the friction resulting from the contact of the belt on a unit of surface $= f\left[\dfrac{T_2}{r}\left(1 + f\dfrac{1}{r}\right)\right]$. This again gives

a new value for the tension of the belt at some point, d (very near c), $= T_2 + f \left[\frac{T_2}{r}(1+f\frac{1}{r})\right] = T_2 \left(1+f\frac{1}{r}\right)^2$. On taking further points in succession, until we take l points and reach the point m on the figure, we have for the tension at $m = T_2\left(1+f\frac{1}{r}\right) = T_1 \therefore \frac{T_1}{T_2}$
$= \left(1+f\frac{1}{r}\right).$

This equation is the well-known one of the hyperbolic logarithm. Where hyperbolic logarithm $\frac{T_1}{T_2} = f\frac{l}{r}$ or $\frac{T_1}{T_2} = e^{f\frac{l}{r}}$, where e is the base of hyperbolic logarithms. We can further transform this equation by substituting the ratio of the angle in degrees, for the length of contact on the arc, compared to the radius.

Thus, $\frac{2r\pi}{360} =$ arc of $1°$, let $l = a \left(\frac{2r\pi}{360}\right) \therefore \frac{T_1}{T_2} = e^{f\frac{2\pi a}{360}}$,

and taking the numerical values of e, π, and dividing out the 360,

$$\therefore \frac{T_l}{T_s} = 2.718^{\,0.017456\, f\, a}$$

$$\therefore log.\left(\frac{T_l}{T_s}\right) = 0.4343\,(0.017456\, f\, a), \quad (1.)$$

$$\therefore log.\, T_l - log.\, T_s = 0.00758\, f\, a$$

$$\therefore \frac{T_l}{T_s} = 10.^{\,0.00758\, f\, a}, \quad (2.)$$

$$\therefore f = \frac{log.\left(\frac{T_l}{T_s}\right)}{0.00758\, a} \quad (3.)$$

As we assumed before, $P = T_l - T_s$. $\therefore T_s = T_l - P$, which inserting in equation 2.

$$\therefore \frac{T_l}{T_l - P} = 10^{\,0.00758\, f\, a}$$

$$\therefore P\left(10^{\,0.00758\, f\, a}\right) = T_l\left(10^{\,0.00758\, f\, a}\right) - T_l$$

$$\therefore P = T_l\left(1 - 10^{-0.00758\, f\, a}\right) \quad (4.)$$

The third equation is the one to which we would now call attention. By it, for any given values for the ratio $\frac{T_1}{T_2}$, we can determine the co-efficient of friction, when, by experiment, we have fixed the greatest difference of the two strains without slipping on a pulley with a given arc (measured by a) of contact.

We would here make a very important observation, which forms the key of the whole system of transmission of force by belts. In practice, *all belts are worked at the maximum co-efficient of friction.* A belt may, when new or newly tightened, work under heavy strain and with a small co-efficient of friction called into action; but in process of time it becomes loose, and it is never tightened again until the effort to perform its task is greater than the value of the co-efficient with a given tension of belt, and the belt slips. We run our belts as slack as possible, so long as they continue to drive.

It has been shown * that the value of $T_1 + T_2$, or the sum of the strains upon the two sides of a belt (loose and tight), is a constant quantity; that is, when a belt is performing work, it will become loose on the one side to the exact amount that it is strained on the other, and when at rest, not transmitting force, the tensions will become equal, and their sum be the same as before. It is manifest that the limit of the strength of a belt is found in the maximum tension T_1, and that this strength being known, the effective pull (P) is further limited, with any given arc of contact, by the value f of the co-efficient of friction.

The discussion has so far been limited to the pull exerted by a belt. When we would include the power which belts will transmit, we have only to multiply the pull by some given or assumed velocity to transform our equations into work performed.

By means of the third equation, we will now deduce a value for the co-efficient of friction as given by the experiments.

All the experiments were with the arc of contact $= 180° = a$,

$$\text{which, substituting } f = \frac{\log. \frac{T_1}{T_2}}{1.3644}$$

and the result of 168 separate experiments of Mr. Towne has given, under tensions of T, from 7 to 110 lbs. per inch of width of belt:

$$\frac{T_1}{T_2} = 6.294. \quad \therefore f = \frac{\log. 6.294}{1.3644} = 0.5833. \dagger$$

---

* Bennett's Morin, page 303, and following ¶ 252.
† Bennett's Morin, page 306, ¶ 253, gives $f = 0.573$.

In this case T_1 has in all cases been so much in excess of T_2 as to slip the belt at a defined, slow, but not accelerating motion.

From an examination of the report of the experiments, we think the reader will coincide with our conclusion that $\frac{6}{10}$ of this value of $\frac{T_1}{T_2}$ can be taken as a suitable basis for the working friction or adhesion which will cover the contingencies of condition of the atmosphere as regards temperature and moisture; or

$$\frac{T_1}{T_2} = 3.7764 \; (maximum \; practical \; value) \therefore f = \frac{log. \; 3.776}{1.3644} = 0.42292.*$$

The experiments further show that 200 lbs. per inch of width of belt is the maximum strength of the weakest part — that is, of the lace holes. Taking this, with a factor of safety at one-third, we have the working strength of the belt, or the practical value for $T_1 = 66\frac{2}{3}$ lbs.† The case when belts are spliced instead of laced, a great increase of strength has been shown, the experiments giving 380 lbs. per inch of width, or 125 lbs. safe working strength.

If we insert these values of f and T_1 in (4,)

$$\therefore P = 66\frac{2}{3} \left(1 - 10^{-0.0078 \times 0.42292 \; a}\right)$$

$$\ddagger \therefore P = 66\frac{2}{3} \left(1 - 10^{-0.003206 \; a}\right) \quad (5.)$$

---

* It should be noted that the experiments were made without any appreciable velocity of belt, and throughout this paper no regard has been paid to the effect of velocity or of the dimensions of the pulleys upon the value of the coefficient of friction.

For pulleys less than 12 inches diameter (with the belts of the ordinary thickness of about $\frac{3}{16}$ths inch), and for velocities exceeding about 1000 Fpm, allowance must be made for the effect of centrifugal force in relieving the pressure of the belt on the pulley, for the rigidity of the belt, and for the *interposition of air between the pulley and the belt*. At high speeds, say 3000 feet velocity of belt per minute, the want of contact can be seen, sometimes, to the extent of one-third the arc encompassed by the belt. The writer has proposed to place a deflector or stripper near the belt, to take off the stratum of air moving with it, but has never tried the experiment, although he has little doubt of its giving some advantage.

† Bennett's Morin, page 306, ¶ 253, gives 55.1 lbs. per inch of width as admissible.

‡ This equation (5) is the really important one in practice, and by means of logarithms can be solved for any values of $a°$ readily; but as some of those who may wish to use it may not be at once prepared to use the logarithmic notation,

The largest possible angle for an open belt, without a carrier or tightener, is 180°, as upon either the driving or the driven pulley this cannot be exceeded; but for crossed, or carried, or tightened belts, the angle may be as large as 270°.

We give the following table of results for different arcs of contact (corresponding to $a°$) within the usual limits of practice.

Table I.

Strain transmitted by belts of one inch width upon pulleys, when the arcs of contact vary as the angles of

90°	100°	110°	120°	135°	150°	180°	210°	240°	270°
Lbs. 32.33	Lbs. 34.80	Lbs. 37.07	Lbs. 39.18	Lbs. 42.06	Lbs. 44.64	Lbs. 49.01	Lbs. 52.52	Lbs. 55.33	Lbs. 57.58

If we suppose the pulley to be one foot in diameter, and to run some number, N. of Rpm, we have the power transmitted $= N\pi P$.

---

from want of use or practice, we give an example. Suppose we take an angle of 90°, the negative exponent then becomes $-0.003206 \times 90 = -0.28854$; subtracting this from 1, we have -1.71146. This term thus becomes $10^{-1.71146}$. Now this expression is only the notation for anti-logarithm -1.71146, or in words the number for which -1.71146 is the logarithm. Logarithmic tables give this number $= 0.51505$, and the equation

$$P = 66\tfrac{2}{3} \left(1 - 10^{-0.003206 \times 90}\right) = 66\tfrac{2}{3} \left(1 - 10^{-1.71146}\right)$$

$$= 66\tfrac{2}{3} \left(1 - 0.51505\right) = 66\tfrac{2}{3} \times 0.48495.$$

$$\therefore P = 32.33.$$

And we give the following table for different arcs of contact (corresponding to $a°$) within the usual limits of practice.

Table II.

Power transmitted by belts on pulleys one foot in diameter one Rpm. Arcs of contact of belts upon pulleys corresponding to the angles.

Inches of Width of Belt.	90°	100°	110°	120°	135°	150°	180°	210°	240°	270°
	Foot-lbs	Foot-lbs	Foot-lbs	Foot-lbs	Foot-lbs	Foot-lbs	Foot-lbs	Foot-lbs	Foot-lbs	Foot-lbs
1	102	109	116	123	132	140	*154	165	174	181
2	203	219	233	246	264	280	308	330	348	361
3	305	328	349	369	396	420	462	495	521	542
4	406	437	466	492	528	560	616	660	695	723
5	508	547	582	615	660	701	770	825	869	904
6	609	656	699	738	792	841	924	990	1043	1084
7	711	766	815	861	924	982	1078	1155	1217	1265
8	813	875	932	985	1056	1122	1232	1320	1391	1446
9	914	984	1048	1108	1188	1262	1386	1485	1564	1626
10	1016	1094	1165	1231	1321	1402	1540	1650	1738	1807

The application of Table II. to any given cases of known angle of the arc of contact, width of belt in inches, diameter of pulley in feet, and number of revolutions, is simply to take the figures from the table for the first two, and multiply by the two succeeding conditions, to obtain the foot-lbs. of power transmitted.

We have taken the following examples: — First. Mr. Schenck (of New York) found an 18-inch wide belt running 2000 Fpm, the pulleys being 16 feet to 6 feet, would give 40 horse-power, with ample margin (one-fourth). (*Sic.*)

If we take the distance from centre of the 16-feet pulley to that of the 6-feet to be 25 feet (about the usual way of placing the fly-wheel pulley of an engine in regard to the main line of shafting), we have the arc of contact subtending about 153°. From Table I. the strain transmissible is 45.1 lbs. × 18 × 2000 = 1,623,600 foot-lbs. = 49.2 horse-power.

Second. Mr. William B. Le Van (of Philadelphia) found by indicator that an 18-inch wide belt running 1800 Fpm, the pulleys being 16 feet and 5 feet respectively, transmitted 43 horse-power, with maximum power transmissible unknown. If we take the centre's

---

*A 1-foot pulley, 1 inch wide, running 215 Rpm, gives 1 horse-power.

distance, as before, at 25 feet, we have the arc of contact subtending about 150°.

From Table I. we derive 44.64 lbs. as the strain transmissible × 18 × 1800 = 1,446,336 foot-lbs. = 43.83 horse-power. The same authority found by indicator that a 7-inch wide belt over two 2-feet 6-inch pulleys, 11 feet centre to centre (horizontal), moving 942 Fpm, gave 8 horse-power. From Table I. for 180° angle, we take 49.01 × 7 × 942 = 323,172 foot-lbs. = 9.79 horse-power. This belt was stated to be very tight.

Third. Mr. A. Alexander (*London Engineer*, March 30th, 1860) gives a rule that a 1-inch belt will, at 1000 feet velocity, transmit 1¾ horse-power.

If we take the contact at 180° from Table I., 49.01 × 1000 = 49,000 foot-lbs, we have only 1½ horse-power.

Fourth. Mr. William Barbour (same journal, March 23, 1860) gives as the power a 1-inch belt will transmit with 1000 feet velocity = 0.927 horse-power, when we derive with 180° angle from our tables = 1½ horse-power.

Fifth. A. B. Ex (same journal, April 6, 1860) gives a rule

$$\frac{diameter\ in\ inches \times Rpm \times breadth\ in\ inches}{5000} = N\ HP,$$

ratio of pulleys not to exceed 5 to 1. Changing this rule to

$$\frac{diameter\ in\ feet \times Rpm \times width\ in\ inches}{5000 \div 12 \div 33,000} = N\ foot\text{-}lbs.$$

$$\therefore \frac{Diameter\ in\ feet \times Rpm \times width\ in\ inches}{0.01263} = N.\ foot\text{-}lbs.$$

$\therefore 79.2 \times diameter\ in\ feet \times Rpm \times width\ in\ inches = N.\ foot\text{-}lbs.$

From Table II. the angle of 120° gives 123 in place of 79.2, and it would appear this authority adopts about ¾ the effect we take.

Sixth. Dr. Fairbairn gives (*Mills and Mill Work*, Part II., page 4) a table of approximate width of leather straps in inches necessary to transmit any number of horses-power, the velocity of the belt being taken at 25 to 30 feet per second (1500 to 1800 per minute), one-foot pulley, 3.6 inches wide, gives one horse-power.

Assume 1650 Fpm, contact 180°, we have, from Table I., 1650 × 49.01 × 3.6 × 1 = 29,112 foot-lbs. = 0.87 horse-power.

Seventh. Rankine gives (*Rules and Tables*, page 241) 0.15 as the co-efficient of friction, probably applicable to the adhesion of belts on pulleys, to be used with his formulæ in estimating the power transmitted. Neither experiments nor practice give so small a co-efficient as this.

We could multiply authorities on these points, but think the corroboration of those we quote with our tables sufficient to establish our experimental and estimated co-efficient of friction, $f = 0.423$, as a proper practical basis.

We give the two following cases not only to show the application of the formula 5, but as matters of some interest.

In the construction of one of the forms of centrifugal machines for removing water from saturated substances, the main or basket spindle is driven by cone-formed pulleys, one of which, being covered with leather, impels the other by simple contact.

In the particular instance taken, the iron pulley on the spindle was 6 inches largest diameter, and the leather-covered driving pulley was 12 inches largest diameter; the length of cones on the face was 4 inches, this last dimension corresponding to width of belt in other cases. By covering the leathered pulley with red lead we were able to procure an *impression* on the iron pulley, showing the width of the surfaces of contact when the pulleys were compressed together with the force generally applied when the machine was at work. This width was, at the largest diameters, almost exactly $\frac{1}{4}$ of an inch. From the nature of the two convex surfaces compressing the leather between them, the actual surface of efficient contact cannot be taken at over half this width (the slight error in estimating this contact as straight lines in place of circular arcs may be neglected). This gives the angle subtended by the arc of contact on the iron pulley $= 2\frac{1}{2}°$, taking equation (5).

$$P = 66\tfrac{2}{3}\left(1 - 10^{-0.003206\,a°}\right) = 66\tfrac{2}{3}\left(1 - 10^{-0.008015}\right) = 66\tfrac{2}{3}\left(1 - 10^{-1.991985}\right) = 66\tfrac{2}{3}(1 - 0.98171) = 66\tfrac{2}{3}(0.01829) = 1.3717.$$

Now, the average diameter of the iron pulley in the middle of its 4-inch face is 4.708 inches $= 0.3923$ feet, with a circumference of 1.2326 feet, and it is usual to run, at the least velocity, $1000°$ Rpm; whence the power given by these pulleys $= 1.3717$ lbs. $\times$ 4 inches $\times$ 1.2326 feet $\times$ 1000 revolutions $= 6757$ foot-lbs. $= \frac{1}{5}$ horse-power. As the work performed by one of these centrifugal machines is the acquirement of velocity under the resistance of the friction of the machine and of the air, and the work of expelling the moisture is so insignificant, in comparison, that it may be neglected in estimating. It can, therefore, be taken as probable that the real power demanded to keep the machine in motion is very nearly that given by calculation. It should be stated that the basket belonging to this particular machine is 29 inches diameter and 12 inches deep.

The second special case we instance at present consists in a proposed arrangement for driving a fan which had previously been found to demand an 8-inch belt on a 10-inch pulley to run it 1275 Rpm. (The arc of contact here was 162°, so that the apparent power, with a very tight belt, was $37\frac{1}{2}$ horse-power; but about $\frac{1}{3}$ of this was defective adhesion from running a rigid belt over so small a pulley.) It was thought desirable to avoid the fast-running countershafts, and drive this direct from an engine-pulley fly-wheel, by impingement, so to speak, of the belt on its tight side between the fan-pulley and another larger carrier pulley, against a portion of the periphery of the fly-wheel.

If we suppose the force demanded, measured on the fan-pulley as before, to be 40 horse-power $= 1,320,000$, and the fan-pulley to be 10 inches diameter $\times$ 16 inches wide, and to run 1250 Rpm,

$$\therefore \frac{1,320,000}{1250 \times 1\frac{9}{12} \times 16 \times \pi} = 25.2 \text{ as the pull, P, on each inch of width of}$$

the belt as it comes from the 10-inch pulley. By substituting this value for P in equation (5), and then reducing the equation to find the value for $a°$, we have $a° = 65°$, which is the angle of contact demanded to give the necessary adhesion.

It will be noticed that this angle is independent of the diameter of the fly-wheel pulley, it being only requisite that that diameter should be such as, with the given or assumed number of revolutions, will produce the given velocity. In the case taken for example, the fly-wheel pulley was 16 feet diameter $\times$ 16 inches wide, with 70 Rpm velocity.

As we have before remarked, the sum of the two tensions on the belt is constant, whether the belt is performing work or not; that is,

$$S = T_t + T_t; \text{ but } P = T_t - T_t. \quad \therefore T_t = T_t - P. \quad \therefore S = 2T_t - P.$$

As we assumed in equation (5) T to equal $66\frac{2}{3}$ lbs., we can substitute the value of P as in Table I., in the equation, $S = 2(66\frac{2}{3}) - P = 133\frac{1}{3} - P$, from which it is evident that the sum of the tensions will vary with P or with the angle of contact. It is evident, also, that the load upon the shaft proceeding from the tensions T_2 and T_1 will be the resultant of whatever angle the belt makes with a line joining the centres of the two pulleys, or as the cosine of that angle.

By constructing on paper a pair of pulleys, it will be readily discerned that the angle in question for small pulleys $= 90° \frac{a}{2}$, and for large and crossed ones, $= \frac{a}{2} - 90°$, we can consequently form the following table:

Table III.

STRENGTH OF LACING OF JOINT $66\frac{2}{3}$ POUNDS PER INCH WIDE,

Showing, first, the sum of tensions on both sides of a belt, per inch of width, whether in motion or at rest, when strained to transmit the maximum quantity of power in general practice; and showing, second, the load carried by the shafts and supported constantly by the journals per inch of width of belt, when the arcs of contact vary as the angles of

	90°	100°	110°	120°	135°	150°	180°	210°	240°	270°
	Lbs.	Lbs.	Lbs.	Lbs.	Lbs.	Lbs.	Lbs.	Lbs.	Lbs.	Lbs.
1st,	101.	98.53	96.26	94.15	91.27	88.69	84.32	80.81	78.	75.75
2d,	71.42	75.47	78.85	81.53	84.32	85.67	84.32	78.05	67.59	53.56

When machinery is driven by gearing, the shafts only carry the running wheels and the weight, and when the machines are thrown on, the friction of the lines increases with the work; but with belts and pulleys the load on the line and its frictional resistance is constant, whether the machinery works or lies idle.

Of course, it is not proper to assume that the load produced by the belt on the shaft is exactly that given by the second line in Table III.; but we can be safe in taking those weights as rarely exceeded, because belts begin to fail when they are; and as rarely much less, because few of our machines are not worked up to their belt capacity.

The advantages shown by the figures on all the tables, but especially on the last, in those arcs of contact over 180° where crossed belts are used, have the substantial ground of practice, although many mechanics are unaware of the facts. The writer will instance a case of several heavy grindstones having from main to counter lines 8-inch crossed belts on pulleys 3 feet diameter, running 120 revolutions, only 8 feet centre to centre, where belts have already lasted, day and night use, 3½ years. For the same purpose, 6-inch open belts were formerly used with an average duration of a few weeks only.

Another use of a crossed belt is for long belts, the crossing effectually preventing those waves which generally impair, if they do not destroy, such belts when open.

Besides the actual power transmitted by belts, which it has here been attempted to embrace in a general law — the application of belting — both the manner and the purpose opens a field for discussion far beyond the limits of the present article, the writer hoping that others will take up the subject, so that the published data of the mechanic may more fully include the practice of the workshop and factory.

Note to the above Article.

162. The following addition to the preceding paper may afford some facility in the use of the results. The figures are derived from, and the final results correspond to, the figures given in Table II.

When the arc of contact of a belt upon the least of 2 pulleys which it connects is

90° 100° 110° 120° 135° 150° 180° 210° 240° 270°

then

3900 3600 3400 3200 3000 2825 2570 2400 2280 2190

is the sum of the multiplication together of the inches of diameter of the pulley by the inches of width of the belt, by the number of Rpm which equal one horse-power. Or to make the use of this perfectly clear we give the following rule: To find the horse-power of 33,000 lbs. lifted one foot high in one minute given out by certain belt, pulley, and speed. Multiply the inches of diameter of pulley by the inches width of belt and by number of Rpm, and divide the result by the numbers given in the last line of figures as relating to the contact the belt has on the smaller of its pulleys, and the quotient will be the number of horse-power.

Machine tools, lathes and boring tools, require to be belted 3 times as strongly as the average use in work, to overcome occasional resistance and starting frictions.

On the Adhesion of Leather Belts to Cast-Iron Pulleys, by Henry R. Towne.

163. The following experiments, undertaken at the instance of Mr. Robert Briggs, had for their object to determine, in a satisfactory and conclusive manner, the true value of the co-efficient of friction of leather belts on cast-iron pulleys. This result has, it is hoped, been attained, and in the preceding article Mr. Briggs has discussed at length the theory of the action of belts, and has also given practical formulæ in which the co-efficient of friction employed is that deduced from these experiments, which latter have been made with great care, and may, it is believed, be accepted as reliable. In order, however, that all interested may judge for themselves of the correctness of the deductions made from them, we present herewith a complete tabular record of the experiments, which will also repay examination as exhibiting several interesting and instructive facts connected with the efficiency of leather belts.

The experiments were made with leather belts of 3 and 6 inches width, and of the usual thickness — about $\frac{3}{16}$ths of an inch. The pulleys used were respectively of 12, $23\frac{5}{8}$, and 41 inches diameter, and were in each case fast upon their shafts. They were the ordinary cast-iron pulleys, turned on the face, and, having already been in use for some years, were fair representatives of the pulleys usually found in practice.

Experiments were made first with a perfectly new belt, then with one partially used and in the best working condition, and, finally, with an old one, one which had been so long in use as to have deteriorated considerably, although not yet entirely worn out. The adhesion of the belts to the pulleys was not in any way influenced by the use of unguents or by wetting them — the new ones when used were just in the condition in which they were purchased — the others in the usual working condition of belts as found in machine-shops and factories — that is, they had been well greased and were soft and pliable.

The manner in which the experiments were made was as follows: The belt being suspended over the pulley, in the middle of its length, weights were attached to *one* side of the belt, and increased until the latter slipped freely over the pulley; the final, or *slipping* weight, was then recorded. Next, 5 lbs. were suspended on *each* side of the belt, and the additional weight required upon *one* side to produce slipping ascertained as before, and recorded. This operation was repeated with 10, 20, 30, 40, and 50 lbs., successively, suspended upon both sides of the belt. In the tables these weights, *plus* half the total weight of the belt, are given as the "equalizing weights" (T_2 in the formulæ), and the *additional* weight required upon one side to produce slipping, is given under the head of "unbalanced weights;" this latter, *plus* the equalizing weight, gives the total tension on the loaded side of the belt (T_1 in the formulæ).

The belt, in slipping over the pulley, moved at the rate of about 200 Fpm, and with a constant, rather than increasing, velocity; or, in other words, the final weight was such as to cause the belt to slip smoothly over the pulley, but not sufficient to entirely overcome the friction tending to keep the belt in a state of rest. In this case (*i. e.* with an excessive weight) the velocity of the belt would have approximated to that of a falling body, while in the experiments its velocity was much slower, and was nearly constant, the friction acting precisely as a brake. By being careful that the final weight was such as to produce about the same velocity of the slipping belt in all of the experiments, reliable results were obtained.

It became necessary to make use of a weight such as would produce the positive motion of the belt described above, as it was found impossible to obtain any uniformity in the results when the attempt was made to ascertain the minimum weight which would cause the belt to slip. With much smaller weights *some* slipping took place, but it was almost inappreciable, and could only be noticed after the weight had hung for some minutes, and was due very probably to the imperceptible jarring of the building. After essaying for some time to conduct the experiments in this way, and obtaining only conflicting and unsatisfactory results, the attempt was abandoned, and the experiments made as first described.

In this way, as may be seen, results were obtained which compare together very favorably, and which contain only such discrepancies as will always be manifest in experiments of the kind. It is only by making a great number of trials and averaging their results, that reliable data can be obtained.

The value of the co-efficient of friction which we deduce from our experiments, is the mean of no less than 168 distinct trials.

It will be noticed, however, that the co-efficient employed in the formulæ is but $\frac{6}{10}$ of the full value of that deduced from the experiments, the latter being 0.5853 and the former 0.4229. This reduction was made, after careful consideration, to compensate for the excess of weight employed in the experiments over that which would just produce slipping of the belt, and may be regarded as safe and reliable in practice.

A note is made, over the record of each trial, as to the condition of the weather at the time of making it — whether dry, damp, or wet — and it will be noticed that the adhesion of the belts to the pulleys was much affected by the amount of moisture in the atmosphere. It is to be regretted that this contingency was not provided for, and a careful record of the condition of the atmosphere kept by means of a hygrometer. The experiments indicate, clearly, however, that the adhesion of the old and the partially-used belts was much increased in damp weather, and that they were then in their maximum state of efficiency. With the new belts the indications are not so positive; but their efficiency seems to have been greatest when the atmosphere was in a dry condition.

Experiments were also made upon the tensile strength of belts, with the following results: The weakest parts of an ordinary belt are the ends through which the lacing holes are punched, and the belt is usually weaker here than the lacing itself. The next weakest points are the *splices* of the several pieces of leather which compose the belt, and which are here perforated by the holes for the copper rivets. The strengths of the new and the partially-used belts were found to be almost identical. The average of the trials is as follows:

Three-inch belts broke through the lace-holes with...... 629 lbs.
"   "   "   rivet "   " ...... 1146 "
"   "   "   solid part "   " ...... 2025 "

These give as the strength *per inch of width*.

When the rupture is through the lace-holes............... 210 lbs.
"   "   "   rivet " .... 382 "
"   "   "   solid part............... 675 "

The thickness being $\frac{7}{32}$ inch ($=.219$), we have as the tensile strength of the leather 3086 lbs. per square inch.

From the above we see that 200 lbs. per inch of width is the ultimate resistance to tearing that we can expect from ordinary belts.

The experiments herein described are strikingly corroborative of those already on record, and this gives increased assurance of their reliability; and, although there is nothing novel either in them or in their results, it is hoped that they will prove of interest, and that an examination of them will lead to confidence in the formulæ which are based upon them.

Three-Inch New Belt.

Equalizing Weights, or Initial Tension on each Side of Belt = T_2.	Unbalanced Weight, Trial No. 1.	Unbalanced Weight, Trial No. 2.	Unbalanced Weight, Average of Trials.	Total Weight on Loaded Side of Belt when Belt Slipped = T_1.	T_1/T_2	Equalizing Weights, or Initial Tension on each Side of Belt = T_2.	Unbalanced Weight, Trial No. 1.	Unbalanced Weight, Trial No. 2.	Unbalanced Weight, Average of Trials.	Total Weight on Loaded Side of Belt when Belt Slipped = T_1.	T_1/T_2
ON 12-INCH PULLEY (BETWEEN SEAMS).						ON 12-INCH PULLEY (ON SEAMS).					
	Dmp.	Dry.					Dmp.	Dmp.			
2.94	17	20	18.5	21.44	7.29	2.94	14	14	14.0	16.94	5.76
7.94	43	58	50.5	58.44	7.36	7.94	30	32	31.0	38.94	4.90
12.94	70	86	78.0	90.94	7.03	12.94	48	51	49.5	62.44	4.83
22.94	108	167	137.5	160.44	6.99	22.94	92	87	89.5	112.44	4.90
32.94	124	226	175.0	207.94	6.31	32.94	107	110	108.5	141.44	4.29
42.94	159	311	235.0	277.94	6.47	42.94	134	138	136.0	178.94	4.17
52.94	210	343	276.5	329.44	6.22	52.94	191	188	189.5	242.44	4.58
Mean					6.810	Mean					4.775

Three-Inch New Belt.

Equalizing Weights, or Initial Tension on each Side of Belt = T_2.	Unbalanced Weight, Trial No. 1.	Unbalanced Weight, Trial No. 2.	Unbalanced Weight, Average of Trials.	Total Weight on Loaded Side of Belt when Belt Slipped = T_1.	T_1/T_2	Equalizing Weights, or Initial Tension on each Side of Belt = T_2.	Unbalanced Weight, Trial No. 1.	Unbalanced Weight, Trial No. 2.	Unbalanced Weight, Average of Trials.	Total Weight on Loaded Side of Belt when Belt Slipped = T_1.	T_1/T_2
ON 23⅜-INCH PULLEY.						ON 41-INCH PULLEY.					
	Dmp.	Dmp.					Dry.	Dry.			
2.94	12	16	14.0	16.94	5.76	2.94	13	13	13.0	15.94	5.42
7.94	34	44	39.0	46.94	5.91	7.94	38	39	38.5	46.44	5.85
12.94	54	77	65.5	78.44	6.06	12.94	55	61	58.0	70.94	5.48
22.94	110	135	122.5	145.44	6.34	22.94	......	......	......	......	......
32.94	161	191	176.0	208.94	6.34	32.94	188	176	182.0	214.94	6.53
42.94	207	234	220.5	263.44	6.14	42.94	......	......	......	......	......
52.94	277	282	279.5	332.44	6.28	52.94	289	295	292.0	344.94	6.52
Mean					6.118	Mean					5.960

Three-Inch Belt.

PARTIALLY USED AND IN GOOD ORDER—ON 12-INCH PULLEY.

Equalizing Weights, or Initial Tension on each Side of Belt = T_2.	Unbalanced Weight, Trial No. 1.	Unbalanced Weight, Trial No. 2.	Unbalanced Weight, Trial No. 3.	Unbalanced Weight, Trial No. 4.	Unbalanced Weight, Trial No. 5.	Unbalanced Weight, Average of Trials.	Total Weight on Loaded Side of Belt when Belt Slipped = T_1.	T_1/T_2
	Damp.	Damp.	Wet.	Dry.	Dry.			
3.2	13	14	26	12	11	15.5	18.7	5.84
8.2	34	36	53	32	32	37.4	45.6	5.56
13.2	57	63	77	50	50	59.4	72.6	5.50
23.2	102	109	149	90	99	109.8	133.0	5.73
33.2	123	128	226	144	144	153.0	186.2	5.61
43.2	174	185	299	183	......	210.2	253.4	5.87
53.2	226	265	338	249	......	268.5	322.7	6.07
Mean								5.754

Three-Inch Belt — Partially Used and in Good Order.

ON 41-INCH PULLEY.						ON 23⅜-INCH PULLEY.						
Equalizing Weights, or Initial Tension on each Side of Belt = T_2.	Unbalanced Weight, Trial No. 1.	Unbalanced Weight, Trial No. 2.	Unbalanced Weight, Average of Trials.	Total Weight on Loaded Side of Belt when Belt Slipped = T_1.	T_1/T_2	Equalizing Weights, or Initial Tension on each Side of Belt = T_2.	Unbalanced Weight, Trial No. 1.	Unbalanced Weight, Trial No. 2.	Unbalanced Weight, Trial No. 3.	Unbalanced Weight, Average of Trials.	Total Weight on Loaded Side of Belt when Belt Slipped = T_1.	T_1/T_2
	Dry.	Dry.					Dmp.	Dmp.	Dry.			
3.2	26	25	25.5	28.7	8.97	3.2	23	22	23	22.6	25.8	8.06
8.2	59	55	57.0	65.2	7.95	8.2	53	54	53	53.3	61.5	7.50
13.2	86	82	84.0	97.2	7.36	13.2	86	89	82	85.6	98.8	7.48
23.2	......	......	......	......		23.2	......	......	141	......	......	
33.2	188	194	190.5	223.7	6.74	33.2	194	210	181	195.0	228.2	6.87
43.2	......	......	......	......		43.2	......	..	228	......	......	
53.2	306	323	314.5	367.7	6.91	53.2	285	316	272	291.0	344.2	6.47
Mean					7.586	Mean						7.276

Three-Inch Old Belt.

ON 12-INCH PULLEY.						ON 23⅜-INCH PULLEY.						
Equalizing Weights, or Initial Tension on each Side of Belt = T_2.	Unbalanced Weight, Trial No. 1.	Unbalanced Weight, Trial No. 2.	Unbalanced Weight, Average of Trials.	Total Weight on Loaded Side of Belt when Belt Slipped = T_1.	T_1/T_2	Equalizing Weights, or Initial Tension on each Side of Belt = T_2.	Unbalanced Weight, Trial No. 1.	Unbalanced Weight, Trial No. 2.	Unbalanced Weight, Trial No. 3.	Unbalanced Weight, Average of Trials.	Total Weight on Loaded Side of Belt when Belt Slipped = T_1.	T_1/T_2
	Dmp.	Dmp					Dmp.	Wet.				
2.75	12	13	12.5	15.25	5.55	2.75	16	20		18.0	20.75	7.55
7.75	35	39	37.0	44.75	5.77	7.75	48	70		59.0	66.75	8.61
12.75	61	66	63.5	76.25	5.98	12.75	72	106		89.0	101.75	7.98
22.75	102	109	105.5	128.25	5.64	22.75	157	174		165.5	188.25	8.27
32.75	142	158	150.0	182.75	5.58	32.75	220	232		226.0	258.75	7.90
42.75	203	220	211.5	254.25	5.95	42.75	256	276		271.0	313.75	7.34
52.75	259	273	266.0	318.75	6.04	52.75	301	311		306.0	358.75	6.80
Mean					5.787	Mean						7.778

Six-Inch New Belt.

ON 23⅜-INCH PULLEY.						ON 41-INCH PULLEY.						
Equalizing Weights, or Initial Tension on each side of Belt = T_2.	Unbalanced Weight, Trial No. 1.	Unbalanced Weight, Trial No. 2.	Unbalanced Weight, Average of Trials.	Total Weight on Loaded Side of Belt when Belt Slipped = T_1.	T_1/T_2	Equalizing Weights, or Initial Tension on each side of Belt = T_2.	Unbalanced Weight, Trial No. 1.	Unbalanced Weight, Trial No. 2.	Unbalanced Weight, Average of Trials.	Total Weight on Loaded Side of Belt when Belt Slipped = T_1.	T_1/T_2	
	Dry.	Dmp.										
6.75	30	36	33.0	39.75	5.89	6.75	32	29	30.5	37.25	5.52	
11.75	53	63	58.0	67.75	5.77	11.75	53	50	51.5	73.25	6.23	
16.75	78	81	79.5	96.25	5.75	16.75	72	72	72.0	88.75	5.30	
26.75	125	135	130.0	156.75	5.86	26.75	118	115	116.5	143.25	5.36	
36.75	161	177	169.0	205.75	5.60	36.75	159	176	167.5	204.25	5.55	
46.75	216	223	219.5	266.25	5.70	46.75	209	236	222.5	269.25	5.76	
56.75	265	269	267.0	323.75	5.70	56.75	265	275	270.0	326.75	5.76	
Mean					5.752	Mean					5.640	

CHAPTER VII.

EXPERIMENTS ON THE TENSION OF BELTS, MADE AT METZ IN THE YEAR 1834.

BY ARTHUR MORIN.

Translated by J. W. Hüttinger, Member of the Franklin Institute, Philada.

Determination of the Natural Tension of Belts.

164. There remains, then, nothing more than to indicate how we have been able, for each series of experiments, with the same relative disposition of the axes of rotation, with the same hygrometric state of the belt, to determine the sum of the tensions t and t'.

There are for this several simple means, but we have usually employed the following. When the apparatus was in place, completely set up and put in motion, and when we were thus assured that the belt had sufficient tension to overcome the resistance, we commenced a series of experiments, during the continuation of which we did not change the distance of the axles; the variation of the tension, or the lengthening of the belt, could then not take place, except by a state more or less hygrometric. The apparatus worked thus: the belt did not slip on the pulley, as long as the resistance offered by the journals was not too great; but if, by not oiling, it happened that the resistance increased to such an extent that the belt began to slip, the shaft ceased to turn with continuity, and as the dynamometer showed, nevertheless, a curve of flexion corresponding to the difference of the tension at the moment; we have to determine t and t' the two simultaneous relations:

$$t = t'e^{\frac{f_i S}{R}}, \quad (t-t')R = FL + 0{,}413 + 0{,}0078\, t + 0{,}0082\, t',{}^*$$

by means of which we can determine t and t', and consequently $t + t' = 2\,T.$

It is by observations of this kind that one has usually determined, at each different position of the apparatus in relation to the wheel, the natural tension of the belt. Although the relative displacement of the

---

* Let $t =$ natural tension, $t' = t +$ friction of belt on drum. All the figures are given according to the metric system, and the comma is used for decimal point.

axes would evidently exercise the principal influence upon the natural tension, nevertheless we must assure ourselves if the very perceptible variations, which the hygrometric state of the belt proved, had not also a notable influence upon this tension, and to take it into account, if not with exactness, at least with a close approximation. This belt runs over the drum, $d\ d$, Fig. 69, close to the wheel, which we had taken the precaution to cover with a light roof, that it might not become too wet and get out of centre; but, as it passed very near to the wheel, it was not possible that, during the rapid movement of the latter, to prevent its becoming wet by the throwing off of water which the floats constantly raised. Further, the belt, which at the beginning of each day was dry, or very nearly so, was at the close almost completely wet. There resulted from this two different effects, which, relatively to the speed of the machine and to the employment of belts under like circumstances, nearly balanced each other. The first is, that the belt lengthened in such a manner that its natural tension diminished; the second is, that the ratio of the friction to the pressure of the belt and the pulley increased notably, and consequently the limit at which the belt would slip was being reached by the first effect, and was receding by the second.

It was necessary, then, at each of the observations made, to determine the tension of the belt, to examine if it were dry or wet, in order to take into calculation the proper value of the relation, $f_{\prime}$, of the friction to the pressure.

Special Observations to Determine the Natural Tension.

165. There remain to be given the results of observations made, to apply the preceding methods to different series of experiments, which I shall do in order.

The apparatus was completely set up, and began to work on the 8th of August, and, from that day to the 17th of the same month, inclusive, it occupied the same place; the belt was but little stretched, and it was frequently observed that when it was damp it slipped when the curve of flexion of the spring corresponded to the scale of loss, with an effort of $26^k,50$, brought to bear upon the circumference of the pulley with a radius of $0^m,308$, we have then

$$FL = 26^k,50 \times 0^m,308 = 8,162,$$

in this case. (See Tables II. and III.)

$$f_{\prime} = 0,377, \quad e = 2,71828, \quad \log. e = 0,43429, \quad \frac{S}{R} = \frac{174°}{180}\pi = 3,037.$$

By means of these statements we deduce from the two preceding equations

$$t' = 13^k,67, \quad t = 3,142 t' = 43^k,95, \quad t + t' = 2\,T = 56^k,62.$$

From the 18th of August the apparatus was moved a little further off, the belt was tightened till the 28th, inclusive, the dry belt slipped when the curve of flexion corresponded to the scale with an effort of $36^k,75$, applied to the circumference of the pulley; we have then

$$FL = 36^k,75 \times 0^m,308 = 11,319;$$

we had besides

$$f_i = 0,282 \text{ (see Table II.)},$$

the other statements are the same, and we deduce from the two equations employed

$$t' = 30^k,08, \quad t = 2,354 t' = 70^k,81, \quad t + t' = 2\,T = 100^k,89.$$

From the 3d to the 7th of November, a time at which the abundance of water permitted me to recommence the experiments which the dryness and other occupations had compelled me to suspend, the belt being slack and damp, I found that it slipped when the curve of flexion indicated upon the scale of loss an effort of 17 kilogrammes brought to bear upon the circumference of the pulley; we have then

$$FL = 17^k,00 \times 0,308 = 5,236, \quad f_i = 0,377,$$

from which we deduce

$$t' = 9^k,01, \quad t = 3,142\,t' = 28^k,30, \quad t + t' = 2\,T = 37^k,31.$$

From the 8th November to the 12th, inclusive, when the experiments were terminated, the belt having been overstretched by moving the axles too far off, we determined the natural tension by another means, employed at the beginning and at the end of the day, or, in other words, when the belt was dry and when it was wet, which is shown in No. 37 of Section III. of M. Poncelet's "Course of Mech." (See Art. 178.) The cast-iron shaft being stopped, and the sluice-gate closed in such a way that the least water possible could enter, we immediately placed upon the horizontal front (up-stream) float of the water-wheel the amount of weight necessary to drive the wheel and make the belt slip upon the cast-iron pulley; but as a little water passed continually, which rose to a certain height on the wheel, it was necessary to repeat the experiment in an inverse order by placing the weights upon the horizontal back (down-stream) float, till the wheel moved again. By taking the mean reckoning of the 2 weights we compensated the effect of the water, because it acted in an opposite direction in the 2 cases, and we had nothing to do but to establish the relation of equilibrium between the 2 weights, the friction of the wheel upon its journals, and the difference of the tensions at the moment of slipping upon the cast-iron pulley. Then calling:

S = the weight which turned the wheel.
$R' = 1^m,85$, the mean radius, including the floats.
$R'' = 0^m,514$, the mean radius of the pulley mounted on the shaft, including the half-thickness of the belt.
N_t = the pressure upon the journals of the wheel.
$f'' = 0,08$, the ratio of friction to the pressure for the journals and their bearings, with lard as lubricator.
$\rho = 0^m,03$, the radius of the journals.
$M = 1587$ kilogrammes, the weight of the wheel and its shaft.
t, t', a and a' retain the same significations and values as below.*

We had at the moment of slipping

$$(t-t')R'' + f''N_t\rho = SR', \text{ and } t = t'e^{\frac{f_tS}{R}}$$

$$N_t = \left\{ \begin{array}{c} 0,96\,(M - t\cos.\,a - t'\cos.\,a') \\ + 0,4\,(t\sin.\,a + t'\sin.\,a') \end{array} \right\} = 0,96\,M - 0,396\,t' - 0,293\,t,$$

and consequently

$$(t-t')R'' + 0,96f''M\rho - 0,396f''t'\rho - 0,293f''t\rho = SR.$$

The observations made when the belt was dry have shown that it slipped under the action of a weight of

$$S = 15 \text{ kilogrammes};$$

we had already (see Tables II. and III.)

$$f_t = 0,282, \quad t = 2,354\,t',$$

and we have deduced from the substitution in the above formula

$$t' = 34^k,76, \quad t = 81^k,825, \quad t + t' = 116^k,58.$$

---

* t = the tension of the conducting strip of the belt.
 t' = the tension of the conducted strip.
 $a = 51°$. $a' = 45°$, the angles which these tensions, t and t', make respectively with the vertical.

Other observations, made when the belt was wet, have given
$$S = 18{,}50 \text{ kilogrammes};$$
we had (see Tables II. and III.)
from which we deduced $f_{\prime} = 0{,}377$,
$$t' = 27^{k}{,}85, \qquad t = 87^{k}{,}50, \qquad t + t' = 115^{k}{,}35.$$

These last two series of experiments show that, when the belt is stretched tight enough, the difference of tension produced by the hygrometic lengthening is very little; and taking the mean of the values of $t + t'$, obtained under extreme circumstances, between which all the results of the same day necessarily come in, when the belt has passed successively from a state of dryness to that of complete saturation, we see that the difference of this mean $t + t' = 115^{k}{,}96$, for each of the extreme values, is only about $\frac{1}{200}$ of its real value.

This result is sufficient for the calculation of our experiments, since it shows that having determined the sum of the tensions in the case when the belt was dry, or in that when it was wet, it will be allowed, without fear of an appreciable error, to regard it as very nearly constant and equal to the value found in one or the other case, which dispenses with many special observations.

But there are more, and it is easy to see that, in our apparatus, and in consequence of its dimensions and use, a considerable variation in the tension of the belt could have only a very slight influence in many cases.

In effect, the formula (see note on page 237),
comes back to
$$FL + f'N'r' = fNr,$$
$$FL = 0{,}96 fQr - 0{,}96 fpr' - (0{,}915\, t + 0{,}961\, t')(f'r' - fr),$$
in which it is evident that the influence of the terms t and t' will be as much less as f and r shall differ less than $f'r'$, and it would be zero if we had $f' = f$ and $r' = r$, which is, however, evident *a priori*. Now, in many experiments, r does not differ from r' but by $\frac{1}{8}$, and for all cases where the surfaces are oily we have, very nearly, $f = f'$.

This observation shows that, if it be necessary to take the results of experiments on the tension of the belt into calculation, the slight variations which it can show, in consequence of its hygrometric state, are without notable influence, and that it is sufficient to have determined the natural tension for each position at some time during the series of experiments, as I have done. As to the rest, one can assure himself directly that in admitting, in the natural tension, a variation $\frac{1}{6}$ to $\frac{1}{8}$, which much exceeds what observation has shown, the value of the ratio f given by the formula below * would not change $\frac{1}{30}$, which may be disregarded by comparison with the real difference which resistance offers.

The result of values found at different times for the natural tension of belts which we have employed for the calculation of the ratio of friction to the pressure.

From the 8th to the 17th of August, included, the formula

$$f = \frac{FL + 0,866}{(0,96Q + 53,10)r}.$$

From the 18th to the 28th of August, included, the formula

$$f = \frac{FL + 1,220}{(0,96Q + 94,63)r}.$$

From the 3d to the 7th of November, included, the formula

$$f = \frac{FL + 0,711}{(0,96Q + 35,00)r}.$$

From the 8th to the 12th of November, included, the formula

$$f = \frac{FL + 1,341}{(0,96Q + 108,77)r}.$$

---

* $Q =$ the total charge of the shaft, including its own weight, that of all the dynamometric apparatus of the pulley of the journals, and of the discs.
$p = 50\ k$, the weight of the pulley.
$F =$ the tension of the spring.
$L =$ its lever-arm, in relation to the axle of the shaft.
$R = 0^m,308$, the exterior radius of the pulley, including the half thickness of the belt.
$r =$ the radius of the journals.
$r' = 0^m,06$, the radius of the eye of the pulley.
$f =$ the ratio of the friction to the pressure for the journal and the box.
$f' =$ the same ratio for the iron eye of the pulley and the cast-iron shaft. We have found it equal 0,144.
$N =$ the pressure exercised by the journals of the shaft upon their bearings.
$N' =$ the pressure exercised by the pulley upon the shaft.

It is easy to see, from what we have said above, that between the tensions t and t', the tension F of the spring, and the friction of the pulley upon the shaft, one will have the relation

$$(t - t')\ R = FL + f'N'r';$$

and that between the friction of the journals upon their boxes, the tension F of the spring, and the friction of the eye of the pulley on the shaft, one will have

$$FL + f'N'r' = fNr;$$

whence we deduce

$$f = \frac{FL + f'N'r'}{Nr}.$$

Verifications of Two Theorems Employed to Prove the Preceding Formulæ.

166. By means of the formulæ which I have just given, it is easy to calculate the results of experiment, but they are based, as we have seen, upon two mechanical theorems, which, although based on principles exempt from all supposition of agreement with the manner in which belts ought to act, appear to me to need the proof of experiment in order that the deductions which I have made might be guarded against all uncertainty. This induced me to make special experiments upon the friction of belts upon wooden drums and cast-iron pulleys, and upon the manner in which their tension varies from wood to iron. I will give an account of them by commencing with the premises.

Experiments upon the Slipping of Belts upon Cast-Iron Pulleys.

167. Three wooden drums, with diameters of $0^m,836$, $0^m,408$, $0^m,100$, were successively employed for these experiments. They were placed horizontally in a fixed position in such a manner that they could not turn, and over them was passed a belt of black curried leather, almost new, but having acquired a certain softness by the use which we had made of it in the previous experiments. Its breadth was $0^m,050$, with a thickness of $0^m,0053$; its rigidity appeared so feeble, that we were justified in disregarding it, in its ratio to the friction on the surface of the drum. The two strips of the belt hung vertically in equal portions on each side of the drum, and to each one of them was attached the platform of a balance to receive the weights. The belt weighed $2^k,295$, each platform of the balance $0^k,229$, consequently the weight of each strip of the belt of equal length was $1^k,376$; the arc embraced was equal to the semi-circumference; we then placed equal weights into each platform, then gradually added to one of them weights enough to make the belt slip on the drum. We see, by this, that the tension t' of the ascending strip was equal to $1^k,376$, plus the weight contained in the corresponding platform, and that the tension t of the descending strip was equal to t' plus the weight added above the first load. Finally, the slipping of the belt took place perpendicularly to the fibres of the wood.

Order of the Following Tables.

168. These details are sufficient to give an idea of the mode of experiment adopted, and nothing remains but to add the table of results:

Table No. I.

Experiments upon the Friction of Belts upon Wooden Drums.

No. of Experiments.	Width of Belt.	Condition of Belt.	Diameter of Drum.	Length of Arc Embraced.	Tension of Ascending Strip.	Tension of Descending Strip or Motor.	Ratio of Friction to Pressure
	M.		M.	M.	Kilogram.	Kilogram.	
1	0,050		0,836	1,313	6,376	30,376	0,497
2	0,050		0,836	1,313	6,376	29,376	0,486
3	0,050	Dry, or	0,836	1,313	6,376	29,876	0,492
4	0,050	somewhat	0,836	1,313	16,376	75,876	0,488
5	0,050	unctuous.	0,836	1,313	16,376	69,526	0,460
6	0,050		0,836	1,313	16,376	68,676	0,458
7	0,050		0,836	1,313	11,376	50,376	0,473
8	0,050		0,836	1,313	11,376	43,376	0,426
						Mean....	0,472
9	0,050		0,408	0,640	6,376	26,876	0,472
10	0,050	Dry, or	0,408	0,640	6,376	31,376	0,458
11	0,050	somewhat	0,408	0,640	6,376	28,676	0,507
12	0,050	unctuous.	0,408	0,640	16,376	63,876	0,479
13	0,050		0,408	0,640	16,376	63,876	0,433
						Mean....	0,462
14	0,050		0,100	0,157	6,376	33,376	0,526
15	0,050		0,100	0,157	6,376	34,376	0,541
16	0,050	Dry,	0,100	0,157	11,376	41,376	0,411
17	0,050	somewhat	0,100	0,157	11,376	44,876	0,438
18	0,050	unctuous.	0,100	0,157	11,376	42,876	0,422
19	0,050		0,100	0,157	16,376	73,376	0,477
20	0,050		0,100	0,157	16,376	76,436	0,490
						Mean....	0,472
21	0,028		0,836	1,313	5,401	32,401	0,570
22	0,028		0,836	1,313	5,401	32,901	0,575
23	0,028	Very dry	0,836	1,313	10,401	51,901	0,512
24	0,028	and rough.	0,836	1,313	10,401	47,401	0,483
25	0,028		0,836	1,313	15,401	62,401	0,446
26	0,028		0,836	1,313	15,401	61,901	0,443
						Mean....	0,504

Observations on the Results contained in the Preceding Table.

169. In comparing with each other the values of the ratio f_t of the friction with the pressure of the belt upon the surface of the oak drum deduced from the formula

$$t = t'e^{\frac{f_t S}{R}} \quad or \; f_t = \frac{2{,}3026}{\frac{S}{R}} \; log. \; \frac{t}{t'},$$

in which $\frac{S}{R}$ expresses the semi-circumference of a circle equal to $\pi = 3{,}1416$, where the logarithms are those given in tables, which brings us back to the form

$$f_t = 0{,}733 \; log. \; \frac{t}{t'},$$

under which it has been employed in the calculations, we see that these values are sensibly constant, and that the particular mean deduced from each of the first three series is the same at $\frac{1}{18}$ nearly, although the extent of the arc embraced, or the diameter of the drum may vary in the proportion of 8 to 2 and to 1, about, and that the tensions may have nearly reached the limits usually given to belts in machines. These three series of experiments plainly confirm, then, the theory adopted and assigned to the ratio f_t of the friction to the pressure for new belts, but soft and even, slipping upon oak drums perpendicularly to the fibres of the wood, the mean value $f_t = 0{,}470$.

This value, deduced from twenty experiments, is much less than was concluded from the experiments of 1831, which gave us the value 0,74 as the ratio of friction of smooth surfaces of curried leather upon oak, the movement being parallel to the fibres of the wood, and after a prolonged contact; but the compressed state of the leather being entirely different in the two cases, it appears to me that this display of results is not surprising.

As to the fourth series of experiments contained in this table, and which relates to the friction of a belt entirely new and very rough, which for more than 8 years dried in a garret, they assign also to the ratio f_t a constant value but a little more than the preceding, which can, without doubt, be attributed to the condition of the rubbing surface of the leather. We will observe, besides, that this belt had only a breadth of $0^m{,}028$, or about half the size of the preceding. This last series confirms, as to belts, the law of the independence of surfaces.

170. Table No. II.

Experiments upon the Friction of Curried Leather Belts upon Cast-Iron Pulleys.

No. of Experiment.	Width of Belt.	Condition of Belt.	Diameter of Pulley.	Length of Arc Embraced.	TENSION		Ratio of Friction to Pressure.	OBSERVATIONS.
					Of Ascending Strip.	Of Descending Strip or Motor.		
	M.		M.	M.	K.	K.		
1	0,050	Dry or somewhat unctuous.	0,610	0,958	6,376	13,476	0,238	This belt was old, and had long been used in a spinning-mill.
2	0,050		0,610	0,958	6,376	16,776	0,308	
3	0,050		0,610	0,958	6,376	15,776	0,288	
4	0,050		0,610	0,958	16,376	29,276	0,301	
5	0,050		0,610	0,958	16,376	40,376	0,282	
6	0,050		0,610	0,958	16,376	37,376	0,262	
						Mean..	0,279	
7	0,050	Dry or somewhat unctuous.	0,610	0,958	6,376	16,000	0,300	This belt was new, and had been employed but a very short time in the experiments upon the friction of journals.
8	0,050		0,610	0,958	11,376	27,876	0,285	
9	0,050		0,610	0,958	11,376	25,876	0,271	
10	0,050		0,610	0,958	16,376	36,376	0,254	
11	0,050		0,610	0,958	16,376	36,376	0,254	
12	0,050		0,610	0,958	26,376	72,876	0,323	
						Mean..	0,281	
13	0,050	Dry or somewhat unctuous.	0,110	0,173	6,376	14,376	0,259	The pulley had been turned, and its breadth was only 0m,03, which reduced the rubbing surface of the belt to 0m,03.
14	0,050		0,110	0,173	6,376	18,376	0,336	
15	0,050		0,110	0,173	11,376	26,876	0,273	
16	0,050		0,110	0,173	11,376	30,876	0,318	
17	0,050		0,110	0,173	16,376	36,876	0,259	
18	0,050		0,110	0,173	16,376	36,876	0,259	
						Mean..	0,284	
19	0,050	Humid and much wet.	0,610	0,958	11,376	30,876	0,317	
20	0,050		0,610	0,958	6,376	19,876	0,361	
21	0,050		0,610	0,958	6,376	19,876	0,361	
22	0,050		0,610	0,958	16,376	51,876	0,366	
23	0,050		0,610	0,958	16,376	57,876	0,401	
24	0,050		0,610	0,958	21,376	90,376	0,458	
						Mean..	0,377	

Experiments upon Curried Leather Belts upon Cast-Iron Pulleys.

171. The experiments upon the friction of curried leather belts upon cast-iron pulleys were made in a like manner as the preceding. The pulleys employed were, first, that of the apparatus described in No. 1 and following, whose breadth of 0^m,10 was double that of the belt. Its surface was slightly convex, and had not been turned after being cast, but it was nearly a true circle, with a diameter of 0^m,610. Secondly, a little pulley, having a diameter of 0^m,110, and a breadth of 0^m,030, and consequently narrower than the belt, which, having a breadth of 0^m,050, projected 0^m,010 at one side or the other: its surface, turned and polished, was slightly convex.

The belt was then used dry, and in an unctuous state, as it was left by the tanner, with two dry pulleys, then entirely saturated with water, and the large pulley also wet. The other data of experiments and the arrangements of the table are shown the same as in preceding experiments, and it is superfluous to enter into further details.

Observations upon the Results contained in the Preceding Table.

172. The examination of the results stated in the preceding table completely confirm those of Table I., and the theory adopted. We see that, in effect, as well as in extent of the arcs embraced, the diameter of the pulleys may vary nearly in the ratio of 6 to 1, the breadth of the belt pressing upon the pulley in that of 2 to 1, and the tension in that of 1 to 3 on the one part, and of 1 to 6 on the other, the value of f_i remained sensibly constant and in the mean equal for the dry belt upon dry pulleys, let

$$f_i = 0{,}282.$$

When the belt is wet, as well as the pulley, the ratio increases and becomes in the mean

$$f_i = 0{,}377.$$

These are the two values which we have adopted at the beginning of this essay and following for the calculation of the tensions of the belt in our apparatus.

By recapitulating the results of these two series of experiments upon the friction of belts upon wooden drums, or upon cast-iron pulleys, one can see that we are justified in concluding therefrom that the ratio of this resistance to the pressure is:

1st. Independent of the breadth of the belt, and of the developed length of the arc embraced, or of the diameter of the drum, or what amounts to the same thing, independent of the surface of contact.

2d. Proportional to the angle subtended by the belt at the surface of the drum.

3d. Proportional to the logarithm, Naperian or hyperbolic, of the ratio of the tensions of the two strips of the belt.

Experiments upon the Variation of the Tension of Belts: Description of the Apparatus.

173. Let us now pass to the experiments which have had for their end the verification of the law of the variation of tension of belts, upon which are partly founded the formulæ employed in the calculation of the experiments upon the friction of journals (or axles).

To make them, I placed vertically above the axle of the waterwheel and of the pulley, mounted on its shaft, a cylindrical oak drum, with a diameter of $0^m,836$, and whose axis was about 3 mètres from that of the wheel. Around this drum, $d'\ d'$, and of the pulley, $d\ d$ (Fig. 69), I passed a belt; but, instead of its being one single piece, it was in two parts joined, near each end, by one of the dynamometers of the power of 200 kilogrammes, with a movable plate and style which are described in Article 179. These dynamometers were easily placed in such positions that that of the descending strip was near the upper drum, and that of the ascending strip near the lower drum, in such a manner that the belt could move over an extent of nearly 2 mètres without a risk of the dynamometers entangling themselves upon the drums.

A thread, rolled several times around the circumference of one of the grooves of the plate of each of the dynamometers, and fastened at the other end to a fixed point, compelled this plate to turn when the apparatus was in motion; and, if the displacement was only in proportion to the extension of the belt, we turned the plate with the hand to obtain a complete trace of the curve.

This trace was besides obtained, as told in the account given, by a style with a tube and capillary orifice incessantly pressed by a spring upon the sheet of paper which the plate carried.

The belt being passed over the two drums, the tension of the strips of the belt was varied, in either direction, by suspending from the exterior circumference of the upper drum a plate, p, which we loaded with weights. The natural or primitive tension was increased by bringing the ends of the belt near to each other, or by shortening it before the experiment.

Apparatus for Experiments upon the Variation of Tension of Belts.

Fig. 69.

The machinery being thus prepared for taking observations, before loading the plate, p, we traced the two curves or circles of curvature of each of the dynamometers, in order to find their tension, or that of the belt in repose, and to obtain by their sum the double of the natural tension. One can understand, besides, that, in consequence of the movement which had taken place in one way or another, and which had necessarily put in play the passive resistances of the forces, these two tensions could never be equal; but this is of little consequence, since we only want their sum. This done, we placed upon the plate a weight, which, being suspended from the circumference of the drum by a cord of a diameter equal to the thickness of the belt, had consequently the same lever-arm as the tensions. The strip of belt opposed to this weight tightened, and the strip placed on the same side relaxed, and we traced new lines of curvature of the dynamometers.

One could, besides, for a same natural tension, make a different series of experiments, including the motive weights, under the action of which the belt slipped on one or the other drum; and, as one also had the opportunity to let the two axles turn during a certain time, under the action of the developed tensions, one can see that the experiments would take in the three cases in practice, namely: that of the variation of tensions before the movement was produced, that of this variation during the movement, and finally that of the slipping.

It is not necessary to add that each one of the dynamometers had been tested* separately, and by gradually attaching equal weights to them, which determined the curvature whose trace we had preserved, we had obtained an exact scale of loss,† which served to estimate the corresponding tension at each curve of flexion traced in the experiments.

These preliminary observations, made and repeated with care, have shown that one of the dynamometers increases by flexion $0^m,00292$, and the other $0^m,00323$, for each kilogramme of increase in the weight which was attached to it. It was then easy, in each case, by comparing the diameters of the curves of flexion obtained with

---

* This word, in the original, is *taré*, which means "spoiled, injured, ruined;" but as neither of these would answer here, I have substituted "tested," as coming nearer the technical meaning. The word *tarer* is "to spoil, to injure," but means also to bring an article into equilibrium before weighing it; hence we may decide that the dynamometers were brought into equilibrium—were regulated or graduated before they were used.

† The words "an exact scale of loss," by good authorities, correspond to the original, which are *eschelle de tare exacte*, and, in a technical sense, may be strained to mean a measure of the error by which to make proper allowance on the record.

the scale of loss, to determine the tension of each strip of the belt with sufficient accuracy.

Arrangement of Table No. III.

174. This statement shows how easy these experiments were by the aid of this apparatus, and there remains nothing but to give the results, which are shown in the following table:

The 1st column shows the numbers of the order of experiment.

The 2d column shows the weights suspended from the circumference of the drum, including that of the plate.

The 3d the tension of the strained strip.
The 4th the tension of the slack strip.
⎫ Deduced from the comparison of the curve of flexion with the scale of loss.

The 5th shows the sum of the two tensions, or the double of the natural tension.

Observations upon the Results contained in the Tables.

175. In examining the results contained in Table III., we see that the first line of each series of experiments corresponds to the case where the additional weight p was zero, and where each strip took the tension corresponding to the distance of the axles from each other. These tensions are not equal on account of the inevitable action of passive resistances brought into play, but they differ very little [from each other] in other respects. In proportion as the weight suspended from the drum increases, the tension of one of the strips increases, and that of the other diminishes, but in such a manner that their sum remains constant, as the fifth column of the table shows.

These results, which completely confirm the theory established by M. Poncelet, besides relating to tensions whose sum rose as high as 90 kilogrammes and more, the highest of which showed as high as 77 kilogrammes, and the smallest diminished to 5 kilogrammes, comprise nearly all cases of practice, and show that this theory can, with surety, be applied to the calculation of all machines driven by belts.

We ought then to regard as demonstrated, both by theory and practice, that in the transmission of motion from one axle to another, by the aid of endless belts, that the sum of the tensions remains constant, be it at the moment of passing from rest to motion, or during motion, finally, be it at the moment when the belt slips upon one of the drums.

176. Table No. III.

Experiments upon the Variation of Tension of Endless Belts to Transmit the Motion to Axles of Rotation.

No. of Experiments.	Weight Suspended to Circumference of Drum Q.	Tension Ascending or Strained Strip. t.	Tension Descending or Slack Strip. t'.	Sum of Tensions. $t + t' = 2T$.	Observations.
	K.	K.	K.		
1	0,	17,49	14,89	32,38	
2	20,23	27,24	5,82	32,86	[drum.
3	27,23	28,63	4,62	33,25	The belt slips upon the
4	0,	39,41	26,03	55,44	
5	10,23	34,05	21,23	55,28	
6	20,23	38,82	14,44	55,26	The dynamometers have moved about a meter.
7	30,23	44,42	10,96	55,38	
8	44,23	49,84	9,41	59,25	
9	0,	33,43	28,26	61,69	
10	25,23	44,89	18,83	63,72	Idem.
11	50,23	53,40	9,24	62,64	
12	0,	30,34	26,27	56,71	[drum.
13	52,00	47,06	7,19	54,25	The belt slips upon the
14	0,	48,76	44,85	93,61	
15	25,23	58,97	31,86	90,83	
16	50,23	71,20	21,57	92,77	
17	0,	44,69	40,24	84,93	
18	50,23	69,96	18,49	88,45	Idem.
19	79,23	77,39	19,69	97,05	
20	0,	39,32	38,35	77,67	
21	40,23	61,14	20,03	81,17*	

* Besides the weight Q there was suspended from the mean circumference of the floats, or at a distance of 1ᵐ,85 from the axis of the wheel, a weight of 10ᵏ,229, which broke the equilibrium.

NOTE.—The belt employed in these experiments was very pliable, soft, and little susceptible to polish itself by slipping.

We have deduced these statements from experiments Nos. 3, 13, and 19.

$$f = 0{,}578, \qquad f = 0{,}596, \qquad f = 0{,}544,$$

whose mean $f = 0{,}539$ exceeds by about $\frac{1}{9}$ the value deduced from direct experiment upon the new belt employed in the experiments upon the friction of journals, and $\frac{1}{12.5}$ of that which was deduced from the experiments upon a very rough new belt.

Tension that can, with Security, be Applied to Belts.

177. I will add that the examination of the condition of belts, after these experiments, and after those relating to their slipping on the surface of the drum, has shown that they had undergone no apparent change, although the small belt, with a breadth of $0^m,028$ and a thickness of $0^m,005$, might have supported a strain of 62 kilogrammes, or $2^k,25$ per square millimètre of a section, and that the belt employed in the last experiment, which was very old and very much worn on the edges, with a breadth of $0^m,050$ and a thickness of $0^m,004$, was for some time subjected to a strain of 97 kilogrammes or of $2^k,06$ per square millimètre. We see, then, that in practice belts can be subjected to a strain calculated at the rate of 2 kilogrammes per square millimètre* of their section, and that their thickness being besides usually limited from $0^m,004$ to $0^m,008$ when they are no longer doubled, one can determine the dimension that it would be proper to give them, and beyond which there would be no advantage in increasing their width out of proportion, as is sometimes done.

Experiments upon the Variation of the Tension of Endless Cords or Belts used in Transmitting Motion. — From "Morin's Mechanics," D. Appleton & Co., N. Y., 1860.

178. "We pass now to an experimental proof of the theory given by M. Poncelet, upon the transmission of motion by endless cords or belts, and will first give a description of its nature.

"When a cord or belt surrounds two pulleys or drums, between which it is designed to maintain a conjoint motion, care is taken to give it a sufficient tension, which is usually determined by trial, but which it would be best to calculate, as we shall see hereafter. The *primitive* tension is, at the commencement, the same for both parts of the belt, and this equality, established in repose, is only destroyed

---

* The practical rule given by Coulomb, fixes the limit of the strain that can, with security, be given to a thread of rope-yarn at 40 kilogrammes, which comes very near to $3^k,70$ per square millimètre.

by the friction of the axles, which may act in either direction, according to that of the motion of the pulleys. Let us examine how this motion is transmitted in such a system. Let C (Fig. 70) be the motive drum; C' the driven drum; T_1 the primitive tension common to the parts A A' and B B' of the belt, from the moment when the drum C begins to turn until it commences to turn the drum C'.

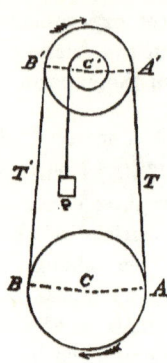

Fig. 70.

"The point A of primitive contact of the part A A' advances, in separating from the point A', in the direction of the arrow; the strip A A' is stretched, and its tension increased by a quantity proportional to this elongation, according to a general law proved by experiment upon traction. At the same time the point B of contact of the part B B' approaches, by the same quantity, towards the point B', so that the portion B B' is diminished by a quantity equal to the increase of that of A A'. If, then, we call T the tension of the driving portion, A A', at the instant of its being put in motion; T' the tension of the driven part, B B'; t the quantity by which the primitive tension, T_1 is increased in the portion A A', and diminished in the part B B', we shall have

$$T = T_t + t, \text{ and } T' = T_t - t,$$

and, consequently,

$$T + T' = 2 T_t.$$

"Then, at any instant, the sum of the two tensions, T and T', is constant, and double the primitive tension.

"Now, it is evident that, in respect to the driven drum, C', the motive power is the tension, T, and that the tension, T', acts as a resistance with the same lever arm, so that the motion is only produced and maintained by the excess, $T - T'$, of the first over the second of these tensions.

"If the machine is, for example, designed to raise a weight, Q, acting at the circumference of an axle with a radius, R', it is easy to see, according to the theory of moments, that, at any instant of a uniform motion of the machine, we must have the relation

$$(T - T') R = Q R' + f N r,$$

N being the pressure upon the journals and r their radius.

"The pressure is easily determined; for, calling a the angle formed by the directions A A' and B B' of the belts with the line of the centres, C C', M the weight of the drum, we see immediately that

$$N = \sqrt{[M + Q + (T - T')\sin. a]^2 + (T + T')^2 \cos.^2 a},$$

an expression which, according to the algebraic theorem of M. Poncelet cited in No. 227,* has for its value to $\tfrac{1}{25}$ nearly; when the first term under the radical is greater than the second,

$$N = 0.96 \, [M + Q + (T - T')\sin. a] + 0.4 \, (T + T')\cos. a.$$

"This value of N being introduced into the formula for equality of moments, we have a relation containing only the values of the resistance, Q, and of the tensions. But as it may be somewhat complicated for application—observing that in most cases the influence of the tensions T and T' upon the frictions will be so small that it may be neglected, at least in a first approximation—we proceed as follows:

"First, neglecting the influence of the tensions upon the friction, we have simply, in the actual case,

$$N = M + Q,$$

and, consequently,

$$(T - T') R = Q R' + f (M + Q) r;$$

whence we deduce

$$T - T' = \frac{Q(R' + fr) + f.Mr}{R} = Q,$$

which furnishes a first value for the difference of tensions, which is the motive power of the apparatus.

"But this is not sufficient to make known these tensions, and it is necessary to determine the primitive tension, T', so that in no case the belt may slip.

"According to the theory of M. Prony, we have, at the instant of slipping, between the tension T and T', the relation

$$T = T' \times 2{,}718 + f\frac{S}{R} T' = K T',$$

the number K being the quantity depending upon the nature and condition of the surfaces of contact as well as upon the angle $\frac{S}{R}$ embraced by the belts upon the drum, C'. These quantities are known, and

---

* See "Morin's Mechanics," p. 266, D. Appleton & Co., 1860.

we may in each case calculate the value of K by this formula, or take it from the table on p. 122, which answers to nearly all the cases in practice.

"By means of this table we have, then, the value of $T = KT'$, and consequently

$$T - T' = (K - 1) T' = Q,$$

Q representing the greatest value which the difference of tensions should attain, to overcome the useful and passive resistances.

"From this relation we may derive the smallest tension to be allowed to the driven portion of the belt, to prevent its slipping.

"We thus have

$$T = \frac{Q}{K-1}.$$

"We should increase this value by $\frac{1}{10}$ at least, to free it from all hazard of accidental circumstances, and to restore the account of the influence of the tensions upon the friction, which was neglected.

"This established, we have

$$T = Q + \frac{Q}{K-1},$$

and consequently

$$T_i = \frac{T+T'}{2} = \frac{1}{2}\frac{K+1}{K-1} Q.$$

"All the circumstances of the transmission of motion will then be determined.

"If these first values of T, T' and T_1 are not considered as sufficiently correct, we may obtain a nearer approximation by introducing them in the value of the pressure N, and thus deduce a more exact value of Q, which will serve to calculate anew T', then T and T_1."

Dynamometer Attachments.

179. "Around this drum and the pulley was passed a belt, which, instead of being in one piece, was in two parts, joined at each end by a dynamometer with a plate and style, of a force of 441 lbs. Moreover, these dynamometers were easily secured in positions, such, that that of the descending portion of the belt was near the upper drum, and that of the ascending near the lower drum. Thus the belt could be moved over a space of 6,56 feet without the risk of the instruments being involved with the drums.

"A thread, wound several times around the circumference of one of the grooves of the plate of each of the dynamometers, and attached by the other end to a fixed point, caused the plate to turn when the apparatus was in motion, and the paper with which the plate was covered received thus the trace of the style of the dynamometer.

"The belt being passed over the two drums, the tensions of the parts were varied at will, in either direction, by suspending at the circumference of the upper drum a plate, p, charged with weights. As to the primitive tension, it was increased by bringing nearer together the ends of the belt, or in diminishing its length before the experiment.

"The apparatus being thus prepared for observations before loading the plate, p, we traced the circles of flexion of each of the dynamometers, so as to have the tensions of the belt at rest, and to obtain by their sum the double of the primitive tension, T_1. We may conceive that these two tensions can never be quite equal; but that is not important, inasmuch as we have to deal only with their sum.

"This obtained, we load the plate with a weight which, being suspended upon the circumference by a cord of a diameter equal to the thickness of the belt, has the same lever-arm as the tensions. That part of the belt opposed to this weight is stretched, and the part on the same side is slackened, and we trace the new curves of the flexion of the dynamometers.

"For this same primitive tension we may make a series of experiments up to the motive weight, under the action of which the belt slides upon either drum."

CHAPTER VIII.

ROPE TRANSMISSION.

The Transmission of Power by Wire Ropes, by W. A. Roebling, C. E., Trenton, N. J.

180. "*The use of a round endless wire rope running at a great velocity in a grooved sheave, in place of a flat belt running on a flat-faced pulley, constitutes the transmission of power by wire ropes.*

"The distance to which this can be applied ranges from 50 or 60 feet up to about 3 miles. It commences at the point where a belt becomes too long to be used profitably, and can thence be extended almost indefinitely.

"In point of economy it costs only $\frac{1}{15}$ of an equivalent amount of belting, and the $\frac{1}{25}$ of shafting. . . .

"The range in the size of wire ropes is small, varying only from $\frac{3}{8}$ to $\frac{3}{4}$ inch diameter in a range of 3 to 250 horse-power.

"In regard to cost, the ropes are the cheapest part of a transmission. For instance, a No. 22 rope — conveying, say 25 horse-power — costs 8 cents per foot, whereas an equivalent belt costs about $1.40 per foot. . . . Their duration is from $2\frac{1}{2}$ to 5 years, according to the speed.

"For the smaller powers it is advisable to take a size larger, for the sake of getting wear out of the rope; although it must be borne in mind that a larger rope is always stiffer than a small one, and therefore additional power is lost in bending it around the sheave. An illustration of that is seen in the case of the 14-feet wheel in the table (page 255), where a $\frac{7}{8}$ rope gives less power than a $\frac{3}{4}$ rope, simply because it is so much stiffer.

"Ropes for this purpose are always made with a hemp core, to increase their pliability.

"It is often required to convey the entire power of a certain shaft which is driven by a belt of a given size. In such a case a simple rule, agreeing with the average result of practice, is, that 70 square feet of belt surface are equal to one horse-power.

"Take, for example, a belt one foot wide running at the rate of 1400 Fpm; then the horse-power $= \dfrac{1400' + 1'}{70'} = 20$; and by referring to the table we find the diameter of the wheel corresponding to this horse-power, and making the same number of revolutions that the belt-pulley does.

Distance of Transmission.

"The wire rope transmission table is arranged for distances ranging from 80 up to 350 or 400 feet in one stretch. For a single stretch extended to, say 450 feet, where no opportunity is presented for putting in an intermediate station, we must use a rope one size heavier; and in a case where there is not sufficient head-room to allow the rope its proper sag, and it has to be stretched tighter in consequence, we must also take a rope one size heavier.

Short Transmission.

"Whenever the distance is less than 80 feet, the rope has to be stretched *very tight*, and we no longer depend upon the sag to give it the requisite amount of tension. Here we must take a rope two sizes heavier than is given in the table, and run at the maximum speed indicated. It is also preferable to substitute in place of the rope of 49 wires a fine rope of 133 wires of the same diameter, which possesses double the flexibility, runs smoother, and lasts longer. In fact, the substitution of a fine rope for a coarse one can be done with advantage in every case in the table where the size admits of it.

Splices.

"Both kinds of rope are spliced with equal facility. The splices are all of the kind known as the long-splice; the rope is not weakened thereby, neither is its size increased any, and only a well-practised eye can detect the locality of one.

Relative Height of Wheels.

"It is not necessary that the two wheels should be at the same level, one may be higher or lower than the other without detriment; and unless this change of level is carried to excess, there need be no change in the size of wheel or speed of rope: the rope may have to be strained a little tighter. As the inclination from one wheel to another approaches an angle of 45°, a different arrangement must be made, as will be shown hereafter.

Table of Transmission of Power by Wire Ropes.

Diam. of Wheel.	No. of Revs.	Trade No. of Rope.	Diam. of Rope.	Horse-Power.	Diam. of Wheel.	No. of Revs.	Trade No. of Rope.	Diam. of Rope.	Horse-Power.
Feet. 4	80	23	$\frac{3}{8}$	3.3	Feet. 9	140	20 19	$\frac{1}{2}$ $\frac{5}{8}$	70. 72.6
4	100	23	$\frac{3}{8}$	4.1	10	80	19 18	$\frac{5}{8}$ $\frac{11}{16}$	55. 58.4
4	120	23	$\frac{3}{8}$	5.	10	100	19 18	$\frac{5}{8}$ $\frac{11}{16}$	68.7 73.
4	140	23	$\frac{3}{8}$	5.8	10	120	19 18	$\frac{5}{8}$ $\frac{11}{16}$	82.5 87.6
5	80	22	$\frac{7}{16}$	6.9	10	140	19 18	$\frac{5}{8}$ $\frac{11}{16}$	96.2 102.2
5	100	22	$\frac{7}{16}$	8.6	11	80	19 18	$\frac{5}{8}$ $\frac{11}{16}$	64.9 75.5
5	120	22	$\frac{7}{16}$	10.3	11	100	19 18	$\frac{5}{8}$ $\frac{11}{16}$	81.1 94.4
5	140	22	$\frac{7}{16}$	12.1	11	120	19 18	$\frac{5}{8}$ $\frac{11}{16}$	97.3 113.3
6	80	21	$\frac{15}{32}$	10.7	11	140	19 18	$\frac{5}{8}$ $\frac{11}{16}$	113.6 132.1
6	100	21	$\frac{15}{32}$	13.4	12	80	18 17	$\frac{11}{16}$ $\frac{3}{4}$	93.4 99.3
6	120	21	$\frac{15}{32}$	16.1	12	100	18 17	$\frac{11}{16}$ $\frac{3}{4}$	116.7 124.1
6	140	21	$\frac{15}{32}$	18.7	12	120	18 17	$\frac{11}{16}$ $\frac{3}{4}$	140.1 148.9
7	80	20	$\frac{1}{2}$	16.9	12	140	18 17	$\frac{11}{16}$ $\frac{3}{4}$	163.5 173.7
7	100	20	$\frac{1}{2}$	21.1	13	80	18 17	$\frac{11}{16}$ $\frac{3}{4}$	112. 122.6
7	120	20	$\frac{1}{2}$	25.3	13	100	18 17	$\frac{11}{16}$ $\frac{3}{4}$	140. 153.2
7	140	20	$\frac{1}{2}$	29.6	13	120	18 17	$\frac{11}{16}$ $\frac{3}{4}$	168. 183.9
8	80	19	$\frac{5}{8}$	22.	14	80	17 16	$\frac{3}{4}$ $\frac{7}{8}$	148. 141.
8	100	19	$\frac{5}{8}$	27.5	14	100	17 16	$\frac{3}{4}$ $\frac{7}{8}$	185. 176.
8	120	19	$\frac{5}{8}$	33.	14	120	17 16	$\frac{3}{4}$ $\frac{7}{8}$	222. 211.
8	140	19	$\frac{5}{8}$	38.5	15	80	17 16	$\frac{3}{4}$ $\frac{7}{8}$	217. 217.
9	80	20 19	$\frac{1}{2}$ $\frac{5}{8}$	40. 41.5	15	100	17 16	$\frac{3}{4}$ $\frac{7}{8}$	259. 259.
9	100	20 19	$\frac{1}{2}$ $\frac{5}{8}$	50. 51.9	15	120	17 16	$\frac{3}{4}$ $\frac{7}{8}$	300. 300.
9	120	20 19	$\frac{1}{2}$ $\frac{5}{8}$	60. 62.2					

Deflection or Sag of the Ropes.

"In the following illustration the upper rope is the pulling rope and the lower one the loose following rope. When the rope is working, the tension, T, in the upper rope, is just double that in the lower rope, hence the latter will sag much lower below a horizontal line than the upper one.

"When the rope is at rest, both ropes will occupy the position indicated by the dotted line, and will have a uniform tension.

"The best way in practice is to hang up a wire in the position the rope is to occupy at rest; that has to be done in any case, in order

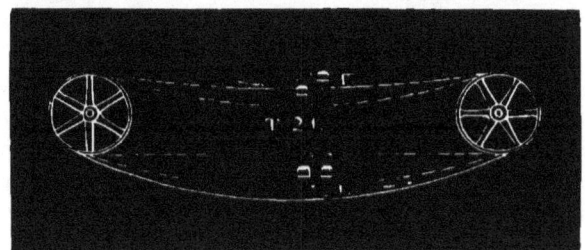

Fig. 71.

to get the length of rope needed. Then hang it so that the deflection, D', below the horizontal line, is about $\frac{1}{35}$ of the whole distance from wheel to wheel. The deflection, D, of the upper running rope will then be about $\frac{1}{45}$ to $\frac{1}{50}$.

"The deflection, D", of the lower working rope is on an average one-half greater than the deflection D' of the rope at rest. This is of importance, as we should know beforehand whether the lower rope is going to scrape on the ground or touch other obstructions; in that case we either have to dig a trench for the lower rope to run in, or else raise both wheels high enough to clear.

"Practically, however, it is not necessary to be so particular about this matter, on account of the stretch in the rope. Wire-rope stretches, comparatively, very little; still, there is some stretch, and it is well to allow for it by stretching the rope a little too tight at first. After running a little it will hang all right. When the rope is very long, it is advisable to take up the stretch at the end of two or three months, as a slack rope does not run so steadily as a tight one.

"Whenever the direction of the motion of the driving wheel is not fixed by other circumstances, it is often advisable to make the lower rope the pulling rope, and the upper the follower, as here shown. In this way obstructions can be avoided, which, by the

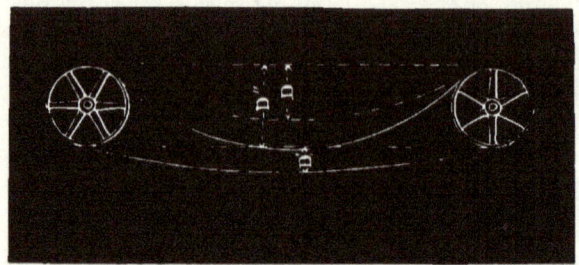

Fig. 72.

other plan, would have to be removed. The ropes will not interfere as long as the difference between the two deflections, D' and D'', is less than the diameter of the wheel.

"These limits are of use whenever, on account of rocks or otherwise, we have to move the wheels closer together, and the question is how far to have them apart with a certain deflection.

The Wheels.

"The bottom of the groove in the wheels is made a little wider, to prevent the filling from flying out. The rope should always run on a cushion of some kind, and not on the iron, which quickly wears it out.

"A variety of material is used for this filling — soft wood, India-rubber, leather, old rope tarred, and oakum.

"To use end-wood the rim has to be constructed on a different plan from that shown here. The objections to it are that it is liable to shrink, and crack and fly out; it is also more severe on the rope. India-rubber is a very good material; strips of an inch square or less can be wedged in very quickly, and will last a long time; the price, however, is 50 to 60 cents per pound, which is rather against it.

"My own practice has been to use leather, also rope and oakum. The leather is cut in sections of the shape shown in Fig. 73, and set in on end around the rim; scraps can be used, cut from old shoes, pieces of leather belting, etc.; they are very thin, and it takes at least a thousand for a 7-feet wheel. When many are wanted, it is worth while to make a die, to cut them out fast. This is the most durable filling that can be made.

"Again, by wedging the groove full of tarred oakum a filling is also obtained, nearly as good as leather, costing less, and not so tedious to put in. Another plan, which I have tried with success, is to revolve the wheels slowly, and let a lot of small-sized tarred ratlin, or jute-yarns, wind up on themselves in the groove; then secure the end, and, after a day or two of running, the pressure of the rope, together with the tar, will have made the filling compact. This makes a cheap filling.

"The double-grooved wheels are filled in the same way.

"The rope will run on such filling without making any noise whatever, and soon wears in a round groove for itself.

"A section of the rim of a 6-feet wheel is here shown with the dimensions marked.

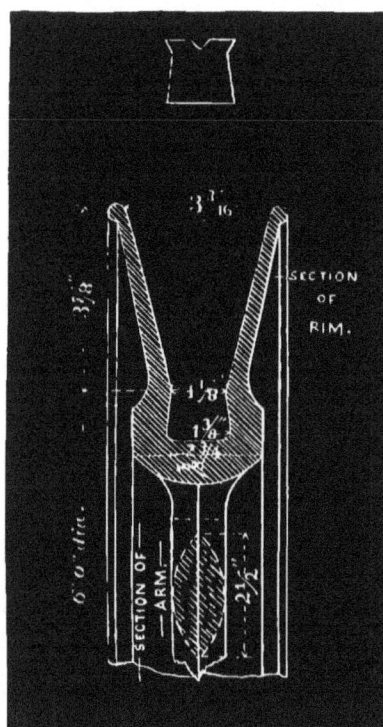

Fig. 73.

"The diameter of the wheel is not reckoned from the outside of the rim, but from the top of the filling, which corresponds to the circle described by the rope. The hub is made of ample size, so as to admit of being bored out for shafts, ranging from 2 to 3½ inches.

. . . "The rope, while running, requires no protection from the weather. If it has to stand still much, pour some hot coal-tar from a can on the rope in the groove of the wheel while running.

"Whenever there is no room for the sag of the rope, and it is inconvenient to raise the wheels higher, or a ditch cannot be dug, it may be supported by a roller in the middle. This supporting roller must be in the centre of the span, and must be at least half the size of the larger wheels." . . .

"Transmissions are in operation a mile in length. The loss of power from friction, etc., or bending of rope, does not amount to 10 per cent. per mile, and need not be taken into account at all for only one station. No slipping of the rope in the groove ever occurs with a proper filling.

"With bearings of a sufficient length under the shaft of the centre wheel, and, by providing them with a self-feeding oil-cup, the axle friction is reduced to a minimum. . . ."

We insert here the general wire-rope table in full, to which reference is made in the preceding notes.

Table of Wire Rope, Manufactured by J. A. Roebling's Sons, Trenton, N. J.

	Rope of 133 Wires.					Rope of 49 Wires.		
Trade Number.	Circumference in Inches.	Diameter.	Ultimate Strength in Tons of 2000 Lbs.	Circumference of Hemp Rope of Equivalent Strength in Inches.	Trade Number.	Circumference in Inches.	Ultimate Strength in Tons of 2000 Lbs.	Circumference of Hemp Rope of Equivalent Strength in Inches.
1	$6\frac{3}{4}$	$2\frac{1}{4}$	74 00	$15\frac{1}{2}$	11	$4\frac{5}{8}$	36 00	$10\frac{1}{4}$
2	6	2	65 00	$14\frac{1}{2}$	12	$4\frac{1}{4}$	30 00	10
3	$5\frac{1}{2}$	$1\frac{3}{4}$	54 00	13	13	$3\frac{3}{4}$	25 00	$9\frac{1}{4}$
4	5	$1\frac{5}{8}$	43 60	12	14	$3\frac{3}{8}$	20 00	$8\frac{1}{4}$
5	$4\frac{3}{8}$	$1\frac{1}{2}$	35 00	$10\frac{3}{4}$	15	3	16 00	$7\frac{1}{2}$
6	4	$1\frac{1}{4}$	27 20	$9\frac{3}{4}$	16	$2\frac{5}{8}$	12 30	$6\frac{1}{4}$
7	$3\frac{1}{2}$	$1\frac{1}{8}$	20 20	8	17	$2\frac{3}{8}$	8 80	$5\frac{1}{2}$
8	$3\frac{1}{8}$	1	16 00	7	18	$2\frac{1}{8}$	7 60	5
9	$2\frac{3}{4}$	$\frac{7}{8}$	11 40	6	19	$1\frac{7}{8}$	5 80	$4\frac{3}{4}$
10	$2\frac{1}{4}$	$\frac{3}{4}$	8 64	5	20	$1\frac{5}{8}$	4 09	4
$10\frac{1}{4}$	2	$\frac{5}{8}$	5 13	$4\frac{1}{2}$	21	$1\frac{3}{8}$	2 83	$3\frac{1}{4}$
$10\frac{1}{2}$	$1\frac{5}{8}$	$\frac{9}{16}$	4 27	4	22	$1\frac{1}{4}$	2 13	$2\frac{3}{4}$
$10\frac{3}{4}$	$1\frac{1}{2}$	$\frac{1}{2}$	3 48	$3\frac{3}{4}$	23	$1\frac{1}{8}$	1 65	$2\frac{1}{2}$
					24	1	1 38	$2\frac{1}{4}$
					25	$\frac{7}{8}$	1 03	2
Ropes from No. 8 to No. $10\frac{3}{4}$ are specially adapted for hoisting-rope.					26	$\frac{3}{4}$	0 81	$1\frac{3}{4}$
					27	$\frac{5}{8}$	0 56	$1\frac{1}{2}$
					$27\frac{1}{2}$	$\frac{1}{2}$		
Copper and steel ropes corresponding to the above sizes are also made.					28	......	Large Sash Cord.	
					29	......	Small	"

"When the power is to be conveyed nearly vertically, no good result is obtained by running the rope, say from A to B, direct, as indicated by dotted lines in the figure below, since it would slip.

"Two carrying sheaves, C and D, must be put up vertically above A, giving a horizontal stretch from C and D to B. This is necessary, in order to maintain the required tension in the rope, which can be obtained in no other way. A, B, and C, and even D, should be of the same size; yet D, which supports the following rope, may be made smaller without damage.

Fig. 74.

"This arrangement must be borne in mind whenever the source of power is located in the cellar, and we want to carry it to an upper story and distribute it horizontally."

Transmission of Motive Forces to a Great Distance.
By A. Achard, C. E., of Geneva, Switzerland.
Published by Dunod, Paris, 1876.

Translated by J. WM. HÜTTINGER, Member of the Franklin Institute, Philada.

181. The transmission by cables is only an extension of the transmission by belts which are universally employed in the interior of shops. Before touching upon the peculiar properties of cables, it is proper to mention what may be said in general about *funicular transmission*.

Funicular transmission consists of 2 shafts, one of which is required to transmit to the other its rotary movement, by 2 mounted pulleys, one on one shaft, the other on the other, lastly, the funicular organ, either an endless belt or an endless cable, which we will call *cord*, for the sake of brevity, without prejudging its nature. We will look only at the case where the 2 shafts are parallel, and where the mean plan of the 2 pulleys coincides, for, as we shall see further on, this is the only thing which interests us on the subject of cables.

At any instant whatever, one can conceive the cord as divided into 4 parts: the length applied to the conducting pulley, the length applied to the conducted pulley, lastly, the 2 lengths free. The latter are called the *strips* of the cord.

The active organ of transmission is the strip which enrolls itself upon the conducting pulley, and unrolls itself from the pulley moved, in other words, which receives from the first a traction which it transmits to the second. For this reason it is called the *conducting strip;* for the same reason the other is called the *conducted strip*.

Being thus, let A, Fig. 75, be the motor shaft, with its pulley of a radius R, and B the shaft moved with its pulley of a radius R'. A power P acts upon the first tangentially to a circle of a radius r, and a resistance Q acts upon the second tangentially to a circle of a radius r'. The arrows, which give the idea of rotation, show that it is the upper cord that is the conductor or motor.

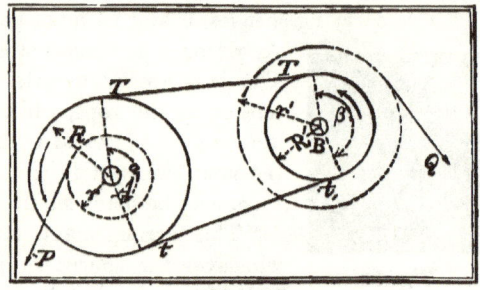

Fig. 75.

In order that transmission may take place, it is necessary that the pulley, A, draw the cord, and that the latter, in its turn, draw the pulley, B. This cannot take place but by reason of the adherence of the cord to the felly, or rim, of the pulleys, and this adherence depends upon 2 elements—the roughness of the parts in contact and the tension of the cord.

Call T the tension of the conducting strip, and t that of the conducted strip. We can imagine the 2 strips cut, and the tension replaced by tractions, respectively equal to T and to t, and acting in regard to each pulley as the tensions do. In this way the conditions of uniform movement can be regarded separately for each pulley. They are, first throwing aside passive resistances:

For the conducting pulley, regarding t as acting in comparison as power, and T as resistance:

$$Pr + tR - TR = o,$$

from which $T - t = P\dfrac{r}{R}$; (1)

For the pulley moved, observing that in respect to the latter, that T is the power, while t acts as the resistance:

$$TR' - tR' - Qr' = o,$$

whence $T - t = Q\dfrac{r'}{R'}$. (2)

The equations (1) and (2) show that the quantities $P\dfrac{r}{R}$ and $Q\dfrac{r'}{R'}$, necessarily equal between themselves, omitting passive resistances, are besides equal to $T-t$. But this is not sufficient to determine T and t; the condition of adherence will supply this.

For which let the force, P, cause the pulley, A, to slip under the cord without moving the latter, it will be necessary that this force agree with the surface of the pulley, that is to say $P\frac{r}{R}$, let it be at least equal to the friction which there may be between the cord and the surface of the pulley, if this slipping had taken place. This friction has for its measure the minimum error which ought to exist between the tension T and t of the 2 strips, for which the strip of tension T draws the other by slipping upon the immovable pulley; for the friction depends only upon the relative movement, which is the same in the two cases. We know, by a known theory, that this error ought to be such that we can have $\frac{T}{t} = e^{fa}$ ($e =$ the base of the natural logarithms, $f =$ the coefficient of friction relative to the two substances in contact, $a =$ the value, ascribed to the radius of the arc of the pulley embraced by the cord). From this equation one can deduce the error $T-t$ in function, be it of t or of T at will. By taking t, we find $T-t = t(e^{fa}-1)$. Consequently, t being provisionally indeterminate, we will find that the pulley, A, will draw the cord if $P\frac{r}{R} < t(e^{fa}-1)$, and will not draw it if $P\frac{r}{R} > t(e^{fa}-1)$.

The same holds good as far as it concerns the pulley moved, the cord will draw it if $Q\frac{r'}{R'} < t(e^{f\beta}-1)$, and will not draw it if $Q\frac{r'}{R'} > t(e^{f\beta}-1)$, ($\beta$ being the arc embraced upon this pulley, and f can take another value, according to the nature of the pulley). It is required then, in order that the strictly necessary value of t be obtained, that the quantity $P\frac{r}{R}$, or its equal $Q\frac{r'}{R'}$, be equal to the smaller of the two quantities $t(e^{fa}-1)$ and $t(e^{f\beta}-1)$. This value is obtained then, by dividing $P\frac{r}{R}$ or $Q\frac{r'}{R'}$ by the smaller of the two quantities $(e^{fa}-1)$ and $(e^{f\beta}-1)$. Having once determined t, T is immediately deduced from it, since $T-t$ is given by (1) and (2).

Algebraically speaking, the solution of the problem comes back to one of the equations (1), and (2) the other equation,

$$\frac{T}{t} = k, \qquad (3)$$

or k is the smaller of the two quantities e^{fa} and $e^{f\beta}$.

It is easy to take account as proof *a posteriori* of the preceding reasoning, that the tensions determined by this, which requires the pulley of least adherence will be, for the strongest reason, sufficient for the pulley of greatest adherence, whatever may be that of the two which is conductor and which is conducted.

In practice, in view of the irregularities of the power to be transmitted, it is advisable to increase the tensions; and in order to keep account of this necessity, we must give to k a co-efficient μ, which is < 1, and which is as much less as the irregularities are great. We will replace, then, the equation (3) by

$$\frac{T}{t} = \mu k. \tag{3a.}$$

After having thrown passive resistances out of account to simplify the reasonings, it is necessary to bring them into the question. They are 2 in number, one of which comes in whatever may be the mode of communicating the movement from one shaft to another, while the second is peculiar to the funicular organ. The first consists in the friction of the bearings. Each shaft is sustained by two supports or rests, holding the bearings which guide its movement of rotation. If we call f' the coefficient of friction, in reference to this contact, the friction of a support will be the product of $\sqrt{\frac{f'}{1+f'^2}}$, by the pressure which the shaft exercises upon it. The quantity $\sqrt{\frac{f'}{1+f'^2}}$, which we design to abridge by f_1, sometimes called *the coefficient of friction of the bearings*. The sum of the frictions relative to the 2 supports will be the product of f_1 by the sum of the pressures which the shaft exercises respectively upon them — a sum which is equal to the resultant F of all the exterior forces applied to the shaft perpendicularly to its direction, and regarded as carried parallel to themselves to a single point of application.

The expression of the friction concerning a shaft will be, then, $f_1 F$, and the momentum of this friction, with reference to the axle of the shaft, will be $f_1 \rho F$, by calling ρ the diameter of the shaft (or of its journal, if there be one).

The second resistance is called *the roughness of the cord*, and consists in this: that, for want of sufficient flexibility of the cable, the strip through which the resistance acts, and which enrolls itself upon the pulley, deviates a little from the direction of the latter, in such a manner as to increase its lever-arm, and consequently its movement.

The same thing does not take place in the other strip, whose unrolling is, on the contrary, favored by the elasticity of the cord.

Among the empiric formulæ proposed to express the roughness, we will choose that which represents the addition of the lever-arm, which it creates by $s\Delta^2$, Δ being the diameter of the cord (or thickness of the belt), and s a coefficient depending upon the material of which it is made.

By introducing these, the equations of uniform movement become:
For the conducting or motor pulley,

$$Pr + tR - T(R + s\Delta^2) - f_i\rho F = 0; \qquad (4.)$$

For the pulley moved,

$$Tr' - t(R' + s\Delta^2) - Qr' - f_i\rho' F' = 0. \qquad (5.)$$

As here we have no longer *a priori* $P\dfrac{r}{R} = Q\dfrac{r'}{R'}$, there are, in reality, 3 unknown quantities, T, t, and P, or rather T, t, and Q, following which we get either Q or P. 3 distinct equations then are necessary. We shall find them by uniting the equation (3 a), which expresses the condition relative to the tensions, to the 2 equations (4) and (5), in which F is a function of P of T of t, and of the weights of the pulley A and of its shaft, and F' a function of Q, of T, of t, and of the weights of the pulley B and of its shaft. The solution of the problem is then possible, *theoretically speaking*. It is for belts only that one can consider the 2 strips as each one being rectilinear and the tension uniform. We shall see that for cables it is different. With the tensions T and t of the 2 strips [of the belt] in motion, there corresponds a tension θ, common to the 2 strips at rest. This tension at rest θ once realized, the putting in motion — in other words, the putting in play of the power and of the resistance — creates of itself the inequality of tension required, and consequently realizes the values T and t.

Such is, briefly, a general *resumé* of the theory of funicular transmission. The following reason will show how we were led to obtain this kind of transmission by metallic cables.

The size to be given to the cable is equal to the quotient obtained by dividing T, the maximum of tension which it has to undergo, by the number which represents the tension compared to unity of a section which the material permits. But the constructor makes the size of T only to a certain limit. In effect, omitting the passive resistances which would uselessly complicate the question, we find, by combining the equations (1) and (3 a),

$$T = \frac{\mu k}{\mu k - 1} \cdot P \frac{r}{R}.$$

Let v be the speed of the cord, and v' that of the point of application of P, we have $\frac{r}{R} = \frac{v'}{v}$. In addition, let P represent the direct action of the first motor (as, for example, if the latter were a hydraulic wheel, mounted upon its shaft A), let it represent an action transmitted by the latter by means of gearing, or otherwise, we have, by calling N the force in horse-power of the first motor, $P = \frac{75\,N}{v'}$.

Then by substituting

$$T = \frac{\mu k - 1}{\mu k} \cdot \frac{75\,N}{v}. \qquad (6.)$$

Thus the power to be transmitted being given, the tension of the cable will be in inverse ratio to its speed. By increasing the speed, the constructor may make the section smaller. By referring to equation (6) it will be easy to take account of the limit of mechanical powers which can be transmitted by a given cable. For this we must, to begin with, take $\frac{\mu k}{\mu k - 1}$ as the value of those cases which require the least tension. Now we know that $k = e^{f a}$, a being the smaller of the two arcs of enrolment. These are the greatest values of f and of a which are to be regarded here. The smaller of the two arcs of enrolment cannot exceed a semi-circumference, and by taking account of the deviation caused by the roughness, we can value it at $0{,}95\varpi$. The maximum of f for belts is about $0{,}30$. Then the value of k will be $2{,}448$. For that of μ, we will admit its maximum to be unity. We will have then $\frac{\mu k}{\mu k - 1} = 1{,}69$.

Of the other part, we can scarcely think of increasing the speed of the belt to more than 25 mètres, nor to run it permanently at a rate of more than $0^{k},25$ per square millimètre of a section. If we wish to find what power we can transmit on these conditions with a belt having a breadth of 250 millimètres, and a thickness of 8 millimètres,*

---

* It is certain that belts of more than 8 millimètres in thickness are often employed. But then the force resulting from the flexion of enrolment acquires a notable value, which diminishes the margin allowed for the effort resulting from the general tension. It would be necessary to admit for this latter a number inferior to $0^{k},25$ in such a way that, in the point of view of power transmitted, little would be gained.

which is already a strong belt, we would have to solve the equation:

$$250 \cdot 8 \cdot 0{,}25 = 1{,}69 \cdot \frac{75\,\dot N}{25},$$

which gives $N = 98\tfrac{1}{2}$ horse-power nearly. ($= 40.97$ ☐' of surface velocity.)

If we had taken account of passive resistances, and had allowed for μ a number < 1, and for the adhesion of the belt to the pulley, less favorable conditions, which use brings sooner or later, we would have obtained a figure sensibly less.

What precedes suffices to show that transmission by belts is only applicable to limited mechanical powers. But even within these limits, the greatness of the distance becomes an obstacle to its employment. Experience has shown that the elasticity of the belt causes oscillations and variations of speed, which increases with the distance, and which, beyond a certain limit, would render the transmission very irregular. Besides, where it would be necessary to make the agent of transmission pass over a great distance in the open air, a belt would be in a bad state of preservation. The search for an agent of transmission, at the same time supple and not too elastic, having sufficient resistance, and not very liable to alter,[*] was then indicated with a view of transmitting great mechanical powers to a distance.

This desideratum has been realized by the invention of metallic cables, due, as we know, to Mr. Ch. F. Hirn, of Colmar.

The Telo-Dynamic System. By C. F. Hirn.

182. "For the transmission of power or 'work' to moderate distances, we have had for ages two main methods employed — the horizontal revolving shaft and the strap pulley — supposing, as is almost always the case, that what we need in the *form* of the trans-

---

[*] As to the question of cost, we can say that, according to average statements, a section of a funicular organ, capable of supporting a continual strain of 100 kilogrammes, arising from the general tension, costs, per running mètre, for

Leather .... francs	3.50.
Caoutchouc .... "	4.25.
Wire (rope) .... "	0.40.

To compare these prices, in a practical point of view, it would be necessary to introduce a coefficient of duration for each substance. But this comparison would be of no account, except between the leather and the caoutchouc; for there would be no competition between the employment of the latter and metallic cables.

mitted work is *rotatory* motion; but these are only capable of transmitting work to extremely short distances, unless with the most serious losses or waste of power by the absorption of work of the motor in torsion and friction of the shaft, and in friction, rigidity, and slip, etc., of the strap."

"For a mere dead pull, such as the alternate strokes needed to work a pump, work is, and has long been, transmitted to very great distances; as by the long lines of 'draw rods' used in mining regions for transmitting the power of a water-wheel by means of a crank on its main axis, *pulling* during half its revolution to raise a heavy weight or 'balance bob' at the remote end of the line of bars; but a system of draw rods cannot economically be employed in producing rotatory motion at great distances.

"In recent days, the idea so clearly seen by Bramah has been realized upon a large scale by Armstrong and others, in the transmission of power by water pressure, or, as it has been called, by 'hydraulic connection;' and Armstrong has even perfected apparatus by which water pressure thus transmitted, through, it might be, miles of pipe, may be converted into rotatory motion."

"Papin saw the possibility of transmitting power through pneumatic connectors, though, if the accounts handed down be reliable, he did not succeed in realizing his notions. There can be no question, however, that power may be transmitted, either by exhaustion or by the condensation of air, for very great distances through tubes, and may be converted into rotatory motion at the remote extremity, or anywhere by the way. . . .

"In both these cases, like that of the hydraulic connector, however, unless the tubes bear a very large proportion in area to the demand for the current, whether of entering or of issuing air that transmits the power, the loss by tube friction becomes very serious. The capital to be sunk in pipes, therefore, is large in relation to the power got, and both this expenditure and the waste of power increase directly with the distance of transmission."

"Meanwhile, however, there exists in actual use, and upon a large scale, another and a simpler apparatus for the transmission of power, in large amount and to very great distances, which has attracted, we may say, no attention as yet in this country. We refer to the system originated by Mr. C. F. Hirn, which has been called that of 'telodynamic transmission,' and some drawings indicative of which were shown as long ago as the Exhibition of 1862."

"Crudely stated, this method appears to the superficial observer to consist in nothing more than in transmitting the rotatory motion of one large grooved pulley, kept revolving by the motor, to another such pulley, at a greater or less distance, by the intervention of an endless wire or steel band, or wire rope, passing over both pulleys; and to the uninstructed observer the whole affair seems nothing more than the old 'belt and pulley,' a mere elongation of that commonplace 'wrapping connector.' The hidden principle involved, however, is something entirely different.

"If we suppose a band of round iron of one inch in diameter to be capable of sustaining a steady pull, without sensible alteration, of 10 tons, and that the bar be pulled with this force endways, so that a point between the motor and the resistance moves at the rate of one foot per second, then it is obvious that the bar itself will be transmitting 'work' at the rate of 10 foot-tons per second. A bar of half its diameter, or $\frac{1}{4}$ its section, can only be strained to 2.5 tons, and at the same rate of 'end-on' motion, can only transmit 2.5 foot-tons of work per second; and so also of a bar $\frac{1}{5}$ of an inch diameter, or $\frac{1}{25}$ of the area of the one-inch bar, it can only transmit .04 ton; and, at the rate of one foot per second, .04 foot-tons of work. But suppose that the half-inch bar moves end-on at the rate of 4 feet per second, and that the $\frac{1}{5}$-inch diameter wire moves at the rate of 25 feet per second, then, as work is made up of pressure, times velocity, all 3 bars, much as they differ in section and in absolute strain upon each, will transmit the same number of foot-tons per second: *i. e., shall all be capable at the resisting end of delivering forth equal quantities of motive power in equal times.* If, therefore, we increase the velocity of motion of the wrapping connector, which is intended to transmit a given amount of motive work in a given time, we may reduce its section, because we have reduced the strain upon it, and hence its total weight in the inverse ratio of the increased velocity. We may, in fact, to put an extreme illustration, reduce the one-inch round bar to an iron wire, as fine as a human hair, and yet (theoretically) get out of it at the resisting end our 10 foot-tons per second."

"Now, this is just the principle which distinguishes Mr. Hirn's method from any common belt and pair of pulleys, and which he has shown can be carried into practical use with great advantage.

"At the motor, be it steam-engine or water-wheel, he places a tolerably large cylindrical-grooved rimmed iron pulley, revolving in a vertical or horizontal plane, to which he communicates rotation at a determinate and considerable speed. Round this he passes a thin wire, or a thin wire rope (which latter in practice he prefers), and this is led away to almost any reasonable distance (the limit is measurable by *miles*), where it is passed over another similar pulley, and returns back as an endless cord to the pulley whence it started. If the distance be more than a few hundred feet, or the intervening surface differ in level, etc., both limbs of the cord are supported and guided at intervals by guide pulleys, as few as possible, leaving the cords in the intervals between these to sag down into such catenary as the strains upon it and its own surplus strength may determine and admit. The periphery of the driving pulley may have an angular velocity as great as possible; the only limit, in fact, is that the speed shall not be likely to destroy the pulley by centrifugal force. The speeds that have been actually employed in the examples to which we are about to refer vary from 10 to 30 yards per second, at the circumference of the pulley. The pulleys themselves have been made of cast-iron and of steel, and they have but one peculiarity of construction, and that is a highly important one. At the bottom of the acute V-shaped groove, going round the circumference, a little trough is formed, dove-tailed, in section, which is filled with a ring of softened gutta-percha let into it, and united at the returned ends. Against this, as the bottom of the V groove, the wire rope of transmission alone bears. It forms a seat for itself, and does not touch the sides of the V groove or other metallic parts of the pulley. The same arrangement is adopted for the receiving pulley for the power at the remote end, and for all guide or supporting pulleys that may be necessary.

"The wire ropes are thus found to hold perfectly, and not to wear sensibly, for long periods.

"Previously to devising this form of pulley, Mr. Hirn had had much difficulty from the wear of the wire ropes, and in other ways, and had tried, in vain, wood, copper, and other linings for the pulley grooves.

"Now, assigning a peripheral velocity to the motor pulley of 30 yards per second, it is obvious, upon the principles we have already stated, that for each horse-power that we require to transmit, we must visit a strain upon the material of the cord, moving at that rate, of $\frac{33000}{90}$ = say 366 lbs.; and taking the breaking strain for average iron wire at 67,000 lbs. per circular inch, and the safe strain at about $\frac{1}{2}$ that, or say 33,000 lbs. per circular inch, then $\frac{33000}{366}$ = 90 nearly, or *a wire of $\frac{1}{90}$ of a circular inch in area will transmit a horse-power per second.*

"The wire must have surplus strength, however, also, for the *loss* of power absorbed by the sources of loss in this method of transmission. These are: 1, the resistance of the air to the rotation of all the pulleys; 2, the rigidity of the wire rope in circumflexure of the 2 main pulleys, and through the change of angular direction at either side of supporting or guide pulleys; 3, the resistance of the air, by friction, to the passage of the wire cord itself through it; 4, the friction of the axles of all the pulleys.

"Where the distances of transmission are moderate, *i. e.*, within a few hundred yards, the actual result of experiments upon the large scale, as stated by Mr. Hirn, show that all these together amount to about $2\frac{1}{2}$ per cent. of the power transmitted. Where supporting and guide pulleys are required, there is to be added to this $2\frac{1}{2}$ per cent., which represents a *constant* resistance due to the motor and transmitting pulleys and rope merely, an *additional* resistance which *varies* with the distance, and has been found, with the usual amount of supporting or guiding pulleys needed for long distances, to amount to about 504 foot-pounds for each 1100 yards in length of the double cord. Thus we see that for short distances the transmitting wire need not be larger than $\frac{1}{85}$ of a circular inch in area per horse-power; and for even such an extreme distance as upwards of 12 *miles*, we need only add $\frac{1}{4}$ to its total area, on account of all resistances due to uselessly consumed power.

"Enough has been said to show how attenuated may be the transmitting cord, and how light it may be.

"With 2 pulleys, each of 12 feet diameter, making 100 Rpm, and with a wire cord of $\frac{2}{3}$ of an inch in diameter, Mr. Hirn has found that 120 horse-power can be transmitted to a distance of 150 yards, with only a waste of power or useless effect of $2\frac{1}{2}$ horse-power.

"The first attempt at practical application of this method was made as long ago as 1850, at Logelbach, near Colmar, at the ancient calico print-works of MM. Haussmann, established in 1772, but shut up from 1841. It was proposed to make the great concern into a weaving factory for cotton, but the immense scattered mass of buildings seemed to forbid the possibility of utilizing them, and yet placing the motive power at any one point. Shafting, as a matter of cost and of waste by distance, was out of the question.

"In this emergency Mr. Hirn first tried this method of force transmission, with a riveted steel ribbon or band to each building from the engine-house. The band first tried was about 2 inches wide by $\frac{1}{75}$ of an inch thick, and on wood-faced drums. The success of the *principle* was complete, but much remained to be discovered before the round wire cord and gutta-percha pulley solved all difficulty, and brought the principle to be a practical reality.

"Since the establishment of MM. Haussmann's weaving-mills on this plan, in 1854, M. Schlumberger has transmitted the power of a turbine at Staffelfelden about 90 yards; in 1857, at Copenhagen, 45 horse-power was transmitted to saw-mills at more than 1000 yards distance from the motor; in 1858, at Cornimont, Vosges, 50 horse-power was transmitted to a distance of 1258 yards; in 1859, at Oberursel, near Frankfort-on-the-Main, 100 horse-power was thus carried 1076 yards; and at Emmendingen, Brisgau, 60 horse-power works a spinning-mill at 1312 yards from the motor; while in 1861, Count d'Espremesnil, at Fontaine le Sonet, Department de l'Eure, transmitted a very large power to saw-mills through 1100 yards, and thence a part to a farther distance of 546, or to a total of 1646 yards, to drive other machinery.

"400 *and upwards* of practical applications have been made of the method already; which has become one of the established and recognized mechanical appliances all through the south-east of France and adjacent Germany, and more particularly in Alsace, the great seat of the French cotton manufacture, and of innumerable connected industries.

"Most of the machinery has been constructed by Messrs. Stein & Co., of Mulhausen, who have acquired great experience in its constructive details and management. And, carrying the principle out to its legitimate end and development, a company has been formed since 1862, at Bale, in Switzerland, under the title of 'Compagnie d'Utilization des forces du Rhin Superieur,' whose object it is to establish water-power to an almost limitless extent at the great fall of the Rhine at Schaffhausen, and thence to transmit it off to great distances and in various directions, and let it off for manufacturing uses. Mr. Hirn has shown that power may be practically and profitably transmitted thus for distances of several miles. To take an extreme example, he proves that 120 horse-power may be transmitted 22,000 yards, or about $12\frac{1}{2}$ miles, and that the loss by uselessly expended power in the transmission even to that extreme distance, shall leave, at the very least, 90 horse-power available at the remote end. On the other hand, he shows that if 200 horse-power be transmitted by ordinary horizontal shafting, 50 per cent. of the motor will become uselessly absorbed within a distance of 1650 yards, and that to obtain 100 horse-power at the remote end of a horizontal shaft $12\frac{1}{2}$ miles long, we must apply at the motor end a power of 788,400 horse-power:* in other words, that while the limit of shaft transmission practically does not exceed 200 or 300 yards, that of telo-dynamic transmission need not, in practice, stop at 5 miles, or even more. Power may thus be transmitted economically and surely, in any direction, over hills and valleys, across rivers, into the depths of coal pits or mines."—*Practical Mech. Jour., March, 1867, p. 358.*

Telo-Dynamic Transmission.

183. We give only those parts which add new facts to the statements already made in an article on this subject.

"The wheels are made as light as is consistent with strength, not only for the sake of reducing the inertia of the moving mass, and the friction on the axes to a minimum, but for the more important object of diminishing the resistance of the air. It can hardly be doubted that an abandonment of spokes entirely, and making the pulley a plain disc, would improve essentially the performance, could discs be made at once strong enough to fulfil the required function and light enough not materially to increase the friction. It will be seen further on that the resistance of the air, which Mr. Hirn admits to be equal to the sum of the other resistances, is, in fact, more than double all the rest put together.

"This figure, which is a half size cross section of the rim, C, of the wheel, represents the form of the groove into which the armature, B, of gutta-percha is compacted, and upon which the wire rope, A, rests. The dove-tail enlargement of this groove at the base is necessary, not merely to secure the gutta-percha against displacement by ordinary causes, but to prevent its being detached by centrifugal force. Mr. Hirn assumes 30 mètres per second to be the velocity which it is expedient ordinarily to give to the circumference of the wheel; but he has carried this occasionally as high as 40 mètres.

Fig. 76.

"At 30 mètres, the centrifugal force generated at the circumference of the smaller pulleys of 2 mètres in diameter, will be between 90 and 100 times the force of gravity; and at 40 it will be nearly 170 times gravity. That is to say, as the circumference of such a wheel measures a little over 20 feet around, if each foot of this circumference weighs one pound, the whole will be dragged in all directions with this last velocity by forces which unitedly will amount to nearly

---

* Compare with Webber's tests of shafting, on page 20.

1¾ tons. It is on this account that Mr. Hirn suggests that the limit of 30 mètres had better not be overpassed, higher velocities endangering the destruction of the wheel.

"The invention of Mr. Hirn was first applied in the transmission of moderate powers to moderate distances.

"Instead of a cable there was used in the beginning a band of steel, having a breadth of about $2\frac{1}{2}$ inches, and a thickness of $\frac{1}{25}$ of an inch. This presented two inconveniences. In the first place, on account of its considerable surface it was liable to be agitated by the winds; and, secondly, it soon became worn and injured at the points where it was riveted.

"It served, however, very well, for 18 months, to transmit 12 horsepower to a distance of 80 mètres.

"A cable was then substituted, and this, first introduced in 1852, is still in good condition.

"These applications have been made for the most part in France, and in the department in which the invention originated, but there are some notable exceptions."

"In the great government manufactory of gunpowder at Okhta, in Russia, which was destroyed in 1864 by explosion, it was determined in the reconstruction of the works to erect the buildings at such a distance from each other that the explosion of one of them should not involve, as happens usually in such cases, the ruin of all the rest.

"This new manufactory, which went into operation in 1867, is composed of 34 different workshops or laboratories, to which motive power is transmitted from 3 turbines, having a total force of 274 horse-power along a line nearly a mile in length.

"Several establishments in Germany employ it for distances varying from 350 to 1200 mètres. An officer of the Danish navy has made one application of it on a line of 1000 mètres, and at the mines of Falun, in Sweden, more than 100 horse-power is transmitted by it to a distance of 5000 mètres.

"The cost of the machinery and its erection is estimated at 5000 francs per kilomètre, exclusive of the necessary constructions at the termini, which will require an additional expenditure of 25 francs per horse-power. In the case of 100 horse-power carried 10 kilomètres, the total expense will therefore amount to 52,500 francs.

"For its practical value, this invention, simple as it appears, is one of the most important that has presented itself in the Exposition, and the jury have shown that they so regard it, by awarding to the inventor the distinction of a grand prize."—*F. A. P. Barnard, LL.D., U. S. Commissioner to Paris Exposition, 1867.*

Wire-Rope Driving.

184. "Among the more recent improvements in the way of *transmitting power* for long distances, is the substitution of belts by endless wire ropes, running at a high speed. Their use bids fair to add immensely to our manufacturing facilities. The distance to which you can thus transfer power ranges from 75 feet to 4 miles. Just where the belt becomes too long for economy, there the rope steps in. In place of a flat-faced pulley, a narrow sheave, with a deep, flaring groove is used, the groove being filled out, or lined, rather, with leather, oakum, India-rubber, or some other soft substance, to save the rope. The essential points are a large sheave, running at a considerable velocity, and a light rope.

"When the distance exceeds 400 feet, a double-grooved wheel is used, and a second endless rope transmits the power 400 feet farther, and so on indefinitely. The loss by friction is about 8 per cent. per mile. A few examples may prove of interest, and give information.

"It is required to transmit 300 horse-power by means of a wire rope. A wheel $14\frac{1}{2}$ feet diameter, making 108 Rpm, is sufficient — the rope running at a rate of 4920 Fpm. Size of rope required, one inch diameter. The distance has nothing to do with it. Again: 'It is desired to transmit, for any distance, as much power as a 12-inch belt will give.' Assuming that the belt travels in the neighborhood of 1300 Fpm, it is about equivalent to 20 horse-power; and a grooved sheave of 7 feet diameter, running 100 Rpm, with a $\frac{5}{8}$-inch rope, will be the proportions required. Again: a 4-feet wheel, running 100 Rpm, with a $\frac{3}{8}$-inch rope, will convey from 4 to 5 horse-power. The cost of the rope is always the smallest item, amounting to a few cents per foot, and not one-tenth the cost of an equivalent amount of belting.

"One is thus enabled, at a small expense, to transmit power in any direction; for instance, to a building lying remote from the main factory buildings, where it is not worth while to put up a separate engine; across rivers, creeks, canals, streets; over the tops of houses; under water; from cellar to roof, etc.

"Frequently an excellent site for water-power remains unimproved for want of suitable building sites in the neighborhood. The water may be conveyed *down stream* by means of expensive canals and flumes; but by a wire rope transmission we can transfer it in any direction, either up stream, across it, or sidewise, up and down grades of one in 8 — in fact, anywhere.

"In many sections of our country, coal is dear and water-power plenty, but not improved, for reasons which may be set aside by the above method. In Europe, over 1000 factories are driven in that way."—*The Manufacturer and Builder, February, 1869, p. 38.*

Note on the Sheaves Used in the Transmission of Power by Wire Ropes.

185. The writer found in his practice that it was not only necessary to have the rims and grooves of the wheels truly turned, the wheels balanced for running without vibration in their bearings, but that, also, some certain protection must be furnished to prevent the rope striking against the flanges of the wheels, to which it is constantly liable by the cross action of air currents, by the irregular wearing of the gum filling, and by the vibration of the wheels.

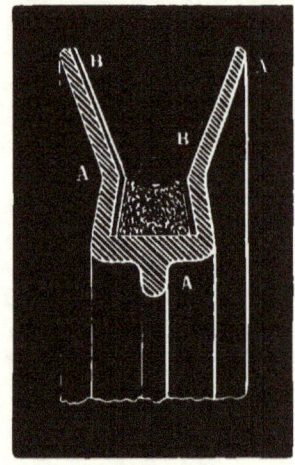

Fig. 77.

In Fig. 77, an efficient and inexpensive method of preserving the rope is shown, by lining the sides of the wheel-groove flanges, A, with leather, B, secured by copper rivets above the gum filling, and held firmly by the filling in the groove.

Rope Gearing.

186. In a paper by Mr. John Ramsbottom, in *Newton's Journal*, Vol. XXI., p. 46, on traversing cranes at Crewe Locomotive Works, dated January 28, 1864, mention is made of the means by which power is communicated from the shop lines of shafting to the gear of the cranes.

"It consists of a $\frac{5}{8}$-inch diameter, soft, white-cotton cord, weighing about $1\frac{1}{2}$ ounces to the foot, running at the rate of 5000 Fpm, in a line with the longitudinal motion of the crane, above the same, and over a 4-feet diameter tightener sheave. This sheave is weighed so as to put a tension on each strand of the cord of 108 lbs., which is found to be the best working strain for keeping the rope steady, and giving the required 'hold' on the main driving pulley.

"The cranes have a span of 40 feet 7 inches, a longitudinal traverse of 270 feet, and the rails are 16 feet above the floor.

"The cord is supported every 12 or 14 feet by cast-iron fixed slippers of plain trough section, $1\frac{3}{8}$-inch wide, with side flanges. These slippers are placed $1\frac{1}{2}$ inches below the working line of the driving side of the cord, so as to allow the driving wheels on the traverser to pass them. They are not oiled, and the friction of the cord in them amounts to $\frac{2}{3}$ of the working load.

"Motion is communicated to the gear of the crane by pressing the cord into grooved cast-iron pulleys. The grooves in the driving pulleys are V-shaped, at an angle of 30°, and the cord does not touch bottom. The guide pulleys have circular grooves, same diameter as the cord, and the pressure pulleys have a circular groove of larger diameter than the cord. The driving pulleys have a diameter equal to 30 times the diameter of the rope. Guards are put on the pulleys to keep the ropes in.

"The driving power of the cord to lift 25 tons is only 18 lbs., irrespective of friction, which is a ratio of 3111 : 1. Light loads are about 800 : 1. In the gib cranes, driven by similar means, the ratio is 1000 : 1, when lifting 4 tons at the rate of 5 feet $1\frac{1}{2}$ inches per minute.

"The actual power required to lift 9 tons, besides the snatch-block and chain, has been found to be 17 lbs. at the circumference of the driving pulley. The crab, when unloaded, requires $1\frac{1}{8}$ lbs. to overcome its friction.

"The cords are soon reduced to $\frac{9}{16}$-inch diameter, and last about 8 months at constant work.

"In an overhead traverser, used in the boiler-shop, lifting 6 tons, 3 years in use, a $\frac{3}{8}$-inch cord was employed, but was afterwards changed for a cord $\frac{1}{2}$-inch in diameter.

"The light driving cord is the only plan compatible with high speeds — a heavy chain, belt, or cord would soon wear out and break by its own weight."

Iron and Steel Ropes Compared.

187. The superintendent of Cedar Point Iron Company, Port Henry, N. Y., says, in reference to the comparative merits of steel and iron wire ropes for hoisting purposes:

"We have used iron and steel wire ropes extensively at our iron ore mines, and our experience in hoisting from 1000 to 1500 tons gross per day awards the superiority by all odds to steel. One steel rope lasts us, on an average, as long as 6 iron ones.

"We attempted to run them on same sheaves as we had used for iron ropes, and found, in some cases, that the steel rope would cut one inch into the cast-iron sheaves in one or two months of use.

"Our sheaves now are lined with wood, which entirely obviates every difficulty in the use of steel, and of course is equally to be commended for iron." — *January, 1873.*

On Rope Gearing for the Transmission of Large Power in Mills and Factories. By Mr. James Durie.*

188. The best means of transmitting power from the prime mover to the various machines in a factory has long been a matter of importance to the engineer and to the manufacturer. Until lately, toothed gearing — either as spur or bevel wheels, or a combination of both — has been almost universally employed for first motions, the smaller powers being taken off pulleys by leather belts. The facility of taking small powers off drums to machines by means of belts, and the absence of noise and vibration, led to the adoption, in the United States of America, of broad leather belts for the transmission of large powers from the prime mover to the shafting in factories; and the success which has followed the adoption of these large belts there, has led to their being adopted by many users of power in this country.

The object of the present paper is to bring before the institution the plan of transmitting large powers by means of round ropes working on grooved wheels, which, in some parts of this country, has been largely adopted as a substitute for toothed gearing. The experience gained by the firm with which the writer is connected, in this mode of transmitting power, has extended over a period of thirteen years; and wherever it has been employed, either to replace toothed gearing in old or for new works, it has always given complete satisfaction.

In this mode of driving, the fly-wheel of the engine is made considerably broader than the fly-wheel of an engine having cogs on its circumference; and, instead of cogs, a number of parallel grooves for the ropes are turned out, the number and size of which are regulated by the power to be taken off the fly-wheel. The power which each of the ropes will transmit depends upon their size, and the velocity of the periphery of the fly-wheel. The ropes employed are of two sizes for large powers, namely, $5\frac{1}{4}$ inches and $6\frac{1}{2}$ inches circumference; another size of rope, $4\frac{1}{4}$ inches circumference, is employed for small

---

* A paper read at the meeting of the Institution of Mechanical Engineers, at Manchester, England, October 25, 1876. From *Iron*, London, Oct. 28, 1876.

powers, but there is no definitely ascertained limit to the size of ropes that may be employed. Where large powers are required, and where large pulleys can be used, it is best to use heavy ropes, and the contrary when the opposite is the case.

The velocity of the periphery of the grooved fly-wheel and pulleys is generally arranged to be between 3000 and 6000 Fpm; and the velocity being settled, and the power of the steam-engine known, the number of the ropes required to transmit the power is then determined, from the experience that has been gained of the amount of power transmitted by ropes in previous cases. It is very essential that the right proportion between the diameter of the ropes and the pulleys is obtained; if the diameter of the pulley is too small, the rope, in continually bending over them, is apt to strain the strands and grind the core into dust, and on the size of the pulley in a great measure depends the life of the rope. As a general rule, the circumference of a pulley should not be less than 30 times that of the rope which works on it. In apportioning the distance between the driver and the driven shafts, great latitude may be taken — a distance of from 20 to 60 feet may be taken as a fair space.

The mode of applying a complete system of rope gearing may be seen at a factory for spinning and weaving jute belonging to Messrs. A. & J. Nicoll, at Dundee, and fitted up by Messrs. Pearce Bros., Lilybank Foundry, of that town; this gearing has been working from June, 1870. The engine fly-wheel is 22 feet diameter, and has 18 grooves cut in its circumference; its width is 4 feet 10 inches over all. The engine makes 43 Rpm, and the velocity of the periphery of the fly-wheel is 2967 Fpm. The power of the engine varies from 400 to 425 indicated horse-power; the power transmitted by each of the ropes, which are $6\frac{1}{4}$ inches circumference, is, therefore, about 23 horse-power. The power is transmitted to the ground floor by 5 ropes on to a pulley 7 feet 6 inches diameter, the power required being 115 indicated horse-power; that to the first floor by 4 ropes to a pulley 5 feet 6 inches diameter, the power required being 92 indicated horse-power; and that to the attic by 6 ropes on to a pulley 5 feet 6 inches diameter, the power required being 138 indicated horse-power; 2 shafts being required in this room, the power to the second shaft is transmitted by horizontal ropes, whilst on the other side of the engine shaft 3 ropes transmit 69 indicated horse-power to a weaving shed, the pulley being 7 feet 6 inches diameter. The ropes should never be so heavily loaded as to draw them, even on short spans, to a near approach to a straight line; in this factory each rope, travelling at a velocity of 2967 Fpm, transmits, as above, 23 indicated horse-power, the tension on the rope is, therefore, $\frac{33,000 \times 23}{2967} = 256$ lbs., which is a very long way under the breaking strength of the ropes.

The ropes do not rest on the bottom of the groove, but on its V-shaped sides; these sides are generally made at an angle of about 43° to each other. If the angle at which the grooves are formed is very obtuse, the ropes will slip; if too acute, much friction may be caused by the rope becoming wedged into the groove. As the sum of the tensions upon the two parts of a band is the same, whatever be the pressure under which the band is drawn or the resistance overcome, the returning side of the rope is as much slackened as the working side is tightened. It is, therefore, generally advisable, when it can be so arranged, to have the tight or driving side of the ropes at the bottom, so that the returning side may lap round the top of the pulleys, and consequently obtain extra bearing surface; when the opposite is the case, the ropes fall sooner out of the grooves, and so lessen the bearing surface. It is not always practicable to arrange this; and in the case of taking the power off both sides of the driving pulley, it is obviously impossible.

None of the shafts of this factory are driven from the fly-wheel by less than 3 ropes, the strain on each rope being only, as shown above, 256 lbs.; it may therefore be supposed that a greater weight may temporarily be put upon a rope, in case any of the ropes should require to be tightened up, and this is often done. A rope is taken off at a meal hour, respliced, and put on again at the next stoppage of the engine, thus avoiding any necessity for night work or overtime. Night work should always be avoided, for besides the extra expense incurred, the work is never so well done by artificial light as by daylight.

In the Samnuggur Jute Factory, Calcutta, which is a one-story building, all the machinery, both spinning and weaving, is on the ground floor. The engines are placed near the middle of the building; they are about 1000 indicated horse-power, and make 43 Rpm; the fly-wheel is 28 feet diameter, and its width 6 feet 7 inches; the velocity of the periphery being 3784 Fpm. The ropes are $6\frac{1}{2}$ inches circumference, 18 ropes transmit the power to the right hand or spinning side, and 7 ropes to the left hand or weaving side, making a total of 25 ropes; each rope therefore transmits 40 indicated horse-power. The tension on each rope due to this load at the above velocity is $\dfrac{33,000 \times 40}{3784} = 349$ lbs., which is rather a heavier load than in the previously described factory; all the shafts are also driven by more than one rope, with the exception of some of the line shafts in the weaving shed. Rope gearing was only adopted in this case after the most searching inquiry as to its suitability for working in the warm and humid climate of Calcutta, and its adoption has been attended with very satisfactory results, both in this case and also at the Sealdah Mills in Calcutta.

The drawings show the arrangements adopted when the factories have been specially designed for rope gearing; but this gearing has often been applied to replace toothed gearing in mills already built. The plan then adopted is to put in a new grooved fly-wheel, or to place grooved segments upon the existing fly-wheel, when the speed is sufficiently great to allow of a limited number of ropes being employed, and the width of the wheel-pit is also sufficient for the purpose; but if this plan cannot be adopted, grooved pulleys are put on the second-motion shaft, and the ropes carried to the different stories of the mill. It has even sometimes been necessary to put in a counter shaft, so as to gain speed and get sufficient length between the centres of the shafts on which the pulleys are placed.

A comparison has now to be made between the system of rope gearing and the other 2 systems in use at the present time. The system in most general use is toothed gearing; in this the first driver is the spur wheel fitted on to the crank shaft of the engine, into which is geared the driven pinion of smaller size. In order to insure these 2 wheels working well, it is absolutely essential that the centres of the engine shaft and the shaft of the pinion are rigidly fixed at the correct distance from each other, and that the teeth of the 2 wheels are accurately of the same pitch and size. The first of these objects is obtained by making the engine bed, in a horizontal engine, a strong rigid casting resting on an expensive ashlar foundation, or in a beam engine the foundations are alone depended upon. If these objects are obtained, the wheels ought to work smoothly and without much noise; but how often this desirable object is obtained, it only requires a walk to be taken through the streets of a manufacturing town, to ascertain by the rumbling noise, sometimes heard at a distance of several hundred yards, that all is not as it should be for the safe and economical transmission of the power of the steam-engine. If the factory consists of more than one story, the power has to be taken from the second-motion shaft, by means of an upright shaft and bevel wheels, requiring heavy wall boxes, and strong walls to keep the wall boxes in their places; the whole object in the construction of the factory being to secure a rigid and immovable structure, a matter which is very difficult to attain.

In the case of rope-gearing, the ropes by which the power is transmitted consist of an elastic substance, and their lightness, elasticity, and comparative slackness between the pulleys, are highly conducive to their taking up any irregularity that may occur in the motive power. This accounts for the slight attachments that are required for shafting driven by ropes from a grooved fly-wheel; and it is the same with all the bearings throughout the mill, the shafts in the various flats only requiring a light wall-box, bracket, or the bearing may be carried on a column of the mill. The cost of fitting up a mill with rope gearing is considerably less than with tooth gearing, when the shafts to be driven revolve at a high speed, but the cost is about the same in other cases. It is, however, rather difficult to give exact figures for this comparison, one great saving being in the foundations of the engine, the wall-boxes, and the extra strength of the walls required for upright shafts.

The great advantage of rope gearing, however, is the entire freedom from any risk of a breakdown; when a rope shows symptoms of giving way — and ropes always give symptoms of weakness long before they break — the weak rope can be removed, and another put in its place at any meal hour or evening. The cost of the maintenance of ropes for transmitting 400 indicated horse-power has been found to be £20 per annum, or about £5 per 100 indicated horse-power per annum. This is made up of the cost of renewal of the ropes, and occasional wages for tightening them. Some ropes have been found to run $10\frac{1}{2}$ years; but, as the general rule, the life of a rope may be said to be from 3 to 5 years, though even 5 years have been often much exceeded.

The friction of rope gearing has often been found to be, for high speeds, considerably less than that of toothed gearing; but the writer regrets not being able to give definite information on this point, which is a very important one to those contemplating altering their gearing or building new works. The reason why no definite reason can be given — beyond the universal impression of those who have adopted them, that ropes require less power to drive than toothed gearing — is, that in all cases where rope gearing has been substituted, other alterations have been made at the same time, or the engines were, after the alteration, driven at an extra speed of 10 or 15 Rpm. However, every one who has substituted rope gearing for toothed gearing, also agrees in bearing testimony to the great improvement and steadiness of driving obtained after the alteration, and that they are enabled to turn off a greater amount of yarns from the machinery in the same time. The tendency at the present time, with the introduction of shorter hours of labor and foreign competition, being to increase the speed of shafting and machinery, to be able by this means to increase the speed of the shafts must be of great advantage to those who own old mills, the toothed gearing of which is generally driven as fast as it is safe to drive it.

The ropes used for rope gearing are mostly made of hemp, carefully selected; the qualification of a good rope being that the fibres are as long as possible, and that the rope should be well twisted and laid, and yet be soft and elastic. It is also very important that the ends of the ropes should be united by a uniform splice — the splice should not be of a greater diameter than the other part of the rope; to effect this object the splice is made about 9 or 10 feet long.

The comparison between rope gearing and toothed gearing having been made, it remains to compare ropes and leather belts, which latter have been largely used in the manufacturing districts of Lancashire and Yorkshire, for the transmission of large powers. The writer has not been able to obtain very satisfactory information as to the amount of power absorbed in friction by large belts; in some cases it has been said to be more, and in some less, than with toothed gearing. The most trustworthy information he has obtained is in the case of a 4-story woollen mill, where an upright shaft, with bevel gearing, was replaced by 2 belts, one 22 inches wide, and the other 27 inches wide, the power transmitted being 400 indicated horse-power, and the speed of the belts 3000 Fpm. The driving pulleys are on the second-motion shaft. In this case the power is stated to be the same with the belts as it was before the alteration; but, as in the case of rope-driving, the "turning" is found to be much superior to what it was before the alteration. The width of the pulleys for ropes is generally rather less than for belts transmitting the same power; but there is some difference of practice as to the width of belt used for transmitting a certain power. The cost of hemp ropes is considerably less than of leather belting, the cost of hemp ropes being about 1*s.* per lb. against 3*s.* per lb. for leather belts. The grooved pulleys for ropes cost more than plain pulleys; but, making allowance for this, the total cost of ropes and grooved pulleys for transmitting a given power, does not exceed one-half or two-thirds the cost of leather belting and flat pulleys. The advantage of ropes over belting, however, lies in the power being divided up into a number of ropes, so that, in the case of any one of the ropes showing symptoms of weakness, that rope may be removed by stopping the engine for a few minutes, the remaining ropes continuing to do the work until a stoppage of the engine occurs. In the case of belting, as only one belt is employed to drive one flat of a mill, if anything were to occur to the belt the whole of that flat would be stopped until the belt was repaired.

Judging from the practice adopted, the comparative amount of power transmitted by certain sizes of ropes and widths of double belts, the writer finds that a $6\frac{1}{2}$-inch circumference rope does about the same amount of work at a given speed of say 3000 Fpm as a belt 4 inches broad. This width, however, represents the smallest width adopted as a rule, 5 inches corresponding to the American practice; but, taking a 4-inch belt, the bearing surface of a $6\frac{1}{2}$-inch circumference rope on the sides of the grooves on a 4-feet 6-inch pulley will be half the circumference, or 85 inches, and, allowing the rope half an inch width of bearing on each side, or 1 inch for both sides, the total bearing surface is $85 \times 1 = 85$ square inches, whilst the belt has $85 \times 4 = 340$ square inches, or 4 times the amount. Consequently, in order that a $6\frac{1}{2}$-inch circumference rope may transmit as much power at the same tension as a 4-inch belt, the effective pressure per square inch of the bearing surface of the rope on the pulley must be at least 4 times as great as that of the belt.

In order to obtain some information bearing on this point, a set of experiments have been made by Mr. A. W. Pearce, at Lilybank Foundry, Dundee, the results of which are given in the following table. The experiments were made with the materials at hand; both the pulleys were just as they came from the lathe, and equally smooth, and they were nearly the same size as named in the table. Comparing together Nos. 2 and 5 experiments, it is seen that a 6-inch circumference ungreased rope, with 336 lbs. suspended at one end, and passing over a 4-feet 9-inch grooved pulley, which was at rest, required only 28 lbs. at the other end to prevent slipping; whilst a half-worn good single leather belt, 4 inches wide, with the same weight at one end, and passing over a 4-feet 6-inch pulley, required 113 lbs. at the other end to prevent slipping, or about 4 times as much as with the rope. The bearing surface of the rope would be only about one-fourth that of the belt, and the effective pressure per square inch of the bearing surface of the rope was consequently in this case 16 times as great as that of the belt. In the experiment No. 4 a double leather belt, 6 inches wide and ⅜ inch thick, with the same weight at one end, and passing over a 4-feet 6-inch pulley, required 98 lbs. at the other end to prevent slipping, or 3½ times as much as with the 6-inch circumference rope. The experiments show, however, such a great difference between the results with different sizes of ropes as to make it impossible to come to any definite proportion between the friction of ropes and belts; but they show, as was to have been expected, that ropes have a considerably greater hold on the V-shaped grooves per square inch of bearing surface than flat belts have on pulleys.

Experiments on the Friction of Ropes and Leather Belts on Cast-Iron Turned Pulleys.

No. of Experiment.	Rope or Belt.	Circumference of Rope or Width of Belt.	Diameter of Pulley.	Load Suspended at One End.	Weight Required at Other End to Prevent Slipping.		Remarks.
					Un-greased.	Greased.	
		Inch.	Ft. In.	Lbs.	Lbs.	Lbs.	
1	Rope.	7	4 9	336	56	102	Rope somewhat worn.
2	"	6	4 9	336	28	90	Rope new.
3	"	5¼	4 9	336	14	......	Rope new.
4	Belt.	6	4 6	336	98	133	Double belt ⅜ in. thick.
5	"	4	4 6	336	113	......	Single belt half worn.

NOTE.—The rope pulley used in these experiments was grooved for 6½-inch circumference of rope.

The writer has been informed that in the United States several rolling-mills are driven by means of flat leather belts, and that very satisfactory results are obtained by their use, and he wishes to draw the attention of members engaged in this department of manufacture to the suitability of rope gearing for this purpose. Although he is not aware of any practical example of its having been applied to driving rolling-mills, he is confident that from the slackness with which the ropes can work, and the hold they have on the grooves of the pulleys, they would be admirably adapted for taking up the shock which is thrown upon the gearing of a train of rolls when the iron enters the rolls.

'Mr. Welsh opened the discussion by a mathematical criticism of the principles involved in the paper, arriving at a corroboration of the angle of 40°, approved by Mr. Durie, on the ground that the tangent of half this angle (0.364) ought to be equal to the coefficient of friction, which, taking the mean of the writer's experiments, was 0.399. He enforced his remarks by symbols and formulæ of fearful complexity, inscribed upon the blackboard.

Mr. Paget said that he had been working in the same direction, and had performed over 900 experiments with a view of getting at some general principle. The results, however, were so absurdly discordant that he had abandoned the task. His best results corroborated Mr. Durie's angle of 40°.

Transmission by Wire Ropes Used as Connecting-Rods.

189. On each of the shafts, A A, Fig. 78, are arranged 3 cranks, each 120° apart. These cranks are connected by wire ropes, which

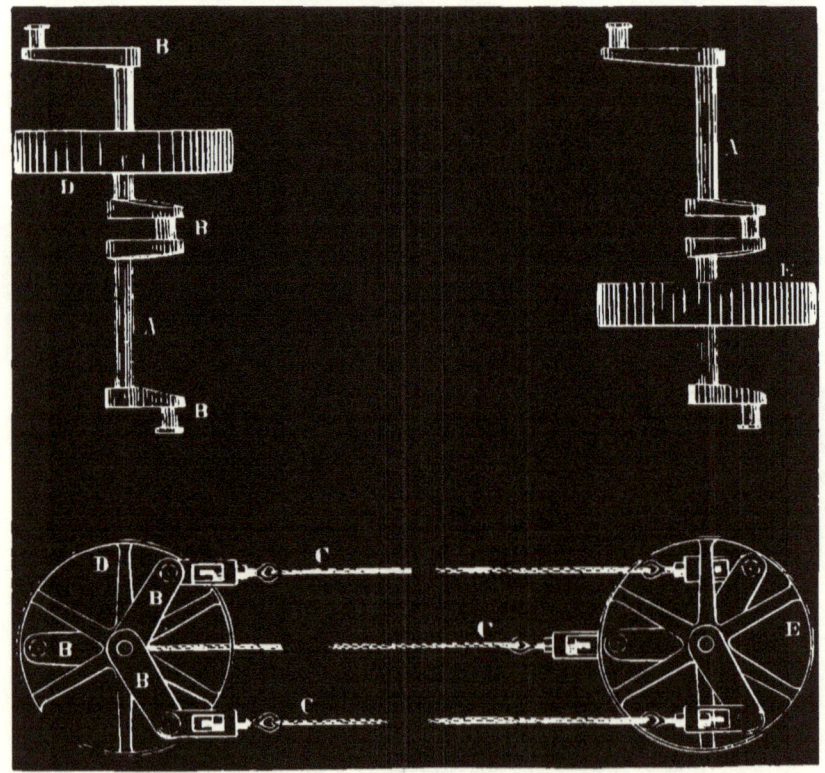

Fig. 78.

may be of considerable length and may also be horizontal, vertical, or inclined, as the location demands.

The pulley, E, may receive the power from the motor, and D may transmit the same to the machinery where it is required.

It is evident that the distance of transmission by this contrivance will be subject to the sag of the ropes; but the rope connections may be multiplied by the use of intermediate shafts, and in that way uniform rotary motion may be transmitted to a considerable distance. The motion must be comparatively slow, however, owing to the severe strain which would be thrown upon the bearings by the surging and swaying of the ropes during the rapid changes of motion to which

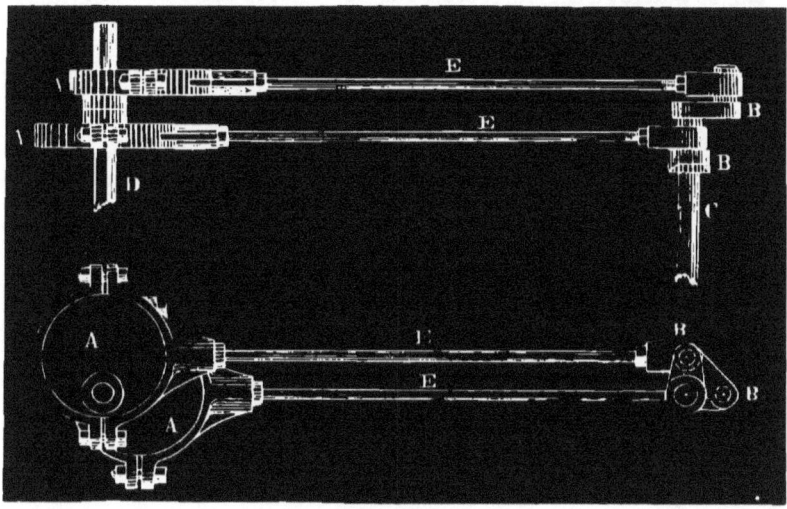

Fig. 79.

they would be liable; but what is lost in velocity may be gained in power transmitted — as that is measured by the strength of the ropes — and it is an easy matter to make them carry heavy burdens safely.

The writer saw this in use at the Whirlpool Rapids, Niagara, for driving the passenger-elevator recently erected at that place.

Transmission of Rotary Motion by Rods.

190. Reciprocating motion by a line of connecting-rods and swinging arms for working pumps and the like is old and in common use at mines and quarries. (See page 267.)

But in order to transmit uniform rotatory motion, limited in distance only by the practical working length of rods, the devices shown in Fig. 79 may be employed.

Two eccentrics, A A, set at right angles on the shaft, D, are connected by rods, E E, to two cranks, B B, which are also secured at right angles on the shaft, C. The motion transmitted is steady and noiseless, is performed in the same direction and in equal time, and is very suitable for valve-gears in steam-engines and the like. The effect would be the same if two cranks were used in place of the eccentrics.

The writer devised and applied this combination of well-known parts to driving the valve cam-shaft and governor of a beam steam-engine built at People's Works, Philadelphia, in the year 1867.

CHAPTER IX.

FRICTIONAL GEARING.

BY E. S. WICKLIN.[*]

191. Frictional gearing is the term applied by Webster to wheels that transmit motion by surface contact, without teeth. Among mechanics and practical men who build and use them, such wheels are usually called "friction gear." When spoken of separately, that is, without reference to their combination, they are called friction "pulleys," especially those with faces parallel to the axes. When made conical, they are termed "bevel friction," and are usually spoken of as "wheels."

This style of gearing is now in use in the lumbering region of the north-west, and is fast gaining favor wherever used. It has some advantages not possessed by other modes of communicating motion, which do not appear to be counteracted by any peculiar disadvantages. As a rule, however, it does not strike the mind of the mechanic favorably when first suggested, but must be seen to be appreciated. The first impression appears to be that the point of contact is too small to possess any considerable amount of adhesive force.

---

[*] JOHN H. COOPER, ESQ., PHILADELPHIA:

Dear Sir: — As you propose to publish, in your "TREATISE ON BELTING," an article on *Frictional Gearing* written by me, which you are at liberty to use, I beg to add a word of explanation.

These papers were prepared some 5 years ago. An extensive experience, however, since that time, with this method of transmitting power, has but confirmed the confidence I then expressed in its efficiency. Pulleys of soft maple, that have been running for 8 years, and that have transmitted 30 horse-power by a single contact, are to-day perfect. Some of these have not perceptibly worn their journals.

Another point it is but just to mention: — A more extended acquaintance with the mills of the north-west has shown me that I had somewhat overestimated the extent to which this frictional transmission is used. Its use, though quite general, is by no means universal.

Truly Yours,
E. S. WICKLIN.

BLACK RIVER FALLS, WIS., March 19, 1877.

It is generally received as a law that friction, or adhesion of contact, is not in proportion to the extent of contact, but to the amount of pressure and the condition of the surfaces. But to many minds, this law appears more as a learned theory than as a practical truth. In fact, there are few, even of our best mechanical thinkers, who do not manifest some surprise when, for the first time, they see with what apparent ease one smooth wheel will drive another equally smooth, by what appears to be a very slight contact; and that the second wheel is not only itself driven, but carries with it ponderous machinery, involving the expenditure often of more than 50 horse-power. Nor is it strange that many minds are unprepared for such results, since most of our mechanics rely, to some extent at least, upon books, in the absence of personal experience; and here books fail.

There are, perhaps, no other means of transmitting motion, about which so little has been written, in proportion to its importance, as frictional gearing. And most that has been written is upon the grooved wheels, which are frictional to their injury, and are, by the unequal motion of the parts in contact, as well calculated to absorb the motive power as they are to destroy each other.

As examples: In the latest edition of Webster's "Unabridged" we find the following definition: "Frictional gearing, wheels which transmit motion by surface friction instead of teeth. The faces are sometimes made more or less V-shaped, to increase or decrease friction, as required."

In a recent work on mill building, perhaps the latest published in this country, a work of large pretensions, the only allusion to this class of machinery is a single sentence, in the chapter on friction, as follows: "Friction also furnishes a convenient medium of communicating and transmitting motion in machinery, as in gigging back the carriage and log in saw-mills; and in some modern mills the whole driving power for both saws and mill-stones is communicated by friction of iron upon iron."

Another late mechanical work gives us the following information upon friction gearing: "The surfaces of the wheels are made rough, so as to bite as much as possible."

The above quotations furnish a sample of what may be learned from books of this very important mode of transmitting motion. And yet, in all the vast lumbering region of the north-west, comprising a large part of 2 or 3 States, and furnishing building and fencing material for several millions of people, there are few mills in which some part of the work is not done by friction gear; and in many mills the whole power, amounting to from 100 to 300 horse-power, is thus transmitted.

The growing popularity and importance of this rather new style of gearing cannot fail to make it a subject of interest to mechanical engineers and manufacturers.

With this view, the writer now proposes to give, in a short series of articles, some observations taken from a practical standpoint, and also the results of a few experiments, made to determine the percentage of adhesive force or traction of these wheels as compared with belted pulleys.

Now a word as to what friction gearing is, where it has become an undoubted success. In large mills, where this gearing is used to transmit power to drive 5 or 6 gangs, one or 2 large circular saws, a muley, gang-edgers, trimmers, slashers, lath-mills, shingle-mills, and more besides; where 20,000 feet of boards may be sawn in an hour. The faces of the wheels are not "made more or less V-shaped, so as to increase or decrease friction as required." Nor is the power "communicated by the friction of iron upon iron." Neither are the surfaces of the wheels "made rough, so as to bite as much as possible." On the contrary, the surfaces are made smooth and straight as possible; one wheel, or pulley, is made of iron, and the other of wood, or of iron covered with wood. So it is seen that the books are wrong, at least so far as applied to the localities where this gearing is most used.

Where it is practicable, this gearing is so arranged that the wood drives the iron. This is done so that the "slip," in starting up machinery while the driving wheels are in full motion, will tend to wear the wooden wheel round rather than to cut it in grooves, which is done to some extent when the wheels are reversed, though this tendency is much less than might be supposed, as in most cases the "bull wheel," used for drawing logs into the mill, is a large wooden wheel, driven by a small one of iron. And these wheels, though started and stopped with the driver in full motion a hundred times a day, work well and last for several years. But for machinery in constant use, the wooden wheel should always drive the iron.

For driving heavy machinery, the wooden drivers are put upon the engine-shaft, and each machine is driven by a separate counter-shaft. 2 or more of these counter-shafts are usually driven by contact with the same wheel, and each is arranged so as to be thrown out from the driver and stopped whenever required, and again started at any moment, without interference with other machinery. This is easily accomplished, as a very slight movement is sufficient for the purpose.

To drive small machinery, these friction drivers are put upon a line-shaft, so as to drive a small counter-shaft, from which the machine is driven by a belt, and stopped and started by throwing out the counter-shaft, and throwing it in again.

To select the best material for driving pulleys in friction gearing has required considerable experience, nor is it certain that this object has yet been attained. Few, if any, well-arranged and careful experiments have been made with a view of determining the comparative value of different materials as a frictional medium for driving iron pulleys. The various theories and notions of builders have, however, caused the application to this use of several varieties of wood, and also of leather, India-rubber, and paper; and thus an opportunity has been given to judge of their different degrees of efficiency. The materials most easily obtained, and most used, are the different varieties of wood, and of these several have given good results.

For driving light machinery running at high speed, as in sash, door, and blind factories, basswood — the linden of the Southern and Middle States — (*Tilia Americana*) has been found to possess good qualities, having considerable durability and being unsurpassed in the smoothness and softness of its movement. Cotton wood (*populus monilifera*) has been tried for small machinery with results somewhat similar to those of basswood, but is found to be more affected by atmospheric changes. And even white pine makes a driving surface which is, considering the softness of the wood, of astonishing efficiency and durability. But for all heavy work, where from 20 to 60 horse-power is transmitted by a single contact, soft maple (*acer rubrum*) has, at present, no rival. Driving pulleys of this wood, if correctly proportioned and well built, will run for years with no perceptible wear.

For very small pulleys leather is an excellent driver and is very durable, and rubber also possesses great adhesion as a driver; but a surface of soft rubber undoubtedly requires more power than one of a less elastic substance.

Recently paper has been introduced as a driver for small machinery, and has been applied in some situations where the test was most severe; and the remarkable manner in which it has thus far withstood the severity of these tests appears to point to it as the most efficient material yet tried.

The proportioning of friction pulleys to the work required, and their substantial and accurate construction are matters of perhaps more importance than the selection of material. The mechanic who thinks he can put up frictional gearing temporarily and cheaply will make it a failure. Leather belts may be made to submit to all manner of abuse, but it is not so with friction pulleys. They must be most accurately and substantially made, and put up and kept in perfect line.

All large drivers — say, from 4 to 10 feet diameter and from 12 to 30-inch face — should have rims of soft maple 6 or 7 inches deep. These should be made up of plank, $1\frac{1}{2}$ or 2 inches thick, cut into "cants," $\frac{1}{6}$, $\frac{1}{8}$, or $\frac{1}{10}$ of the circle, so as to place the grain of the wood as nearly as practicable in the direction of the circumference. The cants should be closely fitted, and put together with white lead or glue, strongly nailed and bolted. The wooden rim, thus made up to within about 3 inches of the width required for the finished pulley, is mounted upon 1 or 2 heavy iron "spiders," with 6 or 8 radial arms. If the pulley is above 6 feet in diameter, there should be 8 arms, and 2 spiders when the width of face is more than 18 inches.

Upon the ends of the arms are flat "pads," which should be of just sufficient width to extend across the inner face of the wooden rim, as described; that is, 3 inches less than the width of the finished pulley. These pads are gained into the inner side of the rim, the gains being cut large enough to admit keys under and beside the pads. When the keys are well driven, strong "lag" screws are put through the ends of the arm into the rim. This done, an additional "round" is put upon each side of the rim to cover bolt-heads and secure the keys from ever working out. The pulley is now put to its place on the shaft and keyed, the edges trued up, and the face turned off with the utmost exactness.

For small drivers, the best construction is to make an iron pulley of about 8 inches less diameter and 3 inches less face than the pulley required. Have four lugs, about an inch square, cast across the face of this pulley. Make a wooden rim, 4 inches deep, with face equal to that of the iron pulley, and the inside diameter equal to the outer diameter of the iron. Drive this rim snugly on over the rim of the iron pulley, having cut gains to receive the lugs, together with a hard wood key beside each. Now add a round of cants upon each side, with their inner diameters less than the first, so as to cover the iron rim. If the pulley is designed for heavy work, the wood should be maple, and should be well fastened by lag-screws put through the iron rim; but for light work it may be of basswood or pine, and the lag-screws omitted. But in all cases the wood should be thoroughly seasoned.

In the early use of friction gearing, when it was used only as backing gear in saw-mills and for hoisting in grist-mills, the pulleys were made so as to present the end of the wood to the surface; and we occasionally yet meet with an instance where they are so made. But such pulleys never run so smoothly nor drive so well as those made with the fibre more nearly in a line with the work. Besides, it is much more difficult to make up a pulley with the grain placed radially, and to secure it so that the blocks will not split when put to heavy work, than it is to make it up as above described.

As to the width of face required in friction gearing: When the drivers are of maple, a width of face equal to that required for a good leather belt (single) to do the same work is sufficient. Or, to speak more definitely, when the travel of the surface is equal to 1200 Fpm, the width of face should be at least one inch for each horse-power to be transmitted, and for drivers of basswood or pine $1\frac{1}{2}$ to 2 inches.

The driven pulleys, as before stated, are wholly of iron. They are similar to belt pulleys, but much heavier, having more arms and stronger rim.

The arm should be straight rather than curved, and there should be 2 sets of arms when the face of the pulley is above 16 inches. For the proportion of these pulleys, a very good rule is to make the thickness of rim $2\frac{1}{2}$ per cent. of the diameter; that is, when the pulley is 40 inches diameter, the rim should be an inch thick.

To secure perfect accuracy these pulleys must be fitted and turned upon the shaft; and when large, should rest in journal boxes in the latter while being turned. If simply swung upon the lathe centres, they are liable to vary while the work is being done. When turned exactly true, round, and smooth, these pulleys must be carefully and accurately balanced. The neglect of this last essential point has worked the destruction of otherwise well-made friction pulleys.

When thus constructed there is a beauty about the movement of this gearing which at once enlists the favor of all who can appreciate the "music of motion," and gives character to its builder. Its efficiency and peculiar advantages will be more fully shown further on.

In the practice of mechanics, we are generally satisfied with an old and familiar principle, without giving ourselves any great trouble to inquire into the comparative degree of its efficiency. But this does not satisfy the requirements of science; nor is it sufficient for the practical mechanic when applied to principles less familiar.

When new modes are introduced as rivals of the old, the question of comparative efficiency is at once raised, and should be met by crucial experiment. But unfortunately for both science and practice, these questions are not generally so met. Too few experiments are made, and those without sufficient care and accuracy to establish principles or remove doubts. No experiment is, however, without some degree of interest, and when all the conditions of a test are known it is not difficult to estimate approximately the value of results. With this view, the conditions and results of a few experiments, made to test the tractive power of smooth-faced friction pulleys, are here given. These experiments, when made, were not meant for publication or for the benefit of science, but to establish rules for private practice. They should be repeated by others before being taken as conclusive.

For the experiments, two pulleys were made in the usual way, one being of wood — soft maple — and the other of iron. Both were accurately and smoothly finished. These pulleys were each 17 inches in diameter, and of 6 inches face, and were put up as shown in the annexed diagram.

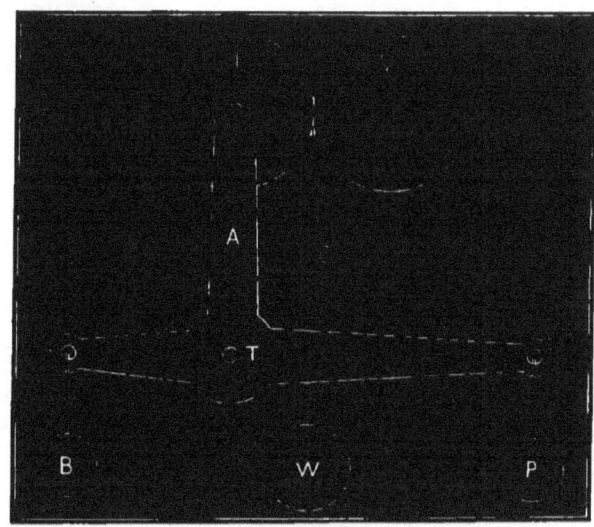

Fig. 80.

A, in the diagram, is a double bell-crank frame, with arms 2 feet long. The ends of the upright arms receive the bearings for the iron pulley, I. The journals of this pulley are $1\frac{1}{2}$ inches in diameter, and 3 inches long, and run in Babbitt boxes. The frame is hung upon journals or trunnions, T, and balanced by the weight, B. W and P are strong packing-boxes, which are filled with scrap-iron to the extent required. The face of the pulley, I, is extended beyond the 6 inches to receive the cord, C, for which purpose a shallow groove is cut in the pulley so as to bring the centre of the cord just to the periphery. The driving pulley, V, is put upon a shaft where it may be made to revolve slowly in the direction of the arrow.

It will be seen that the weight in the box, P, upon the horizontal arm will bring the pulleys together with a pressure just equal to the weight. The wooden pulley being in motion, the pressure, when sufficient, will roll the other pulley and raise the weight, W.

The manner of experimenting was to put a given weight upon the cord, C, and, while the driving pulley was moving, to load the box, P, until the weight, W, was carried up. The machinery was then stopped, when the weight would slowly descend, slipping the iron pulley backwards upon the wood. The weight in the pressure box was now noted; the weight was again raised, and the pressure increased sufficiently to hold the weight from slipping down, and the pressure again noted.

In the following table, the figures on the left show the weights raised. The second column gives the pressure just sufficient to bring the weight up; and the third column shows the weight necessary to raise and hold the weight, without slip.

Weight Raised.	FRICTION PULLEYS.		BELTED PULLEYS.	
	Pressure Required to Just Raise the Weight.	Pressure Required to Raise the Weight without Slip.	Pressure Required to Raise the Weight.	Pressure Required to Raise the Weight without Slip.
Lbs.	Lbs.	Lbs.	Lbs.	Lbs.
10	29	33	30	34
20	58	65	60	69
30	87	96	91	120
40	115	125	121	159
50	143	154	153	199
60	171	185	183	242
70	199	214	213	247
80	225	244	239	332
90	264	289	278	375
100	295	312	310	419
120	354	387	372	487
140	416	438	442	563
160	477	499	524	652
180	538	561	592	731

After these experiments were made and twice repeated with the pulleys, the frame, A, was reversed, so that the weight in the pressure box would tend to separate the pulleys. They were then connected by a 6-inch leather belt, and the experiments repeated with the results given in the fourth and fifth columns of figures.

It will be seen that, in this test, the traction of the friction wheels was greater than that of the belted pulleys, and considerably more than is usually supposed to be obtained from belts upon pulleys of either wood or iron; and that, while there is a marked falling off in the adhesion of the belt as the work increases, that of the friction increases as the labor becomes greater. Also, that the difference in the pressure required to just do the work, and that necessary to do it without loss or slip, advances in an increasing ratio with the work of the belt; but in the friction it is almost constant throughout the whole range of experiments. The figures applied to the friction wheels are the mean results of repeated experiments; those applied to the belted pulleys are each of a single test. It is not thought that these experiments were sufficient to fully establish all that the figures show; but they were enough to prove that smooth-faced wheels possess a much higher tractive power than has been generally supposed. They are given without further deduction or comment.

And now a word as to some of the advantages of friction gearing. Being always arranged with a movable shaft, so that the wheels may be thrown together or apart with the greatest ease, the machine driven by it is started and stopped at any moment while the driving wheel remains in motion. And when stopped, the separation is complete, and may so remain for any number of minutes or months without attention, and may be again started at any moment without the least inconvenience or injury. So slight is the separation required, that it is done almost without an effort. And by it we entirely dispense with the nuisance of loose pulleys, belt shifters, and idle running belts; and with the risk of throwing off and putting on belts. It obviates the delay and labor of shipping and unshipping pinions, and the rattle and bang and frequent breaking of clutches. It is durable, and requires no repairs; it is compact, and economizes room. It does not increase the pressure on journals when the speed is quickened, as is the case with belts running with great velocity, but remains constant at all speeds. And it will transmit any amount of power, from a hundredth part of a horse-power to 100 horse-power, with no greater percentage of loss, and with less pressure on journals than can be done by belts.

It is not contended that this style of gearing should supersede the belt. There are hundreds of situations in which nothing can take the place of belts. The ease with which they can be carried in almost any direction, and to any reasonable distance, will perhaps always place them foremost as a means of transmitting power. But where several machines, that must be run independently of each other and be stopped and started without interference, are driven by the same motor, one connection, at least, should be frictional; and that, if practicable, should be the connection nearest the motor. Where the motions are slow and the occasions for stopping few, this is of less importance; but where the speed is considerable, and the stoppages are frequent, it will be found a very great convenience.

Since the introduction of friction as a means of transmitting motion, it has often been desirable to apply the principle to bevel gearing. Frequently, however, this has been unsuccessful. The failures have resulted either from the want of a correct knowledge of the principles of bevel gearing, or from imperfect workmanship in the application of those principles.

When correctly and substantially built and accurately put up, bevel and mitre friction pulleys, within certain limits, operate just as well as in the other form. True, we cannot in these, as in the cylindrical pulleys, extend the face *ad libitum* without greatly increasing the diameter; and for this reason, when great power is to be transmitted, it is not convenient to use this form of gearing. But in all fast motions, where not more than 10 horse-power is to be transmitted, the bevel friction is one of the best means of connecting at an angle. It may be adapted to almost any change of speed, and set to any angle, either right, obtuse, or acute, and has the same advantages in operation as the other form of friction. And when it is required to reverse the motion at pleasure, it is most conveniently done by setting two bevel pulleys upon one shaft, facing toward each other, and placing one upon another shaft between them, so that it may be brought into contact with either.

In building this gearing, the iron cone, or pulley, is made similar to a bevel pinion, except as to the teeth, instead of which there is a smoothly turned face. The same care should be bestowed upon the accuracy of finish and balance that is required in the other form of friction pulley; but the pulley may be made somewhat lighter in the rim, as the conical form gives additional strength. In making the wooden driver — the iron pulley being furnished — the first point is to determine the exact diameter and bevel, for upon the correctness of these, to a great extent, depends the success of the work.

To obtain these dimensions, place a square across the smaller end of the finished iron pulley, and set a bevel to it, as shown in Fig. 81. This will give the correct bevel for the face of the driver.

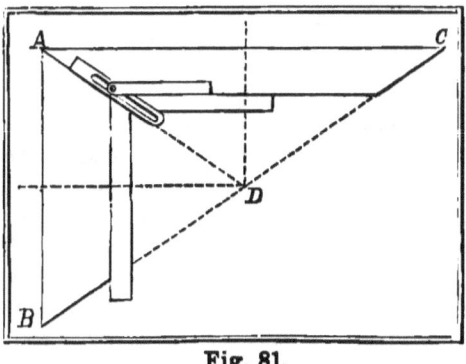

Fig. 81.

Next, upon any plane surface of sufficient size, draw the lines, A B and A C, making the length of the line, A B, just equal to the larger diameter of the iron pulley, and the angle at A a right angle. Then, with the square and bevel, or with a movable T square adjusted to the bevel, draw the lines B C and A D. The distance, A C, is the diameter required for the driver, and the other dimensions are easily obtained.

To obtain the bevels for pulleys to work on shafts placed at acute angles, draw the lines as in the annexed diagram, Fig. 82.

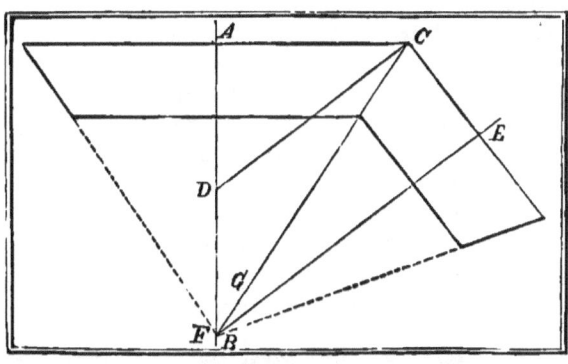

Fig. 82.

First, draw the line, A B, to represent the driving shaft. Then, at a right angle, draw the line A C, making its length equal to half the diameter of the driving pulley. Next, at the angle at which the shafts are to be set, draw the line C D; and at a right angle from this line, draw the line C E, making its length equal to half the required diameter of the other pulley. From the point, E, parallel to C D, draw the line E F, which will represent the other shaft. Now, from the point of a section of this and the line A B, draw the line G C, which will give the bevels for both pulleys.

If not above 2½ feet in diameter, the driver of the bevel pulleys may be built upon a "hub flange"—a disk of iron of about two-thirds the diameter of the pulley, with a hub projecting from one side. The hub should extend half an inch beyond the thickness of the wood to receive an annular disk of smaller diameter, through which the whole may be securely bolted together.

Upon the flange, around the hub, the pulley should be built. The first 2 or 3 inches, to form the back, should be of hard wood put on radially. For the balance, use soft maple. It is, in the present state of our knowledge, the only wood that can be recommended for this form of friction gear. It should be laid on this, as upon all friction drivers, with the grain running tangentially as nearly as possible. And each subsequent course should be made smaller, so as to form the bevel. The layers are put together with glue or white lead, and carefully and thoroughly nailed. The builder should be careful to make the joints perfect, and to put the wood snugly around the hub.

When the wood is built up to sufficient thickness, the other flange should be put on, and the whole bolted together and turned to the exact diameter and bevel required, and the pulley should be balanced with the utmost care.

For a larger bevel driver, it is best to use an iron centre with arms, and a flanged rim something like a car wheel. The diameter of the rim or cylinder should be a few inches less than the smaller diameter of the face of the pulley, and that of the flange something less than the larger diameter. Upon this wheel, the wooden rim is built as directed upon the hub flange, except that the bolts must be put in as the work progresses, so that subsequent layers will cover the heads; and the pulley is finished without the smaller flange.

The diagram (Fig. 83) shows a cross section of this pulley, which will be understood without further explanation.

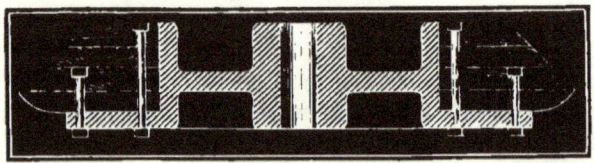

Fig. 83.

In setting up this gearing, it is of the utmost importance that the countershafts line exactly to the centres of the main or line shafts, and at the precise angle for which the pulleys were fitted; and that they are substantially set, so as not to get out of line.

This gearing is thrown on and off, connected and separated, by moving the countershafts endwise in their bearings. This may be done by allowing the end of the shaft to extend through beyond the outer bearing far enough to receive an extra box, one end of which is closed and Babbitted to receive the end pressure. This box is set up by a lever, to which it is pivoted. And by having the end of the shaft grooved where it is embraced by this box, it will be drawn back where the lever is released. In light work, it is as well to make the outer bearing do the whole, by making it both an end and side bearing, and having the box movable in a line with the shaft.

The pressure required to hold these pulleys up to the work is not great, and is easily applied by finishing the end of the shaft, and using a flat bearing of anti-friction metal, the full size of the shaft. Sometimes a steel point, like a lathe centre, is set against the end of the shaft to receive the pressure; but this is a very bad arrangement. It makes the bearing surface too small, and is one of the worst forms of bearing to keep supplied with oil. A flat bearing of wood, especially of hard maple, is very much better than this.

When there is considerable difference in the sizes of bevel pulleys working together, the end pressure is most upon the shaft carrying the larger; but this may frequently be neutralized, upon lines having several of these drivers, by setting them with their faces reversed.

A point that should never be lost sight of, in constructing setting levers for all friction work, is to make them adjustable, so that the pressure may be easily increased if required. This is sometimes done by a ratchet with several notches, into any one of which the lever may be drawn; but it is generally better to have but one catch, and to make the adjustment elsewhere. This may be done by connecting the lever, to the part to be moved, by a rod having adjusting nuts, or by making the fulcrum adjustable by bolt or set screw.

These adjustments should be made by the person having charge of the machinery, not by the operator of each machine. They should be kept tight enough to do the work required; but more than this is a waste of power, and a useless strain upon the machinery.

It may seem unnecessary to give the diagrams of lines for the dimensions of bevel gearing, as these are well understood. But it must be remembered that we have no work on millwrighting at present that gives information on this point of any scientific or practical value, and that our millwrights are not all familiar with the construction of this gearing. Our mills, though superior, are built without rules or uniformity of construction.

INDEX.

Abel, C. D., on belting, 8, 48.
Achard, A., on rope transmission, 260.
Action of belting, 130.
Adhesion, actual, no rule for, 148.
    of belting, 18, 49, 105, 108, 111, 131, 147-149, 161-165.
    of surfaces. 145, 147, 288.
Adhesive, 82, 92, 93, 181, 182, 183.
Advantages of belting, 67.
Albert, Capt., experiment, 13.
Alexander, A., rule for horse-power, 222.
Alexander Bros., on belting, 57, 191.
Allen, Z., on shafting, 132.
"*American Artisan,*" on belting, 59, 60.
Angle of belt, 110, 149, 159, 220.
    of grooved gear, 77, 80, 164, 285.
Animal substances for threads, 90.
Annan, Wm., belt lacing, 189.
Apparatus for experiments upon the variation of tension, Morin's, 244.
Appleton, American Cyclopædia, 84.
    Dictionary of Mechanics, 30.
    Mechanics' Magazine, 48.
    Morin's Mechanics, 248, 250.
Aqua ammonia for belts, 183.
Arc of contact, 18, 111, 115, 131, 152, 153, 214, 218, 221, 225, 226, 242, 243.
Area, sectional, of belt, 103.
Arkwright, 8.
Armengaud, Ainé, on belting, 124, 129, 164.
Armour, J., power in motion, 61.
Arrangement of Table III., 246.
Arresting and imparting motion, 166.
Association, N. E. Cotton, 132, 149.
Athenæum cement, 182.
Atmospheric influence on the adhesion of belts, 53, 101.
Auchincloss at Paris Exposition, on belting, 203.
Average tension or working strain on belting, 13.
Award to N. Y. Rubber Co., 13.

Babbage & Barlow pulley, 21.
Babcock & Wilcox on belting, 96.
Bacon, F. W., on belting, 107, 108, 162, 183.
Baird, H. C., 86.
Baker, 8.
Band links, 169.
    pulleys, 8.

Bands, 98.
Band-saw blades, strength of, 96.
Barbour, Wm., example, 222.
Barlow & Babbage pulley, 21.
Barnard, F. A. P., report, 274.
Bays defined, 8, 137.
Beard, I. H., on belting, 38, 44, 62.
Belting Co., N. Y., 12.
Belts, 59, 60, 100.
    action of, 130.
    adhesion of, 18, 49, 85, 111, 114, 147, 149.
    advantages of, 67.
    and wood friction compared, 295.
    angle of, 110.
    angular, 201, 205.
    average friction of, 18.
    best leather for, 91, 116, 131.
    big, 24.
    binders, 14, 21, 47, 66, 90, 91, 117, 166, 174, 176.
    breaking strain of leather, 116, 131, 147, 229.
    broad, 197, 202.
    bulky, 64.
    canvas, 97.
    care of, 38, 46, 47, 58, 82, 114, 124, 196.
    catgut, 90.
    centrifugal force of, 150.
    Christie, J., on, 71.
    Claudel's formula, 13, 14.
    clean, 39, 46, 196.
    Clissold's patent, 205.
    clutch, 170.
    coarse loose leather, 47.
    compared with gear, 22, 67, 70, 77, 84.
    compound, 92, 202, 203, 208.
    compressed, 125.
    condition of, 11, 37, 43, 73.
    contact of, rule for, 18, 93, 95, 112.
    cotton cord, 275.
    cotton webbing, 71.
    creep of, 75.
    crossed, 49, 50, 88, 98, 152, 226.
    disadvantages of, 67.
    double, 27, 42, 117, 118, 154, 162, 197.
    double and single, 50, 73, 82, 92, 93, 117, 118, 162.
    double edges, 158, 159, 160, 202.
    dressings for, 47, 58, 92, 107, 124, 182, 183, 196.
    driving or main, 92, 147, 154, 156, 157, 158, 159, 160, 197, 202.
    driving power of, 48, 54, 57, 108, 120, 130.

301

INDEX.

Belts, edge-bound, 203.
    edge-laid, 197, 209.
    eel-skin, 47, 90.
    elastic, 150.
    elasticity of, 60.
    examples of, 19, 24, 25, 27, 28, 29, 30, 31, 33, 34, 35, 41, 42, 51, 64, 69, 71, 72, 83, 85, 90, 94, 95, 96, 107, 118, 120, 154, 157, 158, 160, 190, 194, 199, 221.
    experiment, 108.
    fan-pulley, 224.
    fastener, Lincolne's, 186.
    fastenings. See Joinings.
    fat for, 92.
    flat, 68, 98, 99.
    flax, 99.
    flesh-side, 57.
    for blowers, 73, 74, 224.
    for centrifugal machine, 223.
    for circular saws, 70.
    for cooling shafts, 180.
    for dies and stamp-mills, 170.
    for driving fans, 60.
    for elevator, 12.
    for high powers, 101.
    for rolling-mills, 72, 73, 94, 208, 284.
    for saws, 119, 173.
    friction, 86.
    frictional surface, 105.
    gearing, 64.
    good leather for, 54, 59, 61.
    grain side of, to pulley, 19, 49, 50, 54, 56, 57, 113.
    gum, 12, 59, 92, 97, 98, 150, 157, 193, 197.
    gut, 15, 47, 68, 90, 128, 209.
    gutta-percha, 68, 99, 119, 120, 129.
    Haines's, 205.
    half cross, 98.
    Hartig, Dr., on tension of, 13.
    heavy, 96, 117.
    hemp, 68.
    holding power of, 73.
    holes, 180.
    homogeneous, 203.
    hooks, champion, 185.
    " Wilson's, 183, 185.
    horizontal, 46, 92, 148, 149, 153, 196.
    how not to run, 24, 64, 65.
    how to clean, 39.
    how to get adhesion, 39.
    how to get regular speed, 39.
    how to oil, 39.
    how to run, 25.
    Hunter's Print-Works, 157.
    impaired, 89.
    inapplicable, 67.
    inclined, 46, 149, 159.
    in coil, to measure, 83.
    India-rubber, 150.
    inextensible, 203.
    intestines, 15.
    in warm places, 97.
    joinings, 46, 59, 91, 93, 113, 117, 183-189, 196.
    lacings, 50, 82, 91, 94, 113, 131, 183, 187, 188, 189.

Belts, laps, 45, 49.
    large, 12, 18, 63, 94, 96, 154.
    larger than largest, 12.
    largest, 96.
    leather, 60, 99.
    links, 169.
    long, 43, 46, 62, 88, 153, 196.
    longest, 12, 197.
    loose, 46, 47.
    loss of velocity, 88, 89.
    main, 154 to 161, 193-198.
    narrow vs. wide, 155.
    new, 73.
    no definite rule for, 149.
    noiseless, 100, 129.
    paper, 187, 198.
    perpendicular, 47, 62, 82.
    philosophy of, 8, 43.
    pieced out, 50.
    pitch line of, 150.
    pliable, 113, 114.
    ponderous, 64.
    proportion, 43.
    pulleys, etc., 60.
    punched holes for, 93, 113.
    putting on, 82, 113, 148.
    quarter-twist, 84, 99, 172-180.
    quick, 44, 46.
    raw-hide, 99, 190.
    riveted single, 108.
    rosin on, 47, 89, 105, 108.
    round, 49, 68, 168.
    rubber, 12, 59, 92, 97, 98, 150, 157, 193-197.
    run to high part of pulley, 22, 44.
    rules and examples for horse-power of, 9, 13, 17, 18, 19, 26, 27, 28, 29, 30, 31, 32, 33, 34, 35, 37, 41, 42, 48, 49, 50, 51, 57, 68, 69, 71, 74, 83, 85, 93, 95, 107, 108, 112, 116, 118, 120, 125, 131, 152, 153, 156, 157, 159, 161, 196, 207, 208, 214, 215, 221, 222, 226, 254, 266.
    run wrong, 64.
    sag of, 62, 109, 147, 158, 166.
    Sampson's, 207.
    Sanderson's, 204.
    sheet-iron, 151, 190, 204.
    shifter, 51.
    shifting, 98.
    shipping, 19.
    short, 43, 62, 82, 153, 165, 196.
    short-lived, 97.
    single, 27, 93.
    slack, 15, 23, 46, 82, 148.
    sliding of, 45, 54, 55, 60, 66, 73, 85, 89, 99, 101, 104, 106, 130, 155, 227, 234, 235.
    slip of, 49, 54, 55, 60, 66, 73, 85, 89, 90, 101, 104, 106, 130, 155, 227, 234, 235.
    slow speed, 144, 145.
    small, 147.
    smooth, 73, 99.
    Spill's, 204.
    splicing, 50.
    steel, 60, 151, 204.
    strain of, 13, 17, 18, 19, 48, 50, 59, 68, 99, 111, 130, 131.

INDEX. 303

Belts, strength of, 49, 57, 111, 119, 131, 147, 149, 208, 212, 213, 229.
    stretched leather, 50, 125.
    stripped, 158, 159, 160.
    strong, 204.
    studs, Blake's, 184.
    superiority of, 45.
    surface contact, 18.
    tensile strength of, 9, 13, 14, 17, 40, 48, 61, 62, 68, 85, 102, 109, 113, 117, 130, 131, 211, 212, 213, 248.
    theory of Franck, 40.
    thickness of, 91, 99, 101, 117, 197, 219, 248.
    thin, 93, 128.
    tight, 21, 130, 148.
    tightener, 21, 66, 90, 91.
    to measure, 15, 111.
    " " in coil, 83.
    "suit work, 74.
    tool for putting on, 209.
    trapezoidal, Hoyt's, 205.
    triangular, 201, 205.
    twist of, 99, 172–180.
    Underwood's patent, 201.
    unmanageable, 64.
    unpliable, 97.
    varieties of, 190.
    various driving, 208.
    vertical, 47, 62, 82, 149, 153.
    vulcanized rubber, 12, 13, 193.
    water-proof, 200.
    water-proofed, 200.
    Weaver's, 170.
    wet weather, effect of, on, 88.
    well-worked, 125.
    well-worn, 73.
    wide, 18, 98, 202.
    wide *vs.* narrow, 155.
    width of, 111, 125.
    wool, 128, 205.
    work done by, 108.
    works badly, 45.
    woven, 99.
    woven wire, 204.
Bennett's Morin's Mechanics, 214, 218, 219, 248.
Berry, L. H., on twist belt, 173.
Bevan, S., on belts, 118.
Bevan's experiments, 103.
Bevel wheels, 8, 298.
Blake's belt-studs, 184.
Box, on mill work, 7, 119.
Briggs & Towne's experiments, 214–231.
Briggs, Robert, essay on belting, 214.
Bryant & Cogan's edge-laid belts, 209.
Buchanan, 8.
Buckskin covered pulleys, 204.
Buel, R. H., on creep of belts, 75.
Buignet, binders, 90.

Cabourg's machine, 203.
Calculation of sectional area of belting, 103.
Campbell & Co.'s belts, 156.
Canvas belts, 97.

Care of belts, 8, 37, 46, 47, 58, 82, 114, 124, 196.
Carillon's rule for horse-power, 125.
Castor oil dressing, 58, 95, 107, 108.
Catgut belts, 90.
Cedar Point Iron Co., on ropes, 276.
Cements for fastening belts, 181.
    " pulley covering, 95, 181.
Centrifugal force, 70, 101, 150.
Chains, 98.
Champion belt, 12, 197.
    belt-hook, 185.
Chase's cement for belts, 181.
Cheever, J. H., gum belt experiments, 195.
Christie, J., on belts, 71.
Circular saw belts, 119, 173.
Clarke, D. K., tables and rules, 13, 209.
Claudel, on belts, 13, 14.
Cleanliness of belts, 46.
Clement, Benjamin, on belts, 95.
Clissold's belts, 205.
Clutch belts, 170.
Clutches of grooved gear, 79.
Coarse, loose, leather belts, 47.
Coefficient of friction, 42, 121, 152, 214, 218.
    " transformation, 88.
Colophonium for belts, 92.
Combined strap-shifter and stop motion, 51.
    fast and loose pulley, 168.
Communicating motion by belts, 45.
Comparative value of belts, 97, 204.
Comparison of single and double belts, 117, 118.
Compound belts, 92, 202, 203, 208.
Compressed leather, 125.
Condition of belts, 11, 37, 43, 73.
Cone pulleys, Rankine, 48, 58.
Cones of pulleys, 58.
Conestoga Mills' belts, 154.
Connected machinery by belts, 45.
Connecting rods, wire ropes as, 285.
    rods to transmit circular motion, 286.
Connectors wrapping, 11, 67.
Contact angle of, 152.
    of belt with pulley, 18, 26, 73, 114, 119, 126, 127, 149, 151, 152, 214.
Contents, 16.
Convexity of pulleys, 18, 44, 68, 71, 93, 95, 99, 105, 110, 153.
Cooling shaft journals, belts for, 180.
Cotton cord for belts, 275.
    spinning, Leigh's, 22.
    webbing belts, 71.
Couch, A. B., example, 31.
Coulomb's rule, 248.
Covering for pulleys, 45, 105, 106, 110, 150, 189, 204.
Crafts & Filbert's loose pulley, 29.
Crane Bro.'s paper belt, 198.
Curried leather belts, 241, 242.
Curtis, H. W., on belts, 158.

Damp places, use rubber belts in, 83.
Deflection, or sag of belts and ropes, 62, 109, 256.

INDEX.

Designing belt gearing, 131.
Determination of the natural tension of belting, 232.
Die and stamp-mill belting, 170.
Dieterich, D. P., gum belts, 193.
Disadvantages of belting, 67.
Disproportion of belting, 45.
Distance between the shafts for belting, 109, 118.
Double and single belts, 50, 117, 118.
   belts, 87, 117, 162.
   edge belts, 158, 202.
   separate belts, 165.
Draw rod transmission, 267.
Dressing for belts, 47, 58, 92, 107, 124, 181-183, 196.
Driving belts, superiority of, 45.
   main, 154, 156, 157, 159, 160, 193, 197, 207, 208.
   power of belts, 48, 54, 57, 95, 120, 121, 147.
   pulleys, well centred, 100.
Drums, immaterial which are driven, 118.
Duration of belts, 27, 59, 85.
Durie, J., on hemp-rope gearing, 277.
Dynamics, law of, 101.
Dynamometer attachments, 243, 251.

Edge-laid belting, 197, 209.
Edwards's, John, belts, 128.
Edwards's untanned leather, 209.
Edwards's, W. T., belts, 207.
Eel-skin bands and ropes, 47.
   belting, 47, 90.
   lacings, 183.
Effect of air on belting, 53, 101.
   of eyelets, 50.
   of slack on belting, 15, 148.
   of speed on belting, 15, 140, 142, 148, 150.
   of tight belting, 24, 148.
Effective radius of pulleys, 59, 98.
Efficiency of belting, to increase, 161-165.
"*Elements* of Mechanics," Nystrom, 35.
Elephant hide, 203.
Elevator belting, 12.
Empirical formulæ, 14.
"*Encyclopædia* of Arts, Manufrs. and Mach.," 21.
"*Engineer* and Machinist's Assistant," 8.
Engineer on belts, 93.
"*Engineer,* The," 94, 187, 222.
"*Engineering* and Mining Journal," 77.
"*Engineering*" on raw-hide belts, 190.
English belting, leather and iron combined, 204.
   horse-power, 7.
   leather belting, 207.
Equal diameters and variable speed, 150.
European compound leather belting, 202.
Examples of belts, 19, 24, 25, 27, 28, 29, 30, 31, 33, 34, 35, 41, 42, 51, 64, 69, 71, 72, 83, 85, 90, 94, 95, 96, 107, 118, 120, 154, 157, 158, 160, 190, 194, 199, 221.
Experiments on belts, 54, 55, 108, 210-213, 232.
   on the friction of belts on wooden drums, 239.

Experiments on the slipping of belts on cast-iron pulleys, 238, 241, 242.
   on the tension of belts, 232.
   on the variation of tension, 248.
   with belting, 195.
Explanations, 5.
Eyelets, 50.

Face of pulleys, 22, 44, 68, 73, 90, 91, 93, 95, 99, 105, 110, 153.
Fairbairn, on belting, 8, 67, 222.
Furey, J., on steam-engine, 6.
Fast and loose pulleys, 21, 29, 116, 168.
Fastening of belts, 46, 59, 91, 92, 93, 181-189.
Fat for belts, 92.
Filberts & Craft's loose pulley, 29.
Flanges to pulleys, avoid, 91.
Flat face pulley, 110.
Flax belts, 99.
Flaxen thread, 90.
Flexible pieces, 98.
Fly-wheel effects on belts, 46.
Force required to break belts, 9.
   transmission by belts, 214.
   transmitted by belts, 9.
Formulæ for belts, 13, 19, 25, 102, 103, 108, 112, 117, 119, 126, 131, 135, 196, 214-225, 232-237, 240, 248-251, 266.
Fpm defined, 5.
Franck, Prof. L. G., on belts, 40.
Franklin Inst. Journal on belts, 38, 40, 44, 47, 48, 66, 205, 214.
Franklin Sugar Refinery on belts, 97.
French government agents, 12.
   horse-power, 7.
Friction by adhesion of surfaces, 145, 206, 288.
   clutch, 170.
   coefficient of, 152, 214, 218.
   of belts, 18, 45, 85, 105, 106, 114, 121, 122, 126, 147, 284.
   of grain side of belts, 19.
   ratio of, 115, 126.
   relative to arc, 18, 114, 119.
   rollers, 162, 165.
   wheels, 145, 206.
Frictional gearing, Robertson, 60, 77.
   gearing, by Wicklin, 288.
Friction formula, 126.
   of belts, experiments on, 55, 195, 239, 241.
Fuller's earth to remove oil, 89.
Funicular transmission, 260.

Gearing, 134, 135, 136.
   compared with belts, 22, 67, 70, 77, 84, 118.
   for mill, 63, 67.
   grooved clutches, 79.
   horse-power of, 136.
   table, 136.
   traction, Hitchcock's patent, 162.
   treatises on, 119, 132-135.
General statements, 196.
German horse-power, 7.

INDEX. 305

Glues and cements, 181–183.
Good leather for belts, 54, 57, 59, 61, 91, 116.
   method of lacing, 189.
Grain side of belting to pulley, 19, 49, 50, 54, 56, 57, 60, 73, 83, 95, 97.
Gramme defined, 6.
Greater than, character to represent, 5.
Greek letters explained, 5.
Gregory, O., "Mechanics," 207.
Grooved friction gear, 145.
   gear, 77.
   pulleys, 152, 164.
      "   materials to fill, 257, 277.
   speed-ring, 78, 79.
Gum belt (rubber), 12, 59, 92, 97, 98, 150, 157, 193, 197.
Gut (intestines) belts, 15, 47, 68, 90, 128.
Gutta-percha belting, 68, 99, 119, 120, 129.
Gwynne & Co., belt and friction, 165.

Haines's patent belt, 205.
Hair side of belting. See Grain Side.
Hale, Kilburn & Co., belt, 158.
Harmony Mill, Cohoes, 27.
Hartig, Dr., on belting, 13.
Hartman, J. M., belts, 95.
Haswell, on belting, 30, 48.
Headless brass belt-screws, 203.
Heavy belting, 96, 117.
Heilman, P., rule for belting, 126.
Hemp ropes, 62, 115, 122, 152, 153, 277–285.
   thread, 90.
Hepburn & Son's riveted double belts, 117.
Heywood, J., belts, 128.
Hide, meaning of, 191.
High part of pulley, belt runs to, 22, 44.
   velocities, 19, 142.
Hippopotamus hide, 203.
Hirn, C. F., on wire-ropes, 100, 266.
Hitchcock, A., traction gearing, 162.
Hoisting gear at Cincinnati, 51.
Holding power of belts, 73.
Holes for quarter-twist belts, 180.
Hooks, belt, 183, 185.
Horizontal belts, 46, 148, 149, 153, 196.
Horse-power, defined, 6, 7.
   English, French, German, Swedish, Russian, 7.
   of belts and examples, 9, 13, 17, 18, 19, 26, 27, 28, 29, 30, 31, 32, 33, 34, 35, 37, 41, 42, 48, 49, 50, 51, 57, 68, 69, 71, 74, 83, 85, 93, 95, 107, 108, 112, 114, 115, 118, 120, 125, 131, 152, 155, 156, 157, 159, 161, 196, 207, 208, 214, 215, 221, 222, 226, 254, 266.
   of gearing, 136.
   table, 136, 139.
Horsford's, E. N., report, 13.
Howarth belt fastener, 187.
Howe, Dr. H. M., on belting, 97.
How not to run belts, 64.
   to find the length of a belt, 15, 111.
   to use rubber belts, 195.
Hoyt, J. B., on belting, 47, 54, 95, 96, 98, 202, 210, 211, 212.

Hoyt & Co., patent angular belting, 205.
Hunebell, M., on belting, 90.
Hungarian leather lacing, 90.
Hunter's Print-Works' belts, 157.
Hüttinger, J. W., translations, 232, 260.
Hydraulic connection, 267.

Illustrations.
   Figs. 1 and 2, fast and loose pulleys, 21, 22.
   3, main driving belts for each floor, 25.
   4, Crafts & Filbert's loose pulley, 29.
   5, driving pulley carrying two belts, 34.
   6, driving pulley belt at 30°, 35.
   7, strap shifter and stop motion, 52.
   8, cones of pulleys, 59.
   9, how not to run main belts, 65.
   10, grooved speed-ring and belt, 79.
   11, 12, 13, belt joinings, 92.
   14, tension roller, 117.
   15, pulley, belt, and weights, 121.
   16 and 17, pulleys, belts, and weights, 123.
   18, grooved rope-wheels, 153.
   19, main pulley driving three belts, 154.
   20, main pulley driving two belts, 155.
   21, main pulley driving three belts, 156.
   22, indicator card of Campbell's engine, 157.
   23, driving belt at Hunter's Print-Works, 158.
   24, single belt driving three shafts, 159.
   25, driving belts of Patent Metal Co., 161.
   26, Hitchcock's traction gearing, 162.
   27, Parker's patent belting, 163.
   28, contrivance to increase contact, 163.
   29, 30, multigrooved wheels, 164.
   31, driving two shafts from one, 165.
   32, friction wheel and belt, 165.
   33, double separate belts, 166.
   34, Shaw's power-hammer belt, 167.
   35, 36, Shinn's fast and loose pulley, 168, 169.
   37, 38, band links, 170
   39, Weaver's belting, 171.
   40, belt as friction clutch, 171.
   41–50, quarter-twist belts, gears, 172–180.
   51, holes for quarter-twist belts, 180.
   52, Wilson's belt-hooks, 184.
   53, 54, Blake's belt-studs, 185.
   55, champion belt-hook, 186.
   56, machine-riveted strapping, 186.
   57, 58, Lincolne belt-fastener, 187.
   59, connecting the ends of belts, 187.
   60, lacing for paper belts, 188.
   61, good method of lacing, 189.
   62, Alexander Bros.' improvement in wide belting, 191.
   63, edge-laid belt, 198.
   64, 65, Underwood's patent angular belting, 201.
   66, tool for putting on belts, 209.
   67, side of leather tests, 212.
   68, diagram showing tension due to contact, 215.
   69, Morin's apparatus, 244.
   70, diagram, Poncelet's theory, 249.

INDEX.

Illustrations.
   71, 72, diagrams showing sag of wire-rope, 256, 257.
   73, diagram of pulley-groove, 258.
   74, " of how to avoid a high inclination of wire ropes, 260.
   75, diagram of rope transmission, 261.
   76, 77, diagrams of section of pulley-groove for wire-rope, 272, 275.
   78, wire-ropes as connecting-rods, 285.
   79, connecting-rods for rotary motion, 286.
   80, diagram of frictional gearing and belting, 294.
   81-83, diagrams of bevel frictional gearing, 298, 299.
Imparting and arresting motion, 166.
Inclined belting, 46, 149, 159.
Increasing the efficiency of belting, 161-165.
Inextensible belting, 203.
Influence of thickness on belting, 39.
Intestines (gut) for belting, 15, 47, 68, 90, 128, 209.
Iron and steel ropes compared, 276.
   pulleys, 8, 55, 150.

Jessup & Moore, on belting, 96.
Jewell, P. & Sons, on belting, 96.
Joinings of belt, 46, 59, 91, 113, 117, 183, 184, 185, 186, 187, 188, 189, 196.
Journals, to cool, 180.

Kilogramme defined, 6.
Kirkaldy's tests of belting, 14.

Laborde, M. E., on belts, 124, 125.
Lacing belts, 50, 82, 91, 94, 113, 131, 183, 187, 188, 189.
Laps, disposition of, 45, 49.
Large leather belts, 12, 18, 63, 94, 96, 154.
Larger pulleys, 23.
Largest belt in the world, 12, 96.
Law of adhesion, 114, 115, 119, 121, 126, 130, 148, 149, 226, 242.
   of coil adhesion, 116.
Leather belting, compound, 202, 203, 208.
   belting, good, 54, 59, 61.
   belting, noiseless, 100, 129.
   best kind for belting, 91, 116, 131.
   compressed for belting, 125.
   covered pulleys, 55, 82, 85, 104, 105, 106, 110, 189, 204.
   oak-tanned, 57, 58, 91.
   patent tanned, 109.
   pulley, 77.
   side of, tests, 212.
   strength of, 9, 13, 14, 17, 40, 48, 49, 57, 61, 62, 68, 102, 111, 113, 117, 119, 130, 131, 147, 149, 208, 211, 212, 213, 229, 248.
   stretch of, 58, 117.
   thong, 90.
Leffel's "Mechanical News," 151.

Leigh, Evan, on belting, 22-27, 198.
Leonard's "Mechanical Principia," 118, 196.
Leroux, M., on belting, 86.
Less than, character, 5.
Letters, Greek, 5.
Le Van, W. B., on belting, 156, 221.
Lever, J. S., on belting, 165.
Lincolne belt fastener, 186.
Line upon line, 49.
Links, band, 169.
Location of shafts and pulleys, 109.
Lockwood & Co., London, 61.
Logarithm, 5.
Long belting, 43, 46, 62, 88, 153, 196.
Loose belting, 46, 47.
   and fast pulley, 21, 29, 116, 168.
Loss of power by axle friction, 100.
   of power by belting in motion, 104.
   of power by slipping of belting, 104, 105.
   of power in telodynamic transmission, 258.
   of velocity by belting in motion, 88, 104.

Machine riveted strapping, 186.
Machinery, connected by belt, 45.
Machinist, skilful, 93.
MacKenzie, water-proof composition, 181, 182.
Main driving belts, 154-161, 193-198.
Man-power defined, 6.
Mason, John, rawhide belt, 190.
Materials for filling grooves in pulleys for wire ropes, 257.
Means of increasing the adhesion of belts, 105, 161-165.
Measuring belts, 15, 111.
"*Mechanical* News," Leffel's, 151.
   "Principia," 118, 196.
"*Mechanic's* Guide," Scholl, 91.
   "Magazine," London, on belts, 28, 40, 44, 45, 187, 205.
Mechanics of belting, 8.
   Overman's, 151.
Metal, to fasten leather to, 182.
Methods of transmission by belts and pulleys, 154.
Mètre defined, 6.
Millimètre defined, 6.
Mills and Mill-work, Fairbairn, 67.
Modulus of elasticity, 103.
Molesworth's formulæ, 31, 44, 181.
Moore's rule for cement, 183.
Morin's data, 6, 7, 14, 17, 18, 41, 44, 48, 88, 99, 121, 130, 214, 232.
Moseley, 45.
Motion, pulleys in, 122.
Mould on leather, to remove, 182.

N. E. Cotton Association, 132, 149.
New Jersey Zinc Co., belting, 94.
Newton's Journal on gear, 77, 275.
New York Belting Co., 193.
Nicholson, J., pulley, 21.

INDEX. 307

Nine dispositions of the quarter-twist belt, 172.
Nobes & Hunter belts, 208.
Norris & Co.'s belt tests, 14.
North British Rubber Co., 208.
Note on the sheaves used in the transmission of power by wire ropes, 275.
Nothing like leather, 77, 211.
Nystrom, J. W., on belting, 35–37.

Oak-tanned leather belts, 57, 81, 91.
Observations on the results contained in tables, 240, 242, 246.
Oil softens gum, 98, 196, 197.
Old device to increase the friction of cords, 163, 164.
On the adhesion of curried leather belts to cast-iron pulleys, 241, 242.
   the strength of belting leather, 210.
"*Operative* Mechanic and British Machinist," 21.
Oval punch, 50, 113.
"*Overman's* Mechanics," 151.
Ox-hide, strength of, 91, 213.

Page's patent tanned leather belts, 109.
Paper belting, 187, 198.
   pulleys, 15.
Parker's patent belting, 162.
Patent belt fastenings, Hoyt & Co.'s, 188.
   Metal Co.'s belts, 160.
Pelterau, display of belting, 203.
   belts, 203.
People's Works' gear, 180, 287.
Period, 5.
Perpendicular belting, 47, 62.
Perrin's band-saw blades, 96.
Philosophy of belting, 8.
Piecing out belts, 50.
"*Polytechnic* Review" on belting, 30, 38, 96.
   "Centralblatt" on belting, 86.
Poncelet's theorem, 234, 246, 248.
Potter's gut belting, 209.
Poullain Bros.' belts, 203.
Power gained by pressure, 114, 126.
   in motion, Armour, 61.
   lost by friction of axle, 100.
   lost by shafting, 145.
   transmission of, 139, 140, 143–148.
Power-hammer, Shaw's, 166.
Practical Mechanics of belting, 9, 115, 124, 129.
   "Mechanics' Journal" on belting, 39, 204, 227, 272.
"*Praktische* Maschinen Constructeur," 104.
Preller's leather, 209.
Pressure blower belt, 73.
Prevention of sliding of belts, 104.
Prony's experiments, 214.
"*Publication* Industrielle" on belting, 124, 164.
Pulley, 7, 8.
   article on, 150.
   balanced, 142.
   cones, 48, 58, 84.

Pulley contact for belting, 18, 73.
   convex, 44, 91, 95, 99, 105, 110, 153.
   covered with leather, 55, 58, 82, 85, 104, 105, 106, 110, 112, 189, 204.
   covered with raw hide, 15.
   covered with wood, 105, 150.
   covering, 15, 45, 105, 110.
   effective radius, 59, 98.
   face, 22, 44, 68, 73, 90, 91, 93, 95, 99, 105, 110, 153.
   fast and loose, 21, 29, 168.
   flanges, avoid, 91.
   grooved, 152, 153, 163, 164, 168.
   in motion, illustrated, 122, 123.
   of different diameters, 127.
   of equal diameters, 127, 150.
   of equal speed, 150.
   of iron, 8, 55, 150.
   of leather, 77.
   of paper, 15.
   of wood, 55, 91, 150.
   or wheels for wire rope, 257.
   rounding, 44.
   small, 92.
   smooth vs. rough, 125.
   tightening, 14.
   true, 91.
   two, driven by one, 165.
Punching belts, 50, 113.
Punctuation marks used, 5.
Purchasing belts, 111–113.
Putting on belts, 82, 113, 148.

Quarter-twist belts, 84, 99, 172–180.
   holes for, 180.
Quick motion belts, 44, 46.

Radius, effective for pulleys, 59, 98.
Ramsbottom, on rope gearing, 275.
Rankine, on belts, 8, 17, 48, 98, 169, 215, 222.
   on centrifugal force, 101.
Ratio between friction and traction, 130.
   of strains, 122.
Rattier & Co.'s belts, 129.
Raw-hide belts, 99, 190.
   pulleys, 15.
Receipts of Spon, 38.
Regulators for mills, 146.
Relation of small shafting to hollow, 132.
Relative height of wheels, 254.
Report of award to New York Rubber Co., 13.
Requisites for a belt, 124.
Resin on belts, 47, 89, 105, 108.
Resistance to bending of belts, 62.
Results, observations on, 240, 242, 246.
"*Review*, Polytechnic," on belts, 30, 38, 96.
Richards, J., on belts, 69.
Richards, London & Kelley, 96.
Rider, A. K., on belts, 116.
Rigger defined, 7.
Right angles, shafts at, 172–178.

Rise of pulley face, 90.
Riveting belts, 108.
Robertson, friction gear, 60, 77.
Roebling, J. A., on belts, 51.
  W. A., on wire ropes, 253.
Roller for tightening, 14, 21, 47, 66, 90, 91, 117, 166, 174, 176.
Rolling-mill belts, 72, 73, 94, 208, 284.
Rope-gearing, cotton, 275.
  "   hemp, 152, 277-285.
  "   wire, 253-275.
Rope of hemp or wire, 62, 115, 122, 152, 153, 204.
  of steel, 204, 259.
  of wire copper, 259.
  slack of, 256, 257.
  transmission, 253-287.
  vertical, 259.
Rossman, C. R., on belts, 90.
Round belts, 49, 68, 168.
Rounding of pulleys, 22, 44, 68, 71, 90, 93, 95.
Rpm, 5.
Rubber belts, 12, 59, 92, 97, 98, 150, 157, 193, 197.
Rule for belt equal to gear, 70.
  for distance between shafts, 109.
  for piecing out belts, 50.
  for sag, 62.
  for tension by weight of belt, 62.
  of thumb, 18.
  to make a double belt, 191.
Rules
  for ascertaining the horse-power of belts. See Horse-power.
  for belting in coil, 83.
  for belt length, 111.
  for belt width, 88, 111, 112, 125, 127.
  for cements and leather dressings, 47, 58, 92, 107, 124, 181-183, 196.
  for centrifugal force, 101.
  for gearing, 135.
  for shafting, 137.
Running conditions of belts, 43.
  off of belts, 105.
Rupture at lacing, 50.
Russian horse-power, 7.

Safe tension on belts. See Tension.
Safety arrangement, 145.
Sag of belts, 62, 109, 147, 158, 166.
  of ropes, 62, 256.
Sampson's patent belting, 207.
Sanderson's patent belting, 204.
Saw-belts, 119, 173.
  blades, band, strength of, 96, 151.
Scellos, E., belts, 203.
Schenck's example, 221.
Scholl, C. F., on belting, 91.
Sectional area of belting, to calculate, 103.
Separate belts from motor, 25.
  double belts, 165.
Shafting, belts to cool, 180.
  horse-power of, 139.
  in general, 136.
  size of, 132-140.

Shafting tests, 20.
Shafts and pulleys, location of, 109.
  at angle driven by belting, 84, 99, 172-178.
Shaw's trip-hammer, 166.
Sheet-iron belts, 151, 190, 204.
Shinn's, J., fast and loose pulley, 168.
"*Shoe* and Leather Reporter" on belting, 210.
Shoe-pegs for belt-joints, 183.
Short belting, 62, 153, 165, 196.
Side of leather, cut showing best part, 212.
  of leather, means, 191.
Simplicity desirable, 22.
Single and double belts, 50, 73, 82, 92, 93, 97, 109, 117, 118, 162.
Size of band, way to express, 102.
Skilful machinist, 93.
Skin, eel, belting, 47.
Slack belting, 15, 23, 46, 82, 148.
  of wire rope, 256.
Sliding (slip), 49, 54, 55, 60, 66, 73, 85, 89, 99, 101, 104, 106, 130, 155, 227, 234, 235.
Slip experiment, 238.
Slow shafts, 23.
Small pulleys, 92.
Smithsonian report, varnish, 181.
Smooth bands, 12, 193.
  faces desirable, 73.
Spanish white for belting, 89.
Special observations to determine the natural tension of belting, 233.
Speed cones (tapering), 48.
  high, 142, 147, 148, 150.
  of belting, 15, 23, 24, 142, 144, 147, 148, 149.
  slow, 140, 141.
  to regulate, 146, 147.
  variation of, 39, 49.
Speed-ring, grooved, 79.
Spier's, John, sheet-iron belt, 190.
Spill's patent belting, 204, 208.
Splicing belts, 50.
  wire rope, 254.
Spon, E., receipts, 38.
  E. and F. N., 69, 116, 119, 132.
S. S. Spencer on transmission of power, 154.
Steel belting, 60, 151, 204.
  ropes, 204, 259, 276.
Stop-motion and shifter, 51.
Strain on belting, 9, 13, 17, 18, 19, 48, 50, 59, 68, 90, 99, 111, 130, 131, 147, 148, 149, 220.
  on belting, effect of, 89.
Strapping, machine-riveted, 186.
Strap-shifter and stop-motion, 51.
Strength of belting, 14, 208, 210-213, 229.
  of gum, 13.
  of gutta-percha, 119.
  of leather, 14, 49, 57, 111, 119, 131, 147, 149, 208, 210-213, 229.
Stretch of leather, 58, 82, 91.
Stretched leather belting, 50, 54.
Strips of the cord, 235, 260.
Sturtevant, B. F., on belting, 73.
Superiority of driving belt, 45, 144.
Surface velocity, 10, 13.
Swedish horse-power, 7.

INDEX. 309

Sweet, J. E., saw-blades, 96.
Swell of pulley face, 99.

Table for converting belt-strain into surface velocity, 10.
  of belt thicknesses, 101.
  of contents, 16.
  of driving power of belts, 120.
  of experiments, belts on cast-iron pulleys, 241.
  of experiments, belts on wooden drums, 239.
    "   " Briggs & Towne, 230, 231.
    "   " on belts, 55.
    "   " on creep of belts, 76.
    "   " variation of tension, 247.
  of fan belts, 74.
  of formulæ for belts, 37, 112.
  of friction of ropes and belts, 284.
  of friction pulley and belted pulleys, 295.
  of horse-power of gears, 136.
  "   "   " shafts, 139.
  of Kirkaldy's belt tests, 14.
  of metric measures, 6.
  of multipliers per arc of contact, 226.
  of power as per arc of contact, 221, 225.
  of ratio of friction and pressure, 127.
  "   " strains on belts, 122.
  of results of belt tests, 208.
  of side of leather tests, 212.
  of slip of belts, 89.
  of strain as per arc of contact, 220.
  of strength of belting leather, 213.
  "   " saw-blades, 96.
  of transmission by wire ropes, 255.
  of units for horse-power, 7.
  of Webber's shafting tests, 20.
  of width of belts, 87.
  of wire rope sizes and strength, 259.
Tanners' dubbing for belts, 38.
Tapering speed-cones, 48.
"*Technologist*" on belting, 90, 107.
Telodynamic transmission, 100, 253-275.
Tensile strength of belting, 9, 13, 14, 17, 40, 48, 61, 62, 68, 85, 102, 113, 117, 130, 131, 208, 210, 211, 212, 213, 229, 248.
  strength of ox-hide, 91, 212, 213.
Tension on belting, 9, 13, 14, 17, 18, 19, 61, 90, 91, 99, 100, 103, 111, 113, 130, 131, 148, 149, 220, 232-248.
  that can with safety be applied to belting, 248.
  variation of, 243-251.
Tests of band-saw blades, 96.
  of belting, Centennial Exhibition, 213.
  of belting, Kirkaldy, 14.
  of shafting, 20.
Theory of belting, Franck, 40.
Therefore, character, 5.
The science of modern cotton-spinning, 22
Thickness, influence of, on belting, 39.
  of belting, 91, 99, 101, 117, 197, 219, 248.
Thin belting, 93, 128.
Thread of hemp or flax, 90.

Thumb rule, 18.
Thurston, Prof. R. H., on belting, 48.
Tight belting, effects of, 24, 148.
Tightening (Tension) roller, 14, 21, 47, 66, 90, 91, 117, 166, 174, 176.
To drive two pulleys from one, 164.
  fasten leather and cloth — glues and cements, 181-183.
  find the horse-power of belting. See Horse-power.
  find the length of belting, 15, 111.
  find the sectional area of belting, 103.
  find the strength of belting. See Strength.
  find the width of belts, 87, 111, 112, 125, 127, 149.
  increase the efficiency of belting, 161-165.
  make a double belt, 191.
  preserve leather, 181-183.
Tool for putting on belts, 208.
Towne, H. R., experiments, 8, 215, 218, 227-231.
Traction gearing, 162.
  of belts and ropes, 18, 130, 260, 261.
Tractive force not with area of contact, 71.
Transformation, co-efficient of, 88.
Transmission by belts, 128, 129, 131, 143.
  by cords and grooves, 163.
  by gearing, 133.
  by hemp ropes, 122, 152, 153, 277.
  by wire ropes, 100, 204, 253-287.
  by wire ropes as connecting-rods, 285.
  early modes, 139.
  of motive forces to a great distance, 260.
  of rotary motion by connecting-rods, 286.
  telodynamic, Achard, A., 260-266.
  telodynamic, Hirn, C. F., 100, 266-274.
  telodynamic, "Manufacturer and Builder," 274.
  telodynamic, Roebling, 253-260.
Treatise on mechanics, 207.
  on mill-gearing, 119.
  on transmission of force, 214.
  on wire-rope driving, 274.
  on wood machinery, 69.
Treatment of belts, 37, 47, 58, 92.
Twist-belts, two or more pulleys, 172-180.

Ultimate tenacity of leather and raw hide, 100, 131.
Unclassified figures and notes, 13.
Underwood's patent angular belt, 201.
Units for horse-power, 7.
  of measure, 6.
Untanned leather belts, 209.

Value, comparative, of belts, 97, 204.
  of greatest tension of wire rope and belts, 100, 103.
Van Nostrand on belting, 86.
Van Riper, P. V. H., on belting, 81.
Variation of speed, 39, 49, 150.
  of tension-apparatus, 243, 247-251.
Varieties of belts, 190.

INDEX.

Varnish, elastic, 181.
Velocity area of belts, 42.
   high, 142, 148, 149.
   low, 140.
   loss of, by belt in motion, 88, 90, 104.
   loss of, by various causes, 89.
   loss of, remedied, 89.
   of surface, 13.
Verification of two theorems, etc., 238.
Vertical belts, 47, 62, 82, 149, 153.
   ropes, 259.

Waterproof cements, 181–183.
   leather belting, 200.
Watt, Jas., on horse-power, 6.
Weale on belting, 7, 48.
Weaver's belting, 170.
Webber, Samuel, on belting, 17.
Webber's tests of shafting, 20.
Webbing, cotton, belting, 71.
Weight of belting, 99.
Welch, E. J. C., on belting gearing, 131.
Wet weather, effect of, on belting, 88.
Wheels or pulleys for wire-rope, 257.

Wheels, relative height of, 254.
Wicklin on friction-gearing, 288.
Wide belting, 98.
Wider belting for certain machines, 18.
Width of belting, to find, 87, 111, 112, 125, 127, 149.
Williams, Jos., on eel-skin belting, 47.
Willis, Prof., on mechanism, 9.
Wilson's belt-hook, 183.
Wire for stopping mill, 146.
   rope, 62, 100, 204, 253–275, 273, 285.
   rope as connecting-rods, 285.
   rope, driving, 274.
   rope, speed, 100.
   rope, table, 259.
   woven belts, 204.
Wood friction bevel wheels, 298.
   pulleys, 55, 91.
Wool belts, 128, 205.
Work lost, due to stiffness of ropes, 100, 253, 234.
Working strain on belting, 13, 17, 18, 19, 48, 50, 59, 99, 111, 119, 131, 229.
Woven belting, 99.
Wrapping connectors, 67, 98.
Wylde's circle of the sciences, 113.

ERRATA.

On page iv., for "Polytechnic Journal," read Polytechnic Review.

On cut No. 37, complete the circle of wheel C.

On page 119, and in original work also, the expression $\frac{FL}{R}$, in four several places, should take the form of an exponent, thus: $T = t \times (2.718) \frac{FL}{R}$; should read $T = t \times (2.718)^{\frac{FL}{R}}$.

On page 165, fig. 31. The lower pulley is A; the middle, B; and the upper, C.

On page 217, line 5 from top, should read $= \left(1 + f\frac{1}{r}\right)^l$.

In footnote to page 232, for "Let $t =$ natural tension, $t' = t +$ friction of belt or drum," read, "The sum of the two tensions t and t' is constant, and, at some instant of the movement of the drum, equal to the double of the tension proper, or natural T given to each one of the strips of the belt in consequence of moving the two axes of rotation further apart, which is independent of the action of the forces and of the resistances which act upon the system. It is clear that we can determine the constant and natural tension T, and the tensions t and t', relative to the moment of slipping. Knowing, then, for a same position relative to the two axes of rotation and a same state of the belt, the sums of the two tensions, the relation $t + t' = 2T$, or the double of the natural tension."

INDEX TO ADVERTISEMENTS.

AMERICAN OAK-LEATHER CO., LEATHER BELTING,	313
ALEXANDER BROS., LEATHER BELTING,	314
THOMAS WOOD, FAIRMOUNT MACHINE WORKS,	314
E. CLAXTON & CO.'S PUBLICATIONS,	315
Auchincloss's Practical Application of the Slide-Valve and Link-Motion to Stationary, Portable, Locomotive, and Marine Engines,	315
Bilgram's Slide-Valve Gears,	315
Cooper's Treatise on the Use of Belting,	315
Grimshaw on Saws,	316
Hartman and Mechener's Conchology,	316
Hobson's Amateur Mechanic's Practical Handbook,	316
Long and Buell's Cadet Engineer,	316
Morton's System of Calculating Diameter, Circumference, Area, and Squaring the Circle,	316
Overman's Mechanics for the Millwright, Civil Engineer, etc.,	316
Riddell's Carpenter and Joiner Modernized,	317
Riddell's New Elements of Hand-Railing,	317
Riddell's Mechanic's Geometry,	317
Riddell's Lessons on Hand-Railing for Learners,	317
Riddell's Artisan,	317
Roper's Catechism of High-Pressure or Non-Condensing Steam-Engines,	317
Roper's Handbook of the Locomotive,	317
Roper's Handbook of Land and Marine Engines,	317
Roper's Handbook of Modern Steam Fire-Engines,	317
Roper's Engineer's Handy-Book,	317
Roper's Questions and Answers for Engineers,	317
Roper's Use and Abuse of the Steam-Boiler,	318
Sloan's City and Suburban Architecture,	318
Sloan's Constructive Architecture,	318
Spang's Practical Treatise on Lightning Protection,	318
Trautwine's New Method of Calculating the Cubic Contents of Excavations and Embankments by the aid of Diagrams,	318
Trautwine's Field Practice of Laying out Circular Curves for Railroads,	318
Trautwine's Civil Engineer's Pocket-Book,	318
Webber's Manual of Power for Machines, Shafts, and Belts,	318
White's Elements of Theoretical and Descriptive Astronomy,	318
Whitney's Metallic Wealth of the United States,	318

THE LARGEST OAK LEATHER TANNERY IN THE WORLD.

AMERICAN OAK LEATHER CO.

Tanners and Manufacturers of

"NON POROUS" AND OAK TANNED LEATHER BELTING,

DEALERS IN

Rubber Belting, Hose, and Packing,
Calcutta Raw-Hide Lace,
Patent Tanned Lace and Mill Supplies.

SALESROOM,
Cor. Walnut and Second Sts.

TANNERY, { Kennar and Florence Sts., Dalton and McLean Aves.,

CINCINNATI, O.

CHICAGO BRANCH,
212 RANDOLPH STREET.

ST. LOUIS BRANCH,
404 NORTH MAIN STREET.

For Patent Double.  C. H. ALEXANDER, H. W. ALEXANDER, E. P. ALEXANDER.  For Leather Belting'

ALEXANDER BROS
MANUFACTURERS OF PURE OAK TANNED LEATHER BELTING.
410 & 412 NORTH THIRD ST. PHILA.

Patentees and Manufacturers of an Improvement in the Construction of

WIDE DRIVING BELTS.

For Description See Page 191.

DOUBLE-BRACED ADJUSTABLE & SELF OILING HANGER.

Ball and Socket
Self-Oiling Pillow Block.

FAIRMOUNT MACHINE WORKS.
Office, 2106 Wood St., Philadelphia.

THOMAS WOOD,
Manufacture as Specialties

Power Looms, Patent Bobbin or Quill Winding Machines, Plain and Presser Beaming Machines, Plain and Presser Spooling Machines, Reeling, Warp Splitting, Dyeing, Sizing, Scouring, Fulling and Calendering Machines,

WARPING MILLS,
16, 18 and 20 yards Circumference,
WITH IMPROVED HECKS.

SHAFTING,
With Patent
ADJUSTABLE SELF-OILING HANGERS,
8, 10, 12, 15, 18, 20, 24 and 30 in. drop.
Also WALL, POST AND GIRDER HANGERS.
Pulleys, from 4 inches to 10 feet in diameter.
PATENT FRICTION PULLEY.
Pulleys in two parts, any size required.
PATENT HOISTING MACHINES.
Oil Presses for Lard, Fish and Paraffine.

GEO. B. CARPENTER & CO.

DEALERS IN

COTTON DUCK
TENTS-AWNINGS
RAINPROOF COVERS
TWINES AND CORDAGE
→ **MILL SUPPLIES** ←
FLAGS AND BANNERS
BLOCKS AND ROPE
WIRE ROPE

202 to 208 South Water Street, Chicago, Ill.

OFFICE OF

THE HEIM LEATHER BELTING CO.

MANUFACTURERS OF

—PURE—

OAK TANNED LEATHER BELTING

324-326-328 PEARL STREET,
NEW YORK.

APRIL, 1883.

Have lately made a Belt for the U. S. Electric Illuminating Co. of N. Y. City, 125 feet in length, and 60 inches wide, double; and one for the U. S. Warehouse Co., of Brooklyn, N. Y., 112 feet in length, and 50 inches wide, double; and, are now making a Belt 150 feet in length and 72 inches wide, double, which will be, when completed, the longest and widest Belt ever made. The largest Belt known is now running at L. Waterbury & Co.'s Rope Manufactory, Brooklyn, N. Y., and was manufactured by the Heim Leather Belting Co. Prize Medals received at the Centennial Exhibition, Philadelphia, 1876; and at Chili Exhibition, 1875; Berlin, 1877; Paris, 1878; and American Institute, N. Y., 1867 to 1881.

MR. TRAUTWINE'S ENGINEERING WORKS.

The Field Practice of Laying out Circular Curves for Railroads. By JOHN C. TRAUTWINE, Civil Engineer. Eleventh edition, enlarged and rewritten. 12mo, morocco, tuck, gilt edge, $2.50.

A New Method of Calculating the Cubic Contents of Excavations and Embankments by the aid of Diagrams; together with Directions for Estimating the Cost of Earthwork. By JOHN C. TRAUTWINE, Civil Engineer. 10 steel plates. Seventh edition, completely revised and enlarged. 8vo, cloth, $2.

Civil Engineer's Pocket-Book of Mensuration, Trigonometry, Surveying, Hydraulics, Hydrostatics, Instruments and their Adjustments, Strength of Materials, Masonry, Principles of Wooden and Iron Roof and Bridge Trusses, Stone Bridges and Culverts, Trestles, Pillars, Suspension Bridges, Dams, Railroads, Turnouts, Turning Platforms, Water Stations, Cost of Earthwork, Foundations, Retaining Walls, etc. etc. In addition to which the elucidation of certain important Principles of Construction is made in a more simple manner than heretofore. By JOHN C. TRAUTWINE, Civil Engineer. 12mo, 695 pages, illustrated with about 700 wood-cuts. Morocco, tuck, gilt edge. Twentieth thousand, revised and corrected, $5.

GRIMSHAW ON SAWS. History, Development, and Action; Classification and Comparison; Manufacture, Care, and Use of all kinds of Saws. By ROBERT GRIMSHAW. Quarto, 280 pages. This thorough work, impartially written in a clear, simple and practical style, treats the Saw scientifically, analyzing its action and work, and describing, under the leading classes of Reciprocating and Continuous Acting Saws, the various kinds of large and small Hand, Sash, Mulay, Jig, Drag, Circular, Cylinder, and Band Saws, as now and formerly used for Cross-Cutting, Ripping, Scroll-Cutting, and all other sawing operations in Wood, Stone, and Metal, Ice, Ivory, etc., in this country and abroad. With Appendices concerning the details of Manufacture, Setting, Swaging, Gumming, Filing, etc.; Tables of Gauges, Log Measurements from 10 to 24 feet, and from 12 to 96 inches; Lists of all U. S. Patents on Saws from 1790 to 1880, and other valuable information. Elegantly printed on extra heavy paper. Second and greatly enlarged edition, with Supplement, 354 illustrations. Quarto, cloth, $4.

Any of the above books will be sent to any part of the United States or Canada on receipt of list price.

Send money in Registered Letter, P. O. Order, or Check.

E. CLAXTON & CO., Publishers,
930 Market St., Philadelphia, Pa.

SCIENTIFIC AND MECHANICAL BOOKS.

AUCHINCLOSS.—The Practical Application of the Slide-Valve and Link-Motion to Stationary, Portable, Locomotive, and Marine Engines, with New and Simple Methods for proportioning the Parts. By WILLIAM S. AUCHINCLOSS, C.E. Seventh Edition. Revised and Enlarged. 8vo, cloth. **$3.00.**

BILGRAM.—Slide-Valve Gears. A new graphical method for Analyzing the Action of Slide-Valves, moved by eccentrics, link-motion, and cut-off gears. By HUGO BILGRAM, M.E. 16mo, cloth. **$1.00.**

COOPER.—A Treatise on the Use of Belting for the Transmission of Power. With numerous illustrations of approved and actual methods of arranging Main Driving and Quarter Twist Belts, and of Belt Fastenings. Examples and Rules in great number for exhibiting and calculating the size and driving power of Belts. Plain, Particular, and Practical Directions for the Treatment, Care, and Management of Belts. Descriptions of many varieties of Beltings, together with chapters on the Transmission of Power by Ropes; by Iron and Wood Frictional Gearing; on the Strength of Belting Leather; and on the Experimental Investigations of Morin, Briggs, and others. By JOHN H. COOPER, M. E. 1 vol., demi octavo, cloth. **$3.50.**

DANBY.—Practical Guide to the Determination of Minerals by the Blowpipe. By DR. C. W. C. FUCHS. Translated and edited by T. W. DANBY, M.A., F.G.S. **$2.50.**

DRAKE.—A Systematic Treatise, Historical, Etiological, and Practical, of the Principal Diseases of the Interior Valley of North America, as they appear in the Caucasian, African, Indian, and Esquimaux varieties of its population. By DANIEL DRAKE, M.D. 8vo, sheep. **$5.00.**

DWYER.—The Immigrant Builder; or, Practical Hints to Handy-Men. Showing clearly how to plan and construct dwellings in the bush, on the prairie, or elsewhere, cheaply and well with Wood, Earth, or Gravel. Copiously illustrated. By C. P. DWYER, Architect. Tenth edition. Demy 8vo, cloth. **$1.50.**

GENTRY.—The House Sparrow at Home and Abroad. By THOMAS G. GENTRY. Demy 8vo, cloth. **$2.00.**

GIRARD.—Herpetology of the United States Exploring Expedition of the years 1838, '39, '40, '41, and '42, under the command of Commodore Wilkes. With a folio Atlas of over Thirty Plates, elegantly colored from nature, executed under the supervision of Dr. Charles Girard, of the Smithsonian Institution. 4to, cloth. **$30.00.**

GRIMSHAW.—On Saws. History, Development, and Action; Classification and Comparison; Manufacture, Care, and Use of all kinds of Saws. By ROBERT GRIMSHAW. Large Octavo. 234 ILLUSTRATIONS. This thorough work, impartially written in a clear, simple and practical style, treats the Saw scientifically, analyzing its action and work, and describing, under the leading classes of Reciprocating and Continuous Acting Saws, the various kinds of large and small Hand, Sash, Mulay, Jig, Drag, Circular, Cylinder, and Band Saws, as now and formerly used for Cross-Cutting, Ripping, Scroll-Cutting, and all other sawing operations in Wood, Stone, and Metal, Ice, Ivory, etc., in this country and abroad. With Appendices concerning the details of Manufacture, Setting, Swaging, Gumming, Filing, etc.; Tables of Gauges, Log Measurements from 10 to 24 feet, and from 12 to 96 inches; Lists of all U. S. Patents on Saws from 1790 to 1880, and other valuable information. Elegantly printed on extra heavy paper. Copiously indexed. Of immense practical value to every Saw user. **$2.50.**

1

GRIMSHAW.—A Supplement to Grimshaw on Saws, containing additional practical matter more especially relating to the forms of saw teeth for special material and conditions, and to the behavior of saws under particular conditions. One hundred and twenty illustrations. By ROBERT GRIMSHAW. Quarto, cloth. **$2.00.**

GRIMSHAW.—Saws. The History, Development, Action, Classification, and Comparison of Saws of all kinds. With Copious Appendices. Giving the details of manufacture, filing, setting, gumming, etc. Care and use of saws; table of gauges; capacities of saw-mills; list of saw patents; and other valuable information. By ROBERT GRIMSHAW, author of "Modern Milling," etc., etc. Second and greatly enlarged edition, with Supplement, 354 illustrations. Quarto, cloth. **$4.00.**

GRIMSHAW.—Modern Milling. Being the substance of two addresses delivered at the Franklin Institute, outlining the modern methods of Flour-Milling by "New Process" and Rollers. By ROBERT GRIMSHAW. 8vo, 28 illustrations. **$1.00.**

HARTMAN and MECHENER'S Conchology.—Conchologia Cestrica. The Molluscous Animals and their Shells of Chester County, Pa. With numerous illustrations. By WILLIAM D. HARTMAN, M.D., and EZRA MECHENER, M.D. 12mo, cloth. **$1.00.**

HOBSON.—The Amateur Mechanic's Practical Handbook. Describing the different tools required in the workshop, the use of them, and how to use them; also, examples of different kinds of work, etc., with descriptions and drawings. By ARTHUR H. G. HOBSON. 12mo, cloth, **$1.25.**

LONG and BUELL.—The Cadet Engineer, or Steam for the Student. By JOHN H. LONG, Chief Engineer, U. S. Navy, and R. H. BUEL, Assistant Engineer, U. S. Navy, Demy 8vo, cloth. **$2.25.**

MORTON.—The System of Calculating Diameter, Circumference, Area, and Squaring the Circle. By JAMES MORTON. 12mo, cloth. **$1.00.**

MOORE.—The Universal Assistant and Complete Mechanic. A Hand-Book of One Million Industrial Facts, Processes, Rules, Formulæ, Receipts, Business Forms, Tables, etc., in over Two Hundred Trades and Occupations. Together with full directions for the Cure of Disease and the Maintenance of Health. By R. MOORE. A new revised edition, illustrated by 500 Engravings. 12mo, cloth. **$3.00.**

NICHOLS.—The Theoretical and Practical Boiler-Maker and Engineer's Reference Book. By SAMUEL NICHOLS, Foreman Boiler-Maker. 12mo, cloth, extra. **$2.50.**

NYSTROM.—A New Treatise on Elements of Mechanics, establishing strict precision in the meaning of Dynamical Terms, accompanied with an Appendix on Duodenal Arithmetic and Metrology. By JOHN W. NYSTROM, C.E. 8vo, cloth. Price reduced to **$2.00.**

NYSTROM.—A New Treatise on Steam Engineering, Physical Properties of Permanent Gases, and of Different Kinds of Vapor. By JOHN W. NYSTROM, C.E. 8vo, cloth. **$1.50.**

OVERMAN.—Mechanics for the Millwright, Engineer, Machinist, Civil Engineer, and Architect. By FREDERICK OVERMAN. 12mo, cloth. 150 illustrations. **$1.50.**

OWEN.—Report of a Geological Survey of Wisconsin, Iowa, and Minnesota. By DANIEL DALE OWEN. 300 Illustrations. 2 Vols. 4to, cloth. **$7.50.**

2

RIDDELL.—The Carpenter and Joiner Modernized. Third edition, revised and corrected, containing new matter of interest to the Carpenter, Stair-Builder, Carriage-Builder, Cabinet-Maker, Joiner, and Mason; also explaining the utility of the Slide Rule, lucid examples of its accuracy in calculation, showing it to be indispensable to every workman in giving the mensuration of surfaces and solids, the division of lines into equal parts, circumferences of circles, length of rafters and braces, board measure, etc. The whole illustrated with numerous engravings. By ROBERT RIDDELL. 4to, cloth. $7.50.

RIDDELL.—The New Elements of Hand Railing. Revised edition, containing forty-one plates, thirteen of which are now for the first time presented, together with the accompanying letter-press description. The whole giving a complete elucidation of the Art of Stair-Building. By ROBERT RIDDELL, author of "The Carpenter and Joiner Modernized," etc. One volume, folio. $7.00.

RIDDELL.—Mechanic's Geometry; plainly Teaching the Carpenter, Joiner, Mason, Metal-plate Worker—in fact, the artisan in any and every branch of industry whatsoever—the constructive principles of his calling. Illustrated by accurate explanatory card-board Models and Diagrams. By ROBERT RIDDELL. Quarto, cloth. Fully illustrated by fifty large plates. $5.00.

RIDDELL.—Lessons on Hand-Railing for Learners. By ROBERT RIDDELL, author of "New Elements of Hand-Railing," "The Carpenter and Joiner Modernized," etc. 4to, cloth. Third edition. $5.00.

RIDDELL.—The Artisan. Illustrated by forty plates of Geometric drawings, showing the most practical methods that may be applied to works of building and other constructions. The whole is intended to advance the learner by teaching him in a plain and simple manner the utility of lines, and their application in producing results which are indispensable in all works of art. By ROBERT RIDDELL, late teacher of the artisan class in the Philadelphia High School, etc. $5.00.

ROPER.—A Catechism of High-Pressure, or Non-Condens- ing Steam-Engines; including the Modelling, Constructing, and Management of Steam-Engines and Steam-Boilers. With valuable illustrations. By STEPHEN ROPER, Engineer. Sixteenth Thousand, revised and enlarged. 18mo, tucks, gilt edge. $2.00.

ROPER.—Handbook of the Locomotive, including the Construction of Engines and Boilers, and the Construction, Management, and Running of Locomotives. By STEPHEN ROPER. Eleventh edition. 18mo, tucks, gilt edge. $2.50.

ROPER.—Handbook of Land and Marine Engines; includ- ing the Modelling, Construction, Running, and Management of Land and Marine Engines and Boilers. With illustrations. By STEPHEN ROPER, Engineer. Tenth edition. 12mo, tucks, gilt edge. $3.50.

ROPER.—Handbook of Modern Steam Fire-Engines. With illustrations. By STEPHEN ROPER, Engineer. 12mo, tucks, gilt edge. $3.50.

ROPER.—Engineer's Handy-Book. Containing a Full Ex- planation of the Steam-Engine Indicator, and its Use and Advantages to Engineers and Steam Users; with Formulæ for Estimating the Power of all Classes of Steam-Engines; also Facts, Figures, Questions, and Tables for Engineers who wish to Qualify Themselves for the United States Navy, the Revenue Service, the Mercantile Marine, or to Take Charge of the Better Class of Stationary Steam-Engines. Fourth edition, revised and enlarged. By STEPHEN ROPER, Engineer. 1 vol., 16mo, 675 pages, tucks, gilt edge. $3.50.

ROPER.—Questions and Answers for Engineers. This little book contains all the Questions that Engineers will be asked when undergoing an examination for the purpose of procuring licenses, and they are so plain that any engineer or fireman of ordinary intelligence may commit them to memory in a short time. By STEPHEN ROPER, Engineer. $3.00.

3

ROPER.—Use and Abuse of the Steam-Boiler. By Stephen Roper, Engineer. Eighth edition, with illustrations. 18mo, tucks, gilt edge. **$2.00.**

SLOAN.—City and Suburban Architecture. In which are exhibited numerous Designs and Details for Public Edifices, Private Residences, and Mercantile Buildings. By SAMUEL SLOAN. Illustrated with 131 full-page engravings, accompanied by specifications and historical and explanatory text. New edition, revised. Imperial 4to. Cloth. **$10.00.**

SLOAN.—Constructive Architecture. A Guide for the Builder and Carpenter; containing a series of Designs for the construction of Roofs, Domes, and Spires, etc.; and giving choice examples of the *Five Orders of Architecture;* selected from the best specimens of Grecian and Roman art, with the figured dimensions of their height, projection, and profile. To which is added a treatise on Practical Geometry. The whole illustrated by 66 plates. By SAMUEL SLOAN. Quarto. Cloth. **$7.50.**

SPANG.—A Practical Treatise on Lightning Protection. By HENRY W. SPANG. With illustrations. 12mo, cloth. **$1.50.**

TRAUTWINE.—A New Method of Calculating the Cubic Contents of Excavations and Embankments by the aid of Diagrams; together with Directions for estimating the cost of Earthwork. By JOHN C. TRAUTWINE, C.E. 10 steel plates. Sixth edition, completely revised and enlarged. 8vo, cloth. **$2.00.**

TRAUTWINE.—The Field Practice of Laying Out Circular Curves for Railroads. By JOHN C. TRAUTWINE, C.E. Eleventh edition, 1882, revised and enlarged. 12mo, tuck. **$2.50.**

TRAUTWINE'S Civil Engineer's Pocket-Book of Mensura- tion, Trigonometry, Surveying, Hydraulics, Hydrostatics, Instruments and their Adjustments, Strength of Materials, Masonry, Principles of Wooden and Iron Roof and Bridge Trusses, Stone Bridges and Culverts, Trestles, Pillars, Suspension Bridges, Dams, Railroads, Turnouts, Turning Platforms, Water Stations, Cost of Earthwork, Foundations, Retaining Walls, etc. By JOHN C. TRAUTWINE, C.E. 12mo, 695 pages, mor. tucks, gilt edges. Twentieth Thousand. Revised and corrected. **$5.00.**

WEBBER.—A Manual of Power for Machines, Shafts, and Belts. With the History of Cotton Manufacture in the United States. By SAMUEL WEBBER, C.E. This work contains over 1200 tests, up to date, of the Power required by Cotton, Woollen, Worsted, and Flax Machinery, Shafting, and Tools, with Summaries of the Machines and Power used in a number of Cotton Mills on various fabrics; Rules and Tables for strength and speed of Shafting and Belting; Corrected Tables of the Centennial Turbine Tests, at Philadelphia, 1876; Breaking and Twist Tables for Yarn and Roving. Historical Sketch of the growth of the Cotton manufacture in the United States. The whole, with an explanatory preface, forming an octavo volume of 236 pages, neatly bound in cloth. **$3.00.**

WHITE.—The Elements of Theoretical and Descriptive Astronomy, for the use of Colleges and Academies. By CHARLES J. WHITE, A.M. Numerous illustrations, 1 vol., demi 8vo. Fourth edition, revised. **$2.00.**

WHITNEY.—Metallic Wealth of the United States, de- scribed and compared with that of other countries. By J. D. WHITNEY. **$3.50.**

Any of the foregoing books will be sent by mail, post-paid, on receipt of price, by

E. CLAXTON & CO., PUBLISHERS,

930 *Market Street, Philadelphia.*

www.ingramcontent.com/pod-product-compliance
Lightning Source LLC
Chambersburg PA
CBHW030011240426
43672CB00007B/907